PRODUCTION ENGINEERING TE

CW00825491

Other Mechanical and Production Engineering titles from Macmillan

INTRODUCTION TO ENGINEERING MATERIALS: *V. B. John*

MANAGEMENT OF PRODUCTION, Third Edition: *J. D. Radford and D. B. Richardson*

THE MANAGEMENT OF MANUFACTURING SYSTEMS: *J. D. Radford and D. B. Richardson*

MECHANICAL ENGINEERING DESIGN, Second Edition: *G. D. Redford*

MECHANICAL TECHNOLOGY, Second Edition: *G. D. Redford, J. G. Rimmer and D. Titherington*

STRENGTH OF MATERIALS, Third Edition: *G. H. Ryder*

AN INTRODUCTION TO PRODUCTION AND INVENTORY CONTROL: *R. N. van Hees and W. Monhemius*

Production Engineering Technology

J. D. Radford, B.SC. (ENG.), M.I.MECH.E., F.I.PROD.E.

D. B. Richardson, M.PHIL., D.I.C., F.I.MECH.E., F.I.PROD.E., A.M.B.I.M.

Brighton Polytechnic

THIRD EDITION

First edition 1969
Second edition 1974
Reprinted 1976 (with corrections), 1978
Third edition 1980
Reprinted 1982

Published by
THE MACMILLAN PRESS LTD
London and Basingstoke
Companies and representatives throughout the world

Printed in Hong Kong

British Library Cataloguing in Publication Data

Radford, John Dennis
Production engineering technology.—3rd ed.
1. Production engineering
I. Title II. Richardson, Donald Brian
621.7 TS176

ISBN 0-333-29397-5
ISBN 0-333-29398-3 Pbk

Contents

Preface to the Third Edition

The main object in writing this book is to provide a concise treatment of production engineering technology for Degree and Higher National Diploma students.

Although the many aspects of the subject have been separately covered in much greater detail in various books and papers, the authors believe that this is the first time that an attempt has been made to contain the necessary work at this level in one volume.

The third edition has enabled us to include new material and to bring cutting tool nomenclature into line with BS 1296. The chapter 'Polymer Processing' has been contributed by our colleague, Mr R. S. G. Elkin, M.I.Mech.E., M.R.Ae.S.

We should like to thank those who, by their suggestions and advice, have assisted in the preparation of the book, and also Miss Grace Vine, who typed the manuscript.

J. D. Radford
D. B. Richardson

1 Introduction

The shaping of materials before they are incorporated into a product usually occurs in a number of stages. Specific examples of the shaping processes used to produce five different parts are illustrated in Fig. 1.1 (*a*) and an outline of the main groups of shaping processes is shown in Fig. 1.1 (*b*). It will be seen that some parts which have been cast, sintered

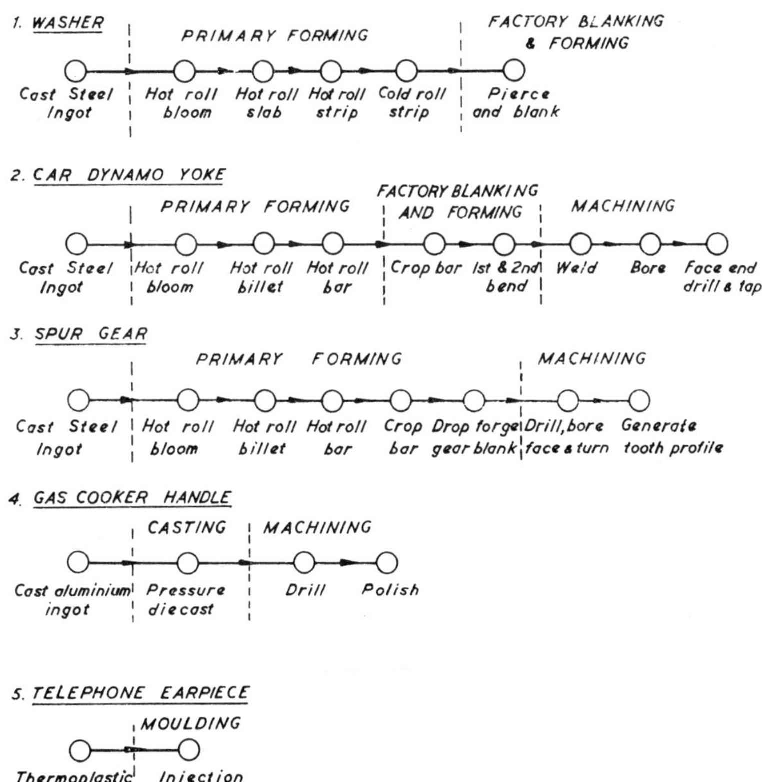

Fig. 1.1 (*a*) Typical shaping process

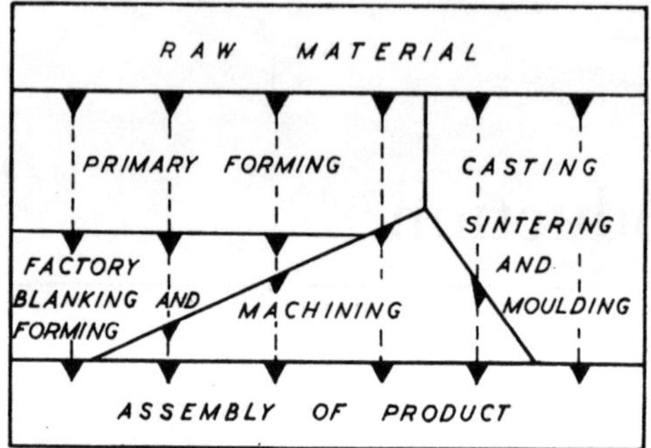

Fig. 1.1 (*b*) Main subdivisions in metal shaping

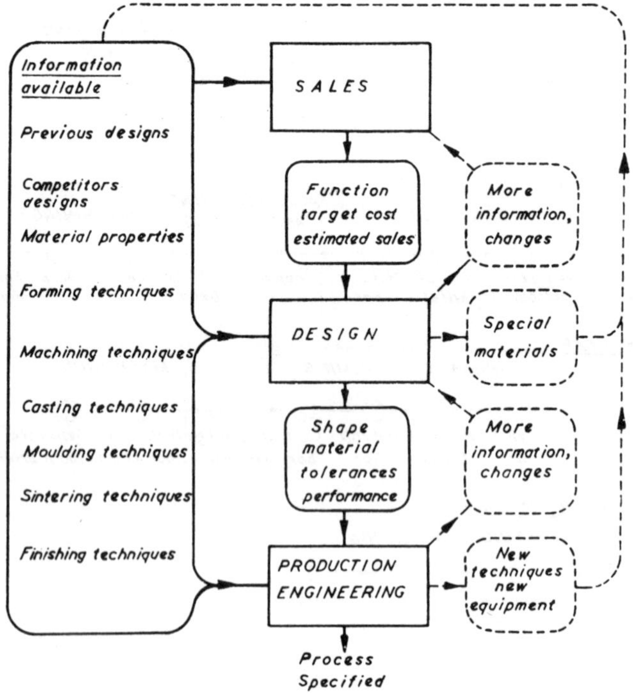

Fig. 1.2 Stages in specification of process

or moulded can be incorporated directly into assemblies without further processes, although usually machining is required. Primary forming operations produce a range of products such as forgings, bar, plate and strip, which is either machined or further formed in the factory. Some factory formed parts, however, still have to be machined before they are assembled.

Within the broad groupings shown in Fig. 1.1 (*b*) lie a very large number of different processes. Some have origins which can be traced back to ancient times, while others are in a very early stage of development. Some are basic techniques which demand considerable experience and skill from those who perform them, while at the other end of the scale there are highly sophisticated processes, often automatically controlled.

The material specified for a part will of course influence the choice of process. Most materials can be shaped by a range of processes, some by a very limited range and others by a range wide enough to embrace most of the known processes. In any particular instance however, there is an optimum sequence of shaping processes. The main factors influencing this choice are the desired shape and size, the dimensional tolerances, the surface finish and the quantity required. The choice must not only be made on the grounds of technical suitability: cost is an important and frequently a paramount consideration. A diagram showing the interaction of factors affecting the choice of process for factory made parts is shown in Fig. 1.2.

Not only must the production engineer know a great deal about methods of materials shaping, but this knowledge must be shared by the designer. New shaping processes are being introduced and existing ones are being developed at such a rapid rate that no book of this type can claim to be completely up to date, nor can any engineer have knowledge in real depth other than in selected fields. A qualitative and partly quantitative account of as many shaping processes as possible has been included so that students entering industry will be able to see current practice as an integrated whole.

2 Manufacturing Properties of Metals

2.1 METAL FORMING PROCESSES

Methods of plastic deformation are used extensively to force metal into a required shape. The processes used are diverse in scale, varying from forging and rolling of ingots weighing several tons to drawing of wire less than 0·025 mm (0·001 in) in diameter. Most large-scale deformation processes are performed hot, so that a minimum of force is needed and the consequent recrystallization refines the metallic structure. Cold working

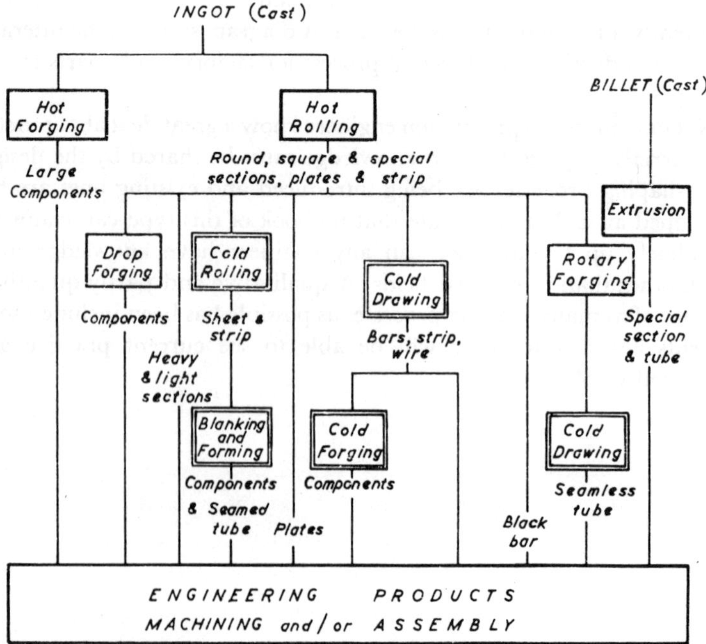

Fig. 2.1 Major metal forming processes: cold operations shown in double frame

is used when smooth surface finish and high dimensional accuracy are required. Although a growing number of components is manufactured completely from a series of deformation processes, metal forming is primarily used to produce such material as bar and sheet which is subsequently machined or pressed into its final shape. A chart showing the major metal-forming processes can be seen in Fig. 2.1.

2.2 YIELDING

To achieve permanent deformation, metal must be stressed beyond its elastic limit. A typical relationship between true stress and logarithmic strain for steel is shown in Fig. 2.2 and the initial yield stress is shown by point A.

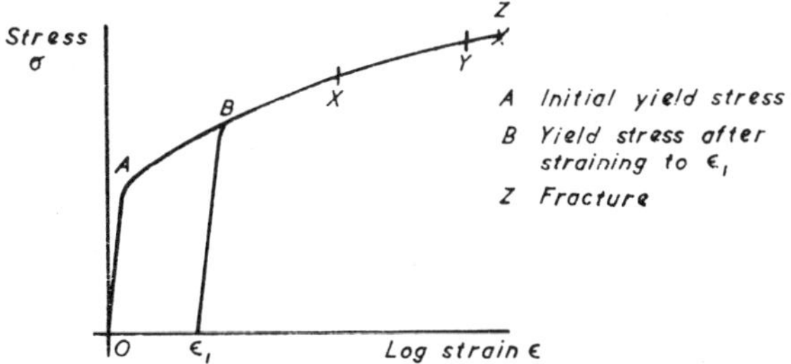

Fig. 2.2 Stress/strain curve for steel

Due to the considerable changes in shape occurring when metal is formed the logarithmic, true or natural strain $\int dl/l$ is preferred to the conventional strain $(l - l_0)/l_0$. The relationship between conventional and logarithmic strains is considered in Chapter 3.2.

The stress system in most metal forming operations is a complex one; hence a knowledge of the stress at which the metal fails in simple tension or compression is of little direct use. The analysis of three-dimensional stresses involves the consideration of three direct stresses and six pairs of shear stresses. In the simple treatment used in this book the stresses are resolved whenever possible into a system containing only three principal stresses. To determine the combination of direct stresses which produces yielding some generally applicable criterion is needed. Two criteria of yielding are commonly used, one proposed by Tresca and the other by von Mises; both are discussed in the next chapter.

2.3 FRACTURE

When metals are deformed below their recrystallization temperature they will work harden due to progressive deformation of the metallic structure making further deformation more difficult. This effect can be observed from the inclination of the stress/strain curve shown in Fig. 2.2. Apart from increasing the yield stress of a material, work hardening reduces its ductility and makes fracture more likely.

Most deforming operations are compressive; this enables the metal to withstand considerably larger strains before fracture than would be possible with tensile deformation. In fact, brittle materials such as cast iron, can be extruded like ductile ones if differential hydrostatic extrusion is used (see Section 6.5.6).

2.4 EFFECT OF TEMPERATURE IN METAL WORKING

Most large-scale processes of ingot and billet reduction and forming are performed at temperatures well above those at which recrystallization occurs. Hot working greatly reduces the yield strength during deformation, but to produce a satisfactory surface finish the product often has to be finished either by descaling and cold working, or by machining. Due to recrystallization, hot working is normally characterized by an absence of strain hardening; however, since the rate of recrystallization is temperature dependent, the working temperature should be sufficiently above the minimum necessary for recrystallization. The rate of straining is also important, for if it is too fast there will be insufficient time for the annealing effect of recrystallization; in fact, when hot worked metal is rapidly strained and then quickly cooled, it will strain harden. On the other hand if the rate of deformation is too slow there will be an undesirable weakening caused by grain growth.

2.5 CONCEPT OF RIGID-PLASTIC MATERIAL

It is convenient in metal working to consider that the material behaves in a rigid-plastic manner (Fig. 2.3). This concept neglects elastic strains as they are very small compared with the total plastic strain which occurs in metal working. The metal is therefore considered rigid up to the stress at which it yields; after yielding it is assumed that no additional stress is needed to increase strain, i.e. no work hardening occurs. This assumption of plastic behaviour is reasonable for hot working processes and it is a fair approximation for cold working when the material has already undergone

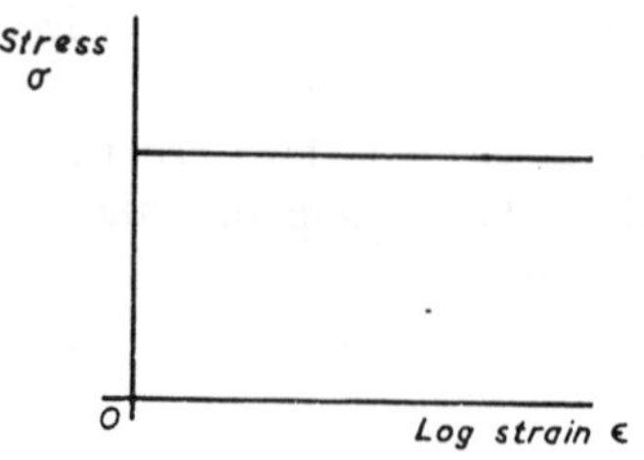

Fig. 2.3 Stress/strain relationship for rigid-plastic material

considerable work hardening and the slope of the stress/strain curve has flattened, (zone XY, Fig. 2.2).

2.6 EFFECT OF FRICTION BETWEEN WORK AND TOOL

In most cold working processes, the coefficient of friction between the plastically deforming material and its constraints is low and Coulomb friction applies. i.e. the frictional force is proportional to the normal force. However, in hot working, the coefficient of friction is high and the yield stress of the material is lower than that for cold working. In consequence the shear flow stress is often reached at the surface of the material and a thin layer of metal adheres to the container or tool. Under these conditions the frictional force is independent of the normal force but depends on the shear flow stress of the metal being formed (see Section 3.10.4).

2.7 EFFECT OF STRAIN RATE

The effect of rapid deformation on yield is as yet imperfectly understood. Strain rate effects in manufacture are inseparable from those due to temperature; in machining and high velocity forming processes there is little heat transfer due to conduction, and the increase in yield stress due to high strain rate is at least partly balanced by thermal softening.

With steel the net effect of strain rate and temperature appears to produce a large increase in the initial yield stress, but at high strains the dynamic increase in yield stress is much less. The resulting stress/strain curve thus indicates a lower rate of strain hardening and approaches that of a rigid plastic material.

Unfortunately, the strain rate and temperature dependence of metals makes accurate quantification of cutting forces from material data impossible.

2.8 HIGH VELOCITY DEFORMATION

Considerable development has occurred in high velocity processes for forming and blanking. Deformation speeds are in the order of 6–300 $\mathrm{m\,s^{-1}}$ (20–1000 ft/s), compared with conventional speeds of up to 2 $\mathrm{m\,s^{-1}}$ (6 ft/s). The main areas of development have been (a) billet forming, (b) blanking and cropping and (c) sheet forming.

The yield stress of steel falls appreciably when preheated above 300°C, thus permitting lower capacity, less expensive forming equipment to be used. However, at very high strain rates the preheat temperatures have to be substantially increased to achieve a similar reduction in yield point. Almost all of the work in high velocity forming reappears as heat in the workpiece and the resultant temperature rise can cause deterioration in metals with a narrow range of working temperatures, such as some of the high strength alloys. Other changes in material properties when subject to rapid deformation will be discussed when the processes are themselves described in later chapters.

2.9 CALCULATION OF DEFORMING LOADS

In the design of machines and tools it is important that the forces necessary to produce a given deformation are known. Most formulae used are derived from a consideration of stresses, work done or metal flow. Where these formulae do not agree with experimental results they often provide a basis for more accurate semi-empirical expressions.

In the uniaxial tensile test, deformation can be assumed to be homogeneous until necking commences. In homogeneous deformation each element keeps its geometrical form: plane sections remain plane and rectangular elements remain rectangular. For homogeneous deformation the applied load F is easily obtained from the expression $F = A \, . \, Y$ where A is the cross-sectional area and Y is the yield stress. The load required to produce plastic flow will vary as deformation proceeds, as both A and Y will change in value. Apart from the work necessary to produce homogeneous deformation, work is also needed to overcome friction and perform redundant work. Friction occurs between the flowing metal and a constraint: this constraint will be the die in wire drawing and extrusion, the rolls in rolling, the dies in forging, or the cutting tool in machining.

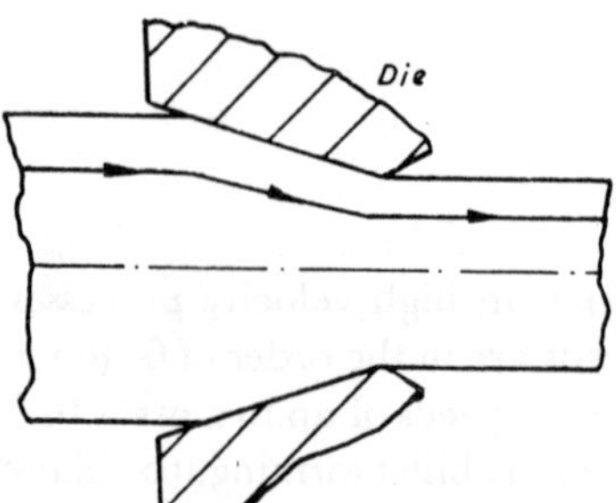

Fig. 2.4 Changes in direction of metal flow in drawing

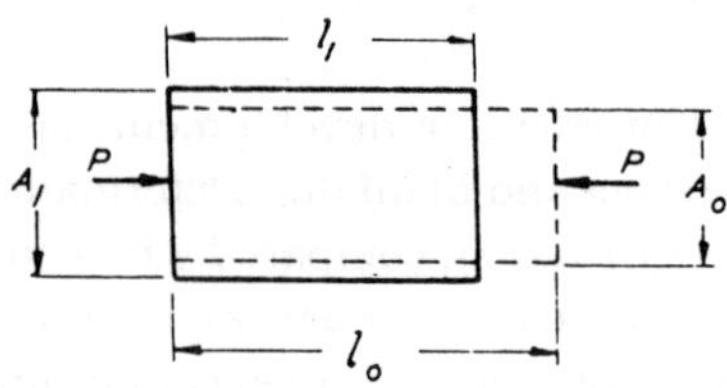

Fig. 2.5 Homogeneous deformation (compressive)

Redundant work is done whenever distortion departs from homogeneous deformation. In practice these departures almost invariably do occur, although their magnitude varies considerably. If wire drawing is considered (Fig. 2.4) it will be seen that, because of changes in direction of flow, the metal shears twice, once at the entry to the die and then again at the die exit. In this example the redundant work done on an element of unit volume is the product of the shear strain and the shear flow stress at the entry and exit of the die.

Summarizing, we may say that the work done in producing deformation is the work necessary to produce homogeneous deformation, plus the work done against friction, plus the redundant work.

2.10 WORK FORMULA METHOD

A specimen homogeneously deformed from length l_0 to length l_1 is shown in Fig. 2.5. At any point in the deformation, $F = YA$, where Y is the yield stress and A the cross-sectional area.

$$\text{Work done} = \int_{l_1}^{l_0} F\mathrm{d}l = \int_{l_1}^{l_0} YA\mathrm{d}l$$

But volume $V = Al$

$\therefore$ Work done/unit volume for homogeneous deformation

$$= \int_{l_1}^{l_0} Y\frac{\mathrm{d}l}{l}$$

For a work hardening material, Y increases with the amount of strain undergone, but for a non work hardening material Y is constant and can be taken outside the integral sign.

Hence work done/unit volume for homogeneous deformation

$$= Y\int_{l_1}^{l_0}\frac{\mathrm{d}l}{l} = Y\ln\left(\frac{l_0}{l_1}\right) = Y\ln\left(\frac{A_1}{A_0}\right) \text{ since } A_1l_1 = A_0l_0 \qquad (2.1)$$

Assuming that deformation is produced by a force F acting on an area A moving through a distance l, then the deforming pressure $p = F/A$.

Work done $= Fl = pAl$,

and work done/unit volume $= p$.

Therefore the term p is numerically equal to the work done/unit volume.

Semi-empirical methods of finding deforming forces often employ the work formula for homogeneous deformation with allowances for friction and redundant work being made by constants and efficiency factors.

2.11 STRESS EVALUATION METHOD

An alternative method of calculating deforming forces is to consider the equilibrium of forces acting on a small element of the deforming metal. By integrating the resulting expression between limits appropriate to the extent of the deformation, the applied stress can be obtained. Friction can be taken into account in this method but not redundant work.

2.12 METAL FLOW METHOD

The third and most comprehensive method of calculating deforming forces is by considering the metal flow in the deformation zone. If there is no metal flow in one of the three principal directions, i.e. plane strain conditions apply, then the behaviour of the plastic metal can be indicated by a slip-line field. Slip-line fields are patterns of orthogonal lines which indicate the direction of maximum shear stress in the deforming metal. Starting from a point at which the stress conditions are known, it is possible by means of the slip-line field to find stresses elsewhere in the plastic zone, and hence the deforming force. Although this method is a tedious one and normally applies only to non work hardening materials the results obtained agree fairly well with experiment. A simplification of the slip-line approach, known as the upper bound method, uses a less complex field and enables deforming forces to be estimated graphically.

3 Basic Plasticity

3.1 It is intended that this chapter should indicate how metal behaves when plastically deformed and so provide a quantitative basis for dealing with metal cutting and forming in the rest of the book. A rigorous mathematical treatment would be inappropriate and space does not permit a complete discussion of this interesting subject. Those seeking a more comprehensive treatment are referred to the bibliography.

3.2 PLASTIC STRAINS

When working below the elastic limit the conventional strain is used, where $e = (l - l_0)/l_0$, l_0 is the original length being considered, and l is the strained length. The much larger strains found in plastic deformation can be more realistically expressed using the logarithmic strain $\varepsilon = \int dl/l$. The greater realism of the logarithmic strain may be appreciated if consideration is given to specimens which are either stretched to double their length or compressed to half their length. Although opposite in direction these strains can be considered to be equivalent deformations. However, only the values of the logarithmic strain indicate this to be so; in fact, to obtain the equivalent conventional compressive strain to a $2l$ extension, the specimen would have to be compressed to zero thickness.

Deformation of specimen of length l	Conventional strain e	Logarithmic strain ε
Extension to $2l$	1	0·693
Compression to $\frac{1}{2}l$	$-\frac{1}{2}$	$-0·693$
Compression to zero thickness	-1	$-\infty$

The relationship between the conventional and logarithmic strains can be obtained by considering the definitions of each.

$$\mathrm{e} = \frac{l - l_0}{l_0} = \frac{l}{l_0} - 1 \qquad \therefore \quad \frac{l}{l_0} = \mathrm{e} + 1$$

$$\varepsilon = \int_{l_0}^{l} \frac{\mathrm{d}l}{l} = \ln\left(\frac{l}{l_0}\right)$$

Hence, $\varepsilon = \ln(\mathrm{e} + 1)$ (3.1)

Because it is frequently easier to measure conventional strain, the logarithmic strain is often derived from it. Apart from being more realistic, logarithmic strains can be added, an operation which is not possible with conventional strains.

$$\varepsilon_{03} = \varepsilon_{01} + \varepsilon_{12} + \varepsilon_{23} = \ln\left(\frac{l_1}{l_0}\right) + \ln\left(\frac{l_2}{l_1}\right) + \ln\left(\frac{l_3}{l_2}\right)$$

$$\varepsilon_{03} = \ln\left(\frac{l_3}{l_0}\right)$$

It is normal in metalworking to ignore the comparatively small elastic strains and to assume that the volume of the material remains unchanged when subjected to plastic strain.

3.3 YIELDING

If the mechanism of metal forming and machining is to be understood, it is necessary to try to predict the stresses and strains occurring at any point in the material. Since both forming and machining produce plastic flow of the material, the stress relationship causing flow is required.

Normally the stress system producing yielding is a complex three-dimensional one containing direct and shear stresses. Consequently a yielding criterion which is of general application and will hold irrespective of the way in which the stresses act is needed. With such a criterion it is possible to associate a complex stress system with an equivalent uniaxial tensile or compressive stress. Two criteria which provide reasonable results for ductile materials have been proposed, one by Tresca, the other by von Mises. They are also sometimes attributed to Guest and Maxwell respectively.

3.3.1 Tresca's criterion of yielding. Tresca proposed that yielding occurs at a point in a material when the maximum shear stress at that point reaches a critical value measurable as the shear stress at which yielding occurs in a uniaxial tensile test.

The tensile stress Y (Fig. 3.1 (a)) is called the equivalent or effective stress and is the value in simple tension which causes yielding. Considering the three-dimensional case shown in Fig. 3.1 (b), the maximum shear stress is caused by the difference between σ_1 and σ_3, the largest and smallest direct stresses acting on the principal planes. The effective stress is then equal to $\sigma_1 - \sigma_3$. It will be noted that, according to Tresca, the intermediate principal stress σ_2 does not affect the maximum shear stress.

Hence, $\tau = \dfrac{Y}{2} = \dfrac{\sigma_1 - \sigma_3}{2}$

$$Y = \sigma_1 - \sigma_3 \tag{3.2}$$

As this criterion ignores the effect of the intermediate principal stress there is some consequent loss of accuracy. However, it is mathematically

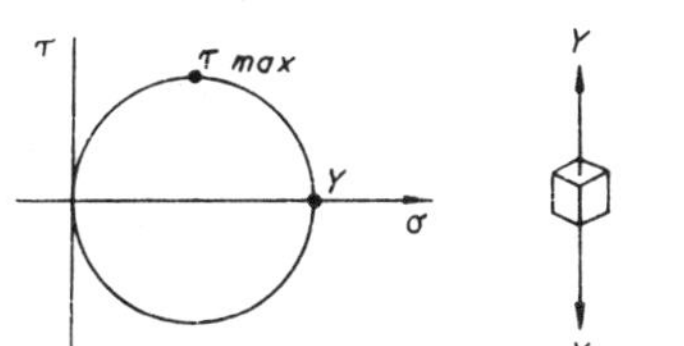

Fig. 3.1 (a) Mohr stress circle for simple tension at yield

Fig. 3.1 (b) Mohr stress for 3-dimensional tensile stress system at yield

simple and is frequently used in its original form or in a modified form as discussed later.

3.3.2 Von Mises' criterion of yielding. This criterion states that when yielding occurs,

$$(\sigma_1 - \sigma_2)^2 + (\sigma_2 - \sigma_3)^2 + (\sigma_3 - \sigma_1)^2 = \text{constant.}$$

It can be shown that this is proportional to the shear stress on the octahedral planes (Fig. 3.2), where

$$\tau_{OCT} = \tfrac{1}{3}[(\sigma_1 - \sigma_2)^2 + (\sigma_2 - \sigma_3)^2 + (\sigma_3 - \sigma_1)^2]^{\frac{1}{2}}$$

A more useful interpretation for metal working calculations is that it is proportional to the shear strain energy, where shear strain energy/unit volume $= 1/12G[(\sigma_1 - \sigma_2)^2 + (\sigma_2 - \sigma_3)^2 + (\sigma_3 - \sigma_1)^2]^{\frac{1}{2}}$ and G is the modulus of rigidity.

Substituting for Y in the case of uniaxial tension, where

$$Y = \sigma_1 \text{ and } \sigma_2 = \sigma_3 = 0,$$

$$Y = \frac{1}{\sqrt{2}} [(\sigma_1 - \sigma_2)^2 + (\sigma_2 - \sigma_3)^2 + (\sigma_3 - \sigma_1)^2]^{\frac{1}{2}} \qquad (3.3)$$

Von Mises' criterion, which takes account of all three principal stresses, gives closer agreement with experimental results than that of Tresca. Fig. 3.1 (*b*) shows that the terms $(\sigma_1 - \sigma_2)$, $(\sigma_2 - \sigma_3)$ and $(\sigma_3 - \sigma_1)$ are each equal to twice the maximum shear stresses in the three Mohr circles.

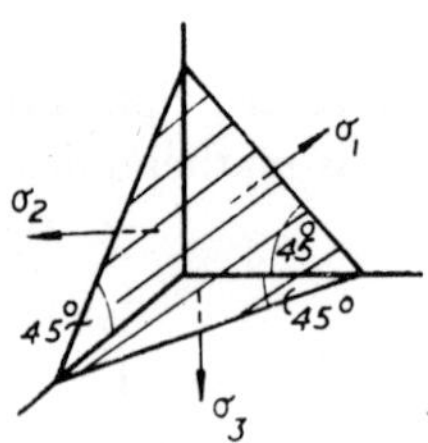

Fig. 3.2 Octahedral plane shown shaded

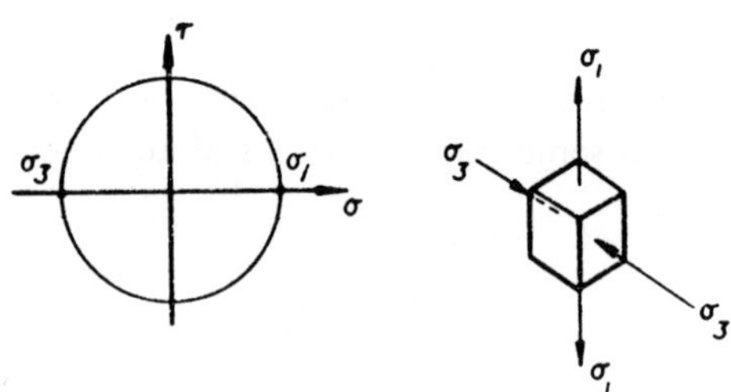

Fig. 3.3 Mohr stress circle for pure shear

From this it is possible to conclude that the mean or hydrostatic stress p does not affect yielding behaviour. In other words, the essential point governing yielding is stress difference and not the absolute value of the stresses.

3.4 COMPARISON OF TRESCA AND VON MISES CRITERIA

3.4.1 Yielding in simple tension. It can be shown that $Y = \sigma_1$ irrespective of the criterion which is used. This is done by simply substituting zero for σ_2 and σ_3 in the equivalent stress formulae.

3.4.2 Yielding in pure shear. The Mohr stress circle for pure shear is shown in Fig. 3.3. The shear stress at which the metal yields is known as the shear flow stress and is denoted by k. In pure shear, $\sigma_2 = 0$ and $\sigma_1 = -\sigma_3$.

Using the Tresca criterion,

$$Y = \sigma_1 - \sigma_3 = 2\sigma_1,$$

and as $\sigma_1 = k$,

$$Y = 2k \qquad (3.4)$$

Using the von Mises' criterion,

$$Y = \frac{1}{\sqrt{(2)}}(6\sigma_1^2)^{\frac{1}{2}} = \sqrt{(3)}\sigma_1$$

and as $\sigma_1 = k$,

$$Y = \sqrt{(3)}k \qquad (3.5)$$

In pure shear it will be seen that the Tresca criterion predicts a higher yield stress than that of von Mises by a factor of $2/\sqrt{3}$, i.e. 15·5% greater.

3.5 EQUIVALENT STRAIN

Because of the large strains common in plastic deformation and the non-linearity of the stress/strain curve, small strain increments ($\delta\varepsilon$) or strain rates ($\dot{\varepsilon}$) are used in analysis rather than absolute strains. The equivalent strain increment $\delta\bar{\varepsilon}$ can be defined in terms of the principal strain increments for the two criteria of yielding. It can be shown that, using Tresca's criterion,

$$\delta\bar{\varepsilon} = \tfrac{2}{3}(\delta\varepsilon_1 - \delta\varepsilon_3) \qquad (3.6)$$

and using von Mises' criterion,

$$\delta\bar{\varepsilon} = \sqrt{[2/9]}[(\delta\varepsilon_1 - \delta\varepsilon_2)^2 + (\delta\varepsilon_2 - \delta\varepsilon_3)^2 + (\delta\varepsilon_3 - \delta\varepsilon_1)^2]^{\frac{1}{2}} \qquad (3.7)$$

3.5.1 Equivalent strain in simple tension. If it is assumed that there is no change in volume when metal is plastically deformed,

$$\delta\varepsilon_1 + \delta\varepsilon_2 + \delta\varepsilon_3 = 0$$

and

$$\delta\varepsilon_1 = -(\delta\varepsilon_2 + \delta\varepsilon_3),$$

but in simple tension

$$\delta\varepsilon_2 = \delta\varepsilon_3$$

$$\therefore \qquad \delta\varepsilon_1 = -2\delta\varepsilon_2 = -2\delta\varepsilon_3$$

If, in the Tresca and von Mises equations for equivalent strain increment, $\delta\varepsilon_2$ and $\delta\varepsilon_3$ are replaced by the appropriate value of $\delta\varepsilon_1$, both equations show that $\delta\bar{\varepsilon} = \delta\varepsilon_1$.

3.5.2 Equivalent strain in pure shear. In pure shear it will be seen from Fig. 3.4 that $\delta\varepsilon_3 = -\delta\varepsilon_1$ and $\delta\varepsilon_2 = 0$, i.e. the Mohr strain circle is centred on zero. Substitution for $\delta\varepsilon_2$ and $\delta\varepsilon_3$ in the equivalent strain increment equations gives,

for Tresca $\delta\bar{\varepsilon} = \frac{4}{3}\,\delta\varepsilon_1$

for von Mises $\delta\bar{\varepsilon} = \frac{2}{\sqrt{3}}\,\delta\varepsilon_1$

Thus the equivalent stress and the equivalent strain increment obtained using the Tresca equations are both greater than those obtained using von Mises' equations by a factor of $2/\sqrt{3}$ for pure shear.

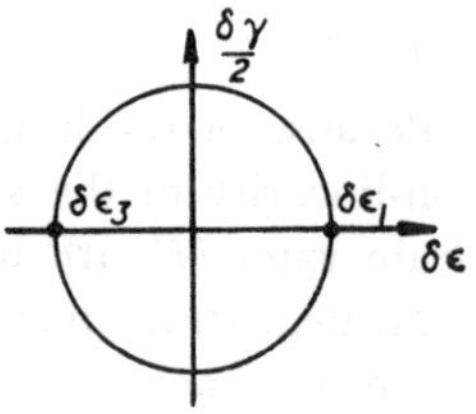

Fig. 3.4 Mohr strain circle for pure shear

3.6 WORK DONE IN PURE SHEAR

The work done in a small increment of plastic flow can be expressed as

$$\delta w = \sigma_1\delta\varepsilon_1 + \sigma_2\delta\varepsilon_2 + \sigma_3\delta\varepsilon_3$$

but in pure shear

$$\sigma_1\delta\varepsilon_1 = \sigma_3\delta\varepsilon_3 \text{ and } \sigma_2\delta\varepsilon_2 = 0$$

$$\therefore \qquad \delta w = 2\sigma_1\delta\varepsilon_1$$

Also, in terms of equivalent values,

$$\delta w = Y\delta\bar{\varepsilon}$$

Substituting the values of equivalent stress and strain obtained from the Tresca equations,

$$\delta w = 2\sigma_1 \times \tfrac{4}{3}\delta\varepsilon_1 = \tfrac{8}{3}\sigma_1\delta\varepsilon_1$$

and from the von Mises' equations,

$$\delta w = \sqrt{3}\sigma_1 \times \frac{2}{\sqrt{3}}\delta\varepsilon_1 = 2\sigma_1\delta\varepsilon_1$$

Von Mises' criterion accurately predicts the work done, whereas Tresca's values for both stress and strain are too great by a factor of $2/\sqrt{3}$ (15·5%) and in consequence the work done is over-estimated by a factor of 4/3 (33·3%).

3.6.1 Modified Tresca criterion. When convenience demands the use of the Tresca equations it is usual to divide the resultant stresses and strains by a factor m, where $1 \leqslant m \leqslant 1{\cdot}155$. This makes the results obtained from the Tresca equations closer to those obtained when using those of von Mises.

3.7 PLANE STRAIN

Fortunately many processes of deformation, such as rolling, forging between flat platens and orthogonal machining, involve negligible strain in one direction, reducing them to two-dimensional problems. This enables a simplified approach to be adopted and allows a field theory to

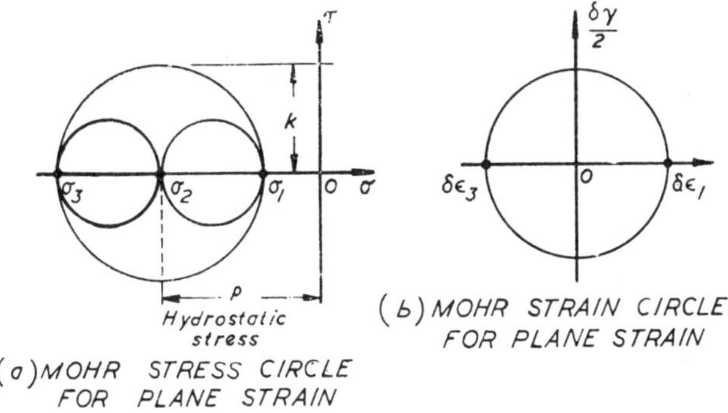

Fig. 3.5

be developed which will account for variations in stress and strain throughout the plastic zone in the metal. Before developing the theory, a number of simplifying assumptions must be made. Elastic strains and work hardening are ignored, i.e. the material is assumed to be rigid-plastic. The material is also assumed to be homogeneous and isotropic, that is, the properties are the same at all points and in all directions.

If the change in shape of an element subjected to plane strain conditions is considered, then by definition of plane strain, there will be no strain in the direction of one of the principal stresses. Assuming constant volume, the change in dimensions in the other two principal stress directions will be equal in magnitude but opposite in sign. That is, if $\delta\varepsilon_2 = 0$ and $\delta\varepsilon_1 + \delta\varepsilon_2 + \delta\varepsilon_3 = 0$, then $\delta\varepsilon_1 = -\delta\varepsilon_3$.

The Mohr strain circle for plane strain (Fig. 3.5) is thus identical to that for pure shear, both being centred on zero. It follows that plane strain deformation is caused by pure shear stress, and metal flow will occur when the maximum shear stress reaches a magnitude k, the shear flow stress. In plane strain, however, a hydrostatic stress p exists, so the centre of the Mohr stress circle is displaced from the origin by an amount equal to p. The hydrostatic stress is the mean of the principal stresses, i.e. $p = (\sigma_1 + \sigma_2 + \sigma_3)/3$. Physically, this means that the stress acting in the direction of zero strain σ_2 and on the planes of maximum shear stress is equal to p in each case.

3.8 SLIP-LINE FIELDS

The variations of stress and strain at points in the plastic zone of material subjected to plane strain conditions can be found by the construction of slip-line fields. The slip lines are lines of maximum shear stress and, since one of these is always accompanied by another at 90° to it, the slip-line field forms a network of straight or curved lines crossing each other at right angles.

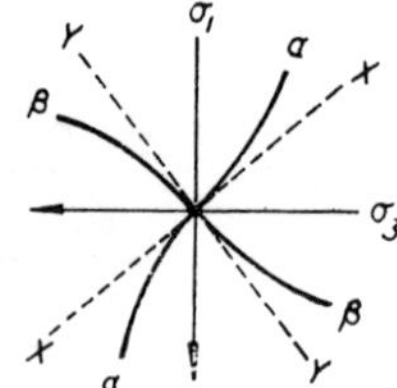

Fig. 3.6 Convention for representing slip lines

One set of slip lines is referred to as α lines and those crossing them as β lines. If the tangent to an α slip line at a point of intersection is considered as an X axis and that to the β line a Y axis, then by convention the direction of the algebraically greatest principal stress σ_1 acting at the intersection should pass through the first and third quadrants (Fig. 3.6). The principal planes are of course at 45° to the planes of maximum shear, and the direction of the principal stresses at the intersection will be 45° to the tangents to the slip lines.

3.9 HENCKY'S EQUATIONS

Although with a rigid-plastic material the shear stress remains constant along a slip line at its flow value k, the hydrostatic stress, and hence the total traction at any point, can vary considerably. Hencky derived equations for calculating the variation of hydrostatic stress at points along a slip line.

Consider a small curvilinear element bounded by slip lines, as shown in Fig. 3.7. The normal stresses, which are also hydrostatic, increase as shown by amounts $(\partial p/\partial \alpha)\delta\alpha$ and $(\partial p/\partial \beta)\delta\beta$. The two curved α lines can be assumed, for a small element, to be concentric circular arcs of radii r_α and $(r_\alpha - \delta\beta)$. If the slip lines turn through an angle $\delta\phi$

$$\delta\alpha = r_\alpha \delta\phi$$

$$\delta\phi = \frac{\delta\alpha}{r_\alpha}$$

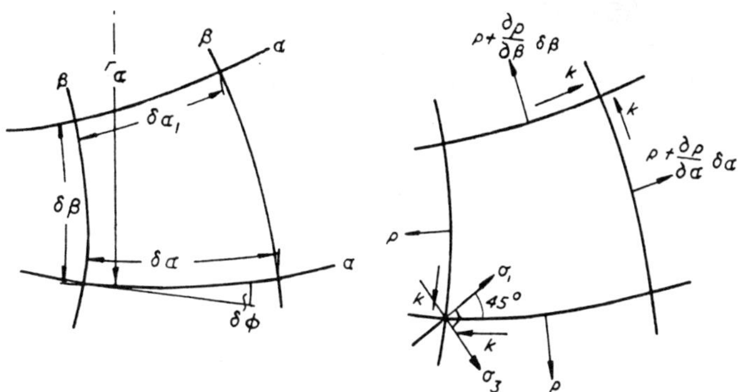

Fig. 3.7 Curvilinear element in slip-line field

Similarly, $$\delta\alpha_1 = (r_\alpha - \delta\beta)\delta\phi$$

Substituting for $\delta\phi$,

$$\delta\alpha_1 = \delta\alpha\left(1 - \frac{\delta\beta}{r_\alpha}\right)$$

Taking moments about the centre of curvature,

$$r_\alpha k\delta\alpha + p\left(r_\alpha - \frac{\delta\beta}{2}\right)\delta\beta = (r_\alpha - \delta\beta)k\delta\alpha\left(1 - \frac{\delta\beta}{r_\alpha}\right) + \left(p + \frac{\partial p}{\partial\alpha}\delta\alpha\right)\left(r_\alpha - \frac{\delta\beta}{2}\right)\delta\beta$$

Ignoring products of infinitesimals,

$$\frac{\partial p}{\partial\alpha} - \frac{2k}{r_\alpha} = 0$$

Substituting $\delta\alpha/\delta\phi$ for r_α and integrating,

$$p - 2k\phi = \text{constant along an } \alpha \text{ line} \tag{3.8}$$

Using a similar approach for the β lines, we obtain the equation,

$$p + 2k\phi = \text{constant along a } \beta \text{ line} \tag{3.9}$$

Knowing the angle through which the slip lines have turned, the change in p can be found. It should be noted that p is considered positive in tension. In the more usual case of a compressive hydrostatic stress, p will be a negative quantity. The slip-line rotation is considered positive when the angle increases in an anti-clockwise direction.

3.9.1 Modified Hencky equations. A modified set of equations was derived by Christopherson, Oxley and Palmer[1] to allow for the effect of work hardening as material passes through the slip-line field. These equations can be derived by considering a diagram similar to Fig. 3.7, but allowing for an increase in shear flow stress as well as hydrostatic stress. The modified equations are

$$p - 2k\phi + \int \frac{\partial k}{\partial \beta}\, d\alpha = \text{constant along an } \alpha \text{ line} \tag{3.10}$$

$$p + 2k\phi + \int \frac{\partial k}{\partial \alpha}\, d\beta = \text{constant along a } \beta \text{ line} \tag{3.11}$$

Although the modified equations allow a greater freedom in plotting a slip-line field by permitting lines of the same family to curve in opposing directions, so far their only application has been to the machining problem.

3.10 SLIP LINES AT METAL SURFACES

3.10.1 Free surfaces. In some forming operations and in machining operations the plastic zone extends to a free surface. As there is no normal

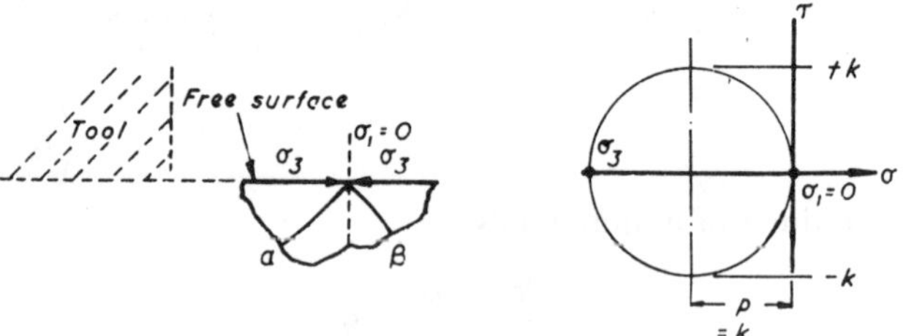

Fig. 3.8 Slip lines at free surface

force on the free surface, this surface is a principal plane, and slip lines must meet it at 45° (Fig. 3.8). The α and β slip lines comply with the convention already described.

3.10.2 Frictionless interface with tool or constraint. If the surface of the tool or constraint is well lubricated there can be no shearing force at the tool surface, and hence this interface is one of the principal planes. From Fig. 3.9 it is seen that k is not now equal to the hydrostatic stress, as the principal stress σ_1 is no longer zero. The slip lines again meet the interface at 45°.

3.10.3 Coulomb friction at interface. In this instance it is assumed that there is a normal stress q exerted by the tool or container on the work, and that the frictional stress at the interface is μq (Fig. 3.10). This frictional stress is balanced by rotating the slip lines from their 45° positions so that the resultant forces on the slip lines oppose the friction

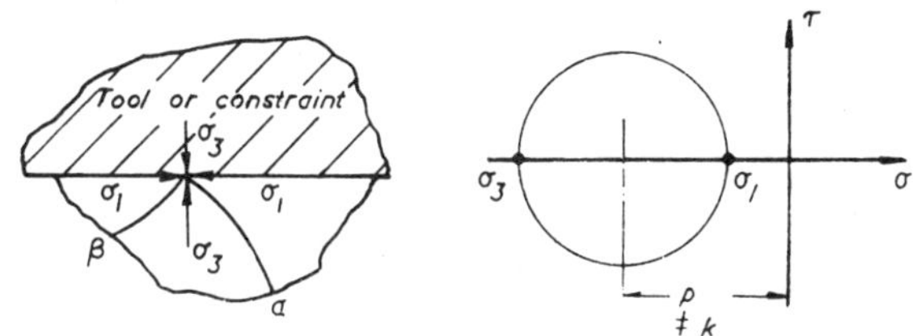

Fig. 3.9 Slip lines at tool face without friction

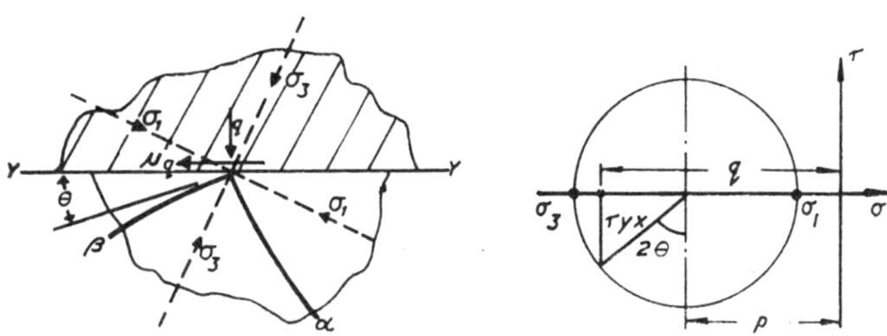

Fig. 3.10 Slip lines at tool surface with Coulomb friction

force. If the plane of the interface YY is inclined at an angle θ to the β slip line, then the shear stress on this plane is found by setting off a radius at angle 2θ from the vertical in the Mohr stress circle.

$$\tau_{yx} = k \cos 2\theta$$

But for equilibrium,

$$\mu q = k \cos 2\theta$$

$$\therefore \qquad \theta = \tfrac{1}{2} \cos^{-1} \frac{\mu q}{k} \qquad (3.12)$$

3.10.4 Sticking friction at interface. During hot working, and in some metal cutting and cold working processes, the friction between work and tool is often so high that the metal sticks to the tool and yielding occurs in the metal just below the interface. For the metal to yield, the tangential stress at the interface must reach the shear flow stress k. The slip lines now rotate so that they are tangential and normal to the interface (Fig. 3.11). In the general case,

$$\tau_{yx} = k \cos 2\theta$$

But for yielding

$$\tau_{yx} = k$$

$\therefore \qquad \cos 2\theta = 1 \qquad$ i.e. $\theta = 0°$ when sticking occurs.

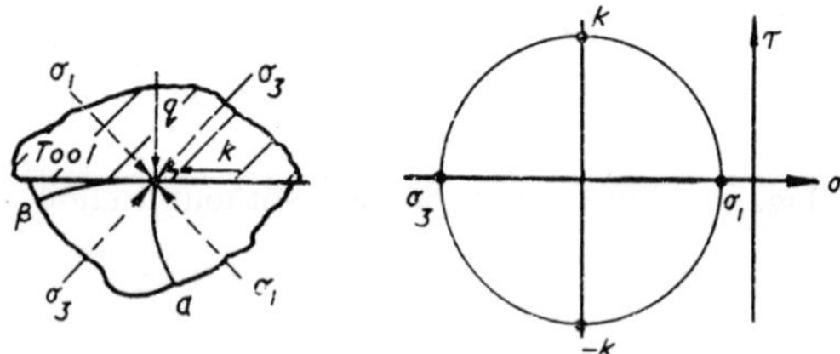

Fig. 3.11 Slip lines at tool surface with sticking friction

It is of interest to note that in metal forming operations with metals observing the von Mises' yield criteria, the value of μ is limited to 0·577 in the plastic zone. With a lubricant present it can be assumed that for almost all metal working operations the tangential stress τ at the interface is proportional to the normal stress q. The value of τ, however, cannot exceed the shear flow stress (k) of the work piece material itself. The normal stress which causes plastic flow is Y, the unaxial yield stress.

Hence for full sticking friction

$$\mu_{max} = k/Y$$

But $\qquad Y = \sqrt{3}k \quad$ von Mises (equation 3.5)

$\therefore \qquad \mu_{max} = 1/\sqrt{3} = 0{\cdot}577$

3.10.5 Singular points. These are points at the surface of the deforming metal where the normal to the surface cannot be specified; they occur, for example at the tip of an extrusion die such as point P in Fig. 3.12. Slip lines can emanate from singular points in any direction into the metal, and these are frequently the centre of a fan of slip lines.

3.10.6 Plane of symmetry. Where a symmetrical operation such as rolling or symmetrical strip extrusion is considered, the slip-line field is identical on each side of the centre line.

The reflexion of the slip lines at the plane of symmetry and the fact that they cross each other at 90° means that they must meet a plane of symmetry at 45° (Fig. 3.12).

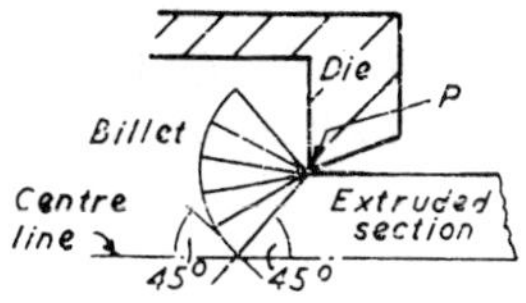

Fig. 3.12 Fan of slip lines emanating from singular point

3.11 VELOCITY DISCONTINUITIES

Metal deformation occurs either as a progressive flow or by block slippages which are produced by velocity discontinuities. At these discontinuities shearing occurs along the slip line, the metal on either side of the slip line having a different velocity. In practice, such discontinuities are mechanically impossible and they consist of very narrow bands of intense shear strain. In some types of deformation the resultant block slippages can be clearly observed, the metal either side of the velocity discontinuity moving as a rigid body.

Unless the metal is to pile up or voids are to occur, the velocity of the metal at any point normal to the slip line must be constant on both sides of the line, so the change in velocity can only be along the slip line. Since shearing does not cause any change of dimension, a velocity discontinuity must be of constant value along a slip line. Where a slip line which is a velocity discontinuity meets an axis of symmetry, the velocity discontinuity will be reflected.

3.11.1 Metal flow inside a slip-line field. Consider two adjacent points A and B along a curved slip line (Fig. 3.13 (*a*)). In order that there shall be no change in length between A and B, this short length can be

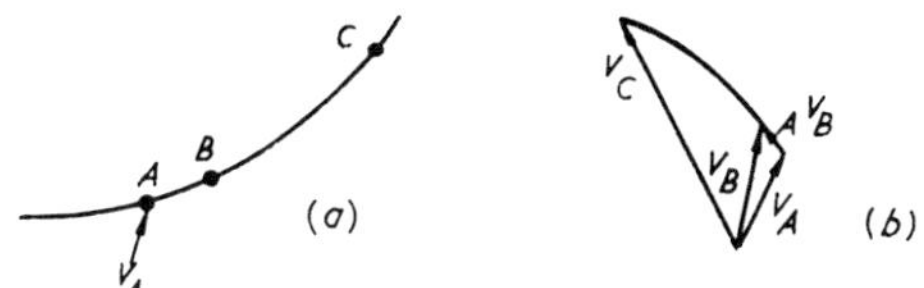

Fig. 3.13 Variation of velocity along a slip line

considered as a rigid link. Then the velocity of B relative to A must be at right angles to AB. Extending this approach to a large number of small steps, it follows that the locus of the end points of the velocities between A and C will be represented by the line in Fig. 3.13 (*b*) which is drawn normal to AC in Fig. 3.13 (*a*).

It is thus possible to construct a velocity diagram or hodograph for any slip-line field, and from the hodograph it is possible to ascertain whether the velocities implied are compatible with the boundary conditions.

3.12 CONSTRUCTION OF SLIP-LINE FIELDS

There is seldom a unique slip-line field solution to any problem, but it is first necessary to propose a field which satisfies the stress conditions at the boundaries. This is then checked by means of a hodograph to see if the velocity conditions at the boundaries are also satisfied. The best field from the boundary conditions can then be selected.

Frequently it is possible to specify some part of a slip-line field, such as a fan of slip lines, and then it is necessary to extend the field so as to meet a boundary. A useful device for doing this employs Hencky's first theorem, which states that any two slip lines of the same family will turn through the same angle when measured from their intersections with two slip lines of the other family.

This is simply proved by applying the Hencky equations to the four points A, B, C and D in Fig. 3.14.

Along α lines $p_A - 2k\phi_A = p_B - 2k\phi_B$

$$p_C - 2k\phi_C = p_D - 2k\phi_D$$

Along β lines $p_C + 2k\phi_C = p_A + 2k\phi_A$

$$p_D + 2k\phi_D = p_B + 2k\phi_B$$

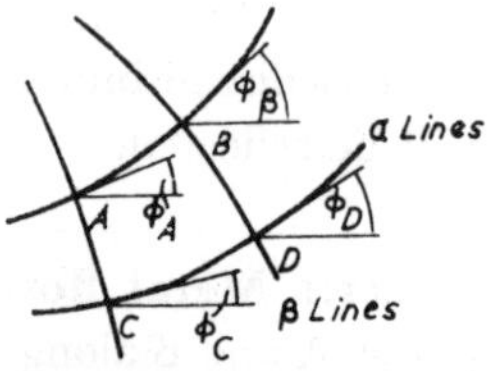

Fig. 3.14 Rotation of slip lines

Substituting for p_A,

$$p_B + 2k(\phi_A - \phi_B) = p_C + 2k(\phi_C - \phi_A)$$

Substituting for p_D

$$p_C + 2k(\phi_D - \phi_C) = p_B + 2k(\phi_B - \phi_D)$$

Subtracting,

$$\phi_A - \phi_B = \phi_C - \phi_D.$$

3.12.1 2:1 plane extrusion. The slip-line field for a perfectly lubricated 2:1 extrusion of flat wide strip is shown in Fig. 3.15. The field consists of two 90° fans radiating from the mouth of the die; one half of the field only has been shown as it is symmetrical about the horizontal centre line. The triangular portion at the top right of the field is a dead metal zone of stationary metal.

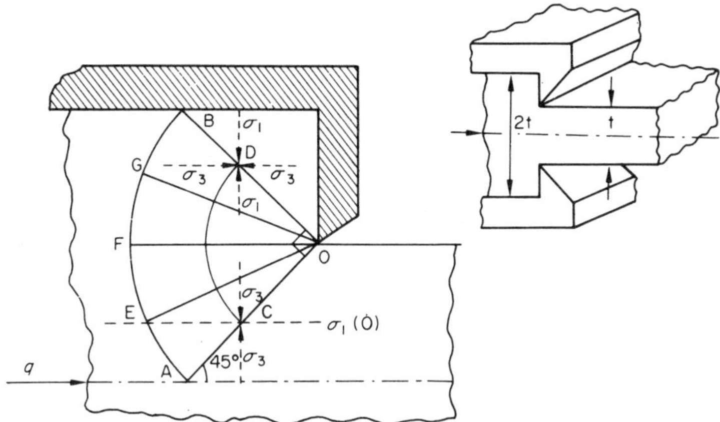

Fig. 3.15 Slip-line field of 2:1 plane perfectly lubricated extrusion

Firstly it is necessary to decide which are the α and which are the β slip lines. This is done by finding a field boundary where the magnitudes of the principal stresses are known. In this instance we consider the principal stresses acting at exit slip line OA. Line OA is at 45°, therefore one principal stress is horizontal and the other vertical; it can be assumed that the extrusion is stress free in the horizontal direction but in the vertical direction there is a compressive stress. Hence σ_1, the algebraically greatest principal stress, acts horizontally and σ_3, the algebraically least principal stress, acts vertically. Applying the convention for deciding which are α and β lines to point C on OA we find that the circular slip line CD is an α line and the radial slip line OA is a β line.

The appropriate Hencky equation can now be selected to find the variation in hydrostatic stress along CD. The Mohr stress circle for slip line OCA is shown in Fig. 3.16 and it will be seen that the value of hydrostatic stress at C (p_C) is equal to $-k$. For an α slip line

$$p - 2k\phi = \text{constant}$$

At C, putting $\phi = 0$

$$p_C = -k$$

As slip line OCA is a straight line, the hydrostatic stress at any point along it is equal to $-k$.

The rotation of the slip line CD from C to D is $\pi/2$ clockwise, clockwise rotation is by convention negative, hence the value of p_D can now be found.

$$p_D - 2k[-\pi/2] = -k$$
$$p_D = -k(1 + \pi) = -4{\cdot}14k$$

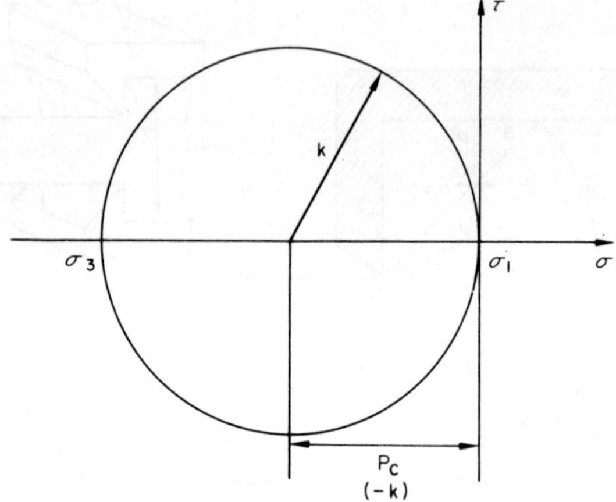

Fig. 3.16 Mohr stress circle for slip line OCA

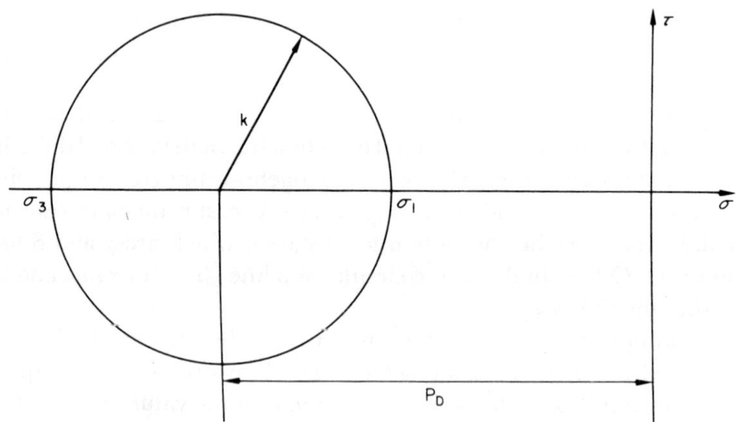

Fig. 3.17 Mohr stress circle for slip line ODB

The Mohr stress circle for point D, and the whole of slip line ODB, is shown in Fig. 3.17. Due to the 90° rotation of the slip-line field σ_3 is now acting horizontally at slip line ODB and σ_1 is acting vertically. The Mohr stress circle is now centred at $\sigma = -4{\cdot}14\,k$ and as its radius is k the value of σ_3 is $-5{\cdot}14\,k$.

The horizontal stress $(-5{\cdot}14\,k)$ is transmitted to the end of the die, through the dead metal zone, producing a force equal to $5{\cdot}14\,k \times$ (projected area of die wall).

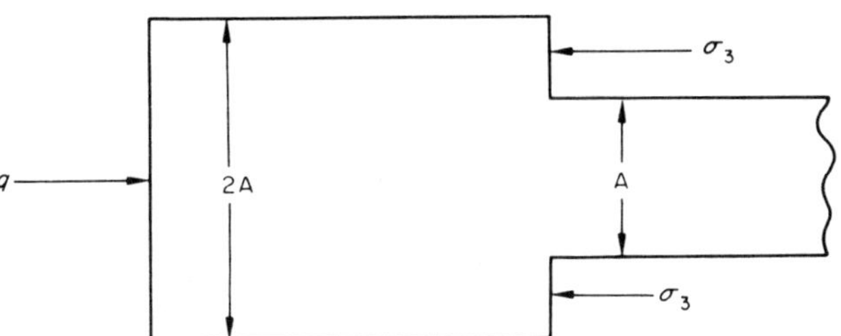

Fig. 3.18 Free body diagram for 2:1 plane extrusion

The stress q exerted by the ram on the billet can be found by considering the free body diagram in Fig. 3.18 and resolving forces horizontally.

$$q \times 2A = 2(5{\cdot}14\,k \times A/2)$$
$$q = 2{\cdot}57\,k$$

The hodograph for a 2:1 plane extrusion can be drawn as follows. All metal to the left of AEFGB, Fig. 3.15, moves horizontally as a rigid body with velocity u, the speed of the ram. It then crosses the velocity discontinuity associated with boundary slip line AEFGB. Consider a particle of metal near the container wall and just below B. First it undergoes sudden shearing in a direction tangential to the curved slip line, i.e. at 45° to the horizontal. It is then constrained to move parallel to the dead metal zone and its absolute velocity can be represented by vector XB_B, Fig. 3.19

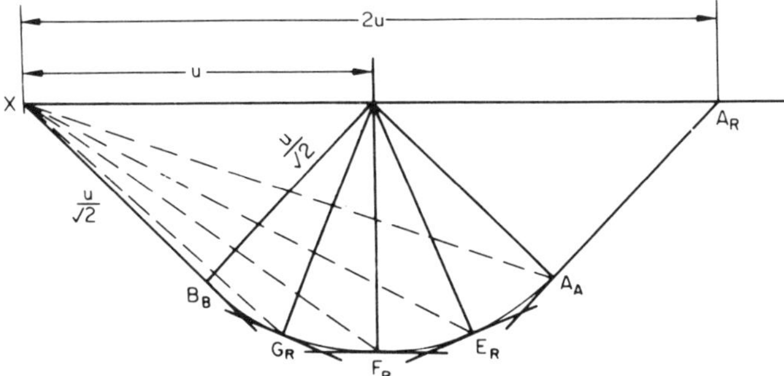

Fig. 3.19 Hodograph for 2:1 plane extrusion

(suffix B indicates metal just below point B). The velocity discontinuity AEFGB is of constant magnitude $u/\sqrt{2}$, however its direction will depend on the tangent to the discontinuity at the point of crossing. By considering metal at the points just to the right of G, F and E and just above A the hodograph can be extended by adding points G_R, F_R, E_R and A_A. The hodograph is completed by considering metal emerging from the plastic zone just above A and crossing the velocity discontinuity AO. This discontinuity is inclined at 45° to the horizontal and the velocity of the rigid metal after leaving AO is horizontal. Point A_R on the hodograph can now be found and $XA_R = 2u$, showing that the slip-line field chosen is compatible with a 2:1 extrusion ratio.

3.12.2 4:1 plane extrusion. The application of a slip-line field to lubricated strip extrusion with square-ended dies giving an extrusion ratio of 4:1 will be considered. This is illustrated in Fig. 3.20. As O is a singular point, it is permitted to draw a fan of slip lines centred on O. For clarity the angle between the radial lines has been made 22½°, although this leads to considerable inaccuracy and a 5° fan would normally be used. At the centre line, being an axis of symmetry, the slip lines will make angles of 45°. To extend the fan, we can consider the point at which the continuation of line OG meets the centre line at 45°. This line will have then turned through 22½°, as will the slip line of the other family at this point. From Hencky's

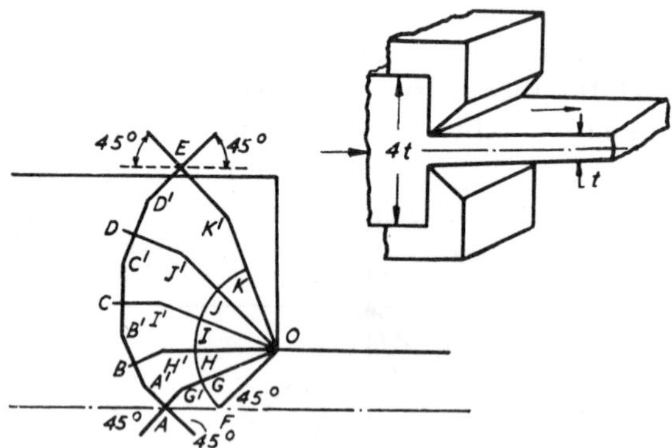

Fig. 3.20 First attempt at slip-line field for 4:1 extrusion ratio

first theorem, the other radial lines will have also turned through the same angle at their point of intersection with the slip line of the other family.

The point A is selected so that $GG^1 = AG^1$, and the point B is similarly selected so that $AA^1 = A^1B$, and $HH^1 = BH^1$. In this way the field can be extended as far as the point E. In this case E does not exactly coincide

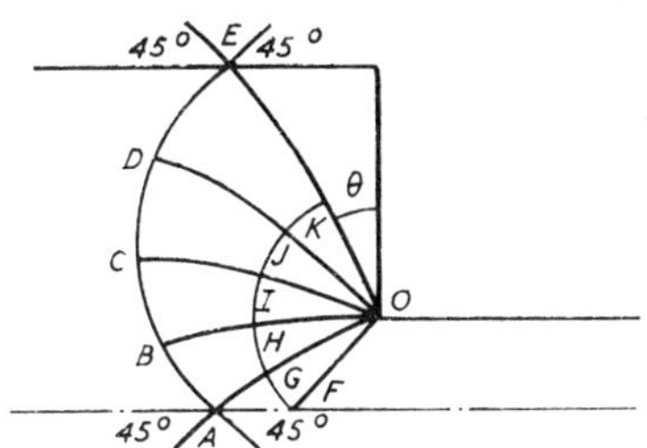

Fig. 3.21 (*a*) Slip-line field for 4:1 extrusion ratio

with the container wall, so a slightly smaller fan angle should be chosen until E does coincide. Since perfect lubrication is assumed, the slip lines meet the container at E at 45°. Having satisfactorily plotted the points of intersection, smooth curves are then drawn through the point to give the field shown in Fig. 3.21 (*a*).

Considering the exit slip line OF, since the extruded strip is assumed to be stress free, the horizontal stress on this slip line is zero. Intuitively, we would expect the vertical stress on this line to be compressive. The Mohr circle is shown in Fig. 3.21 (*b*) for the line OF, where $p = -k$. By definition the circular slip line FK must be an α line, since the zero horizontal stress is algebraically greater than the compressive vertical stress.

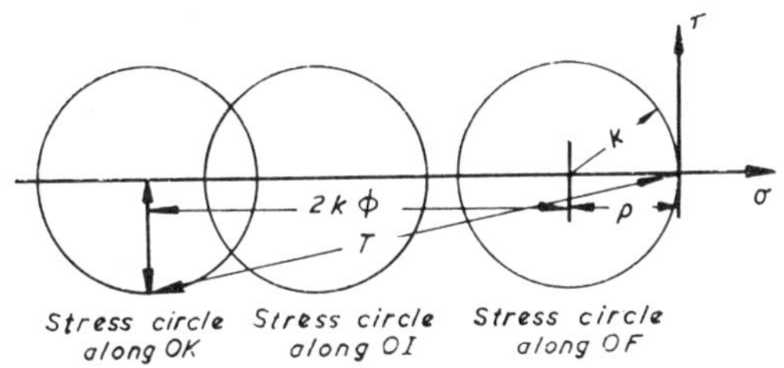

Fig. 3.21 (*b*) Mohr stress circles for radial lines around fan *OFK*

The Hencky equation for an α line is $p - 2k\phi = $ constant. In this case the rotation ϕ of the slip line is clockwise (i.e. negative); p therefore becomes more compressive by an amount $2k\phi$ as we move from F to K. At any point between O and K the total stress T is equal in magnitude to $[(k + 2k\phi)^2 + k^2]^{\frac{1}{2}}$.

Proceeding along the β line KE, ϕ increases (anticlockwise) by an amount $(\pi/4) - \theta$. Since $p + 2k\phi = $ constant along a β line, p decreases in value (it becomes more compressive) as we approach E. From a knowledge of the angle turned through, the hydrostatic stress and hence the total stress can be evaluated at each point. Since we are interested in the horizontal force on the end of the die, it is necessary to find the horizontal components of these stresses and integrate them along the total length of the slip line (Fig. 3.22) as described by Alexander.[2]

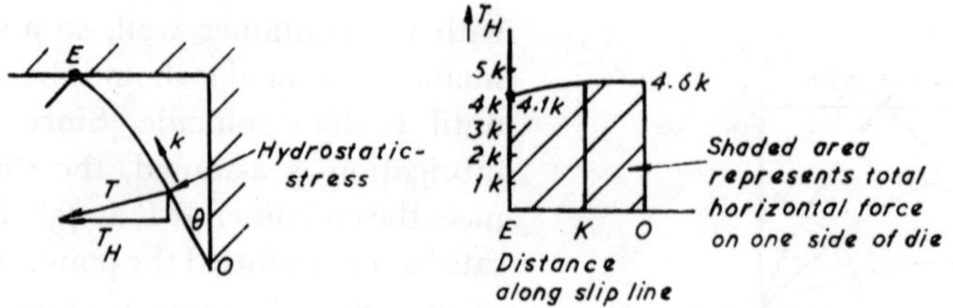

Fig. 3.22 Representation of stresses and forces acting on end of die

To find the total extrusion force it is necessary to double this value, since the other side of the extrusion die must also be considered. Where friction occurs between the surface of the metal and the container the extrusion force will, of course, be increased by an amount equal to the friction force.

The hodograph can be constructed as follows. All metal to the left of line AE in Fig. 3.21 (*a*) moves as a rigid body at a constant speed *u*, the speed of the ram. Shearing occurs along EO, leaving a stationary wedge of 'dead metal' in the corner of the die. Immediately beneath the point E, because of the constraint offered by the dead metal zone in the corner of the die, the metal will flow in the direction of the boundary line. The continuity requirement demands that the normal velocity either side of the α boundary slip line is constant, at this point equal to $u\sqrt{2}$ at 45° to the horizontal. The velocity discontinuity is therefore also $u\sqrt{2}$ and the point E_B, (Fig. 3.23) can be found. The discontinuity is constant in magnitude along the slip line, but it will turn through 90° between E and A. Hence the velocities immediately to the right of D, C, B and above A will be represented by the discontinuity vectors D_R, C_R, B_R and A_A.

Next the velocity of the metal at points along the α slip line joining F and K will be considered. The velocity discontinuity along the slip line

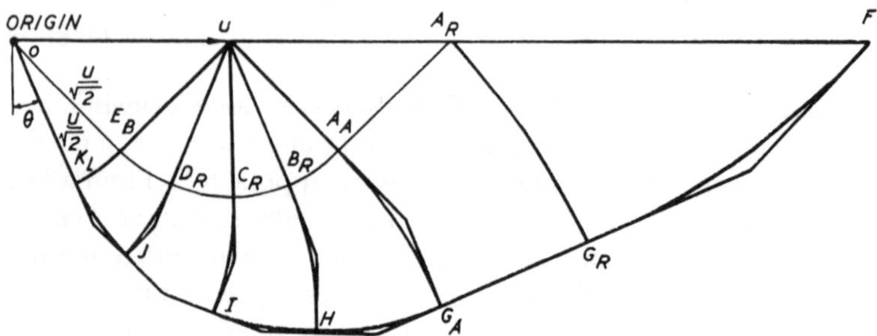

Fig. 3.23 Hodograph for 4 : 1 extrusion

EKO is represented by the narrow fan $OE_B\ K_L$ of magnitude $u/\sqrt{2}$, OK_L representing the velocity of the metal just to the left of K. Velocities of points J, I, H and G_A along slip line EKO can be found because the corresponding parts of the slip-line field and the hodograph are normal to one another. A third velocity discontinuity, also of magnitude $u/\sqrt{2}$ occurs along the slip line AGO. This discontinuity is represented on the hodograph by the fan bounded by $A_A - A_R$ and $G_A - G_R$. It now remains to locate the velocity of *F*, which is found by drawing the hodograph from G_R normal to GF in the slip-line field.

For the velocities to be compatible, it is necessary to show that OF is in fact four times *u*, since for plane strain extrusion with a reduction ratio of 4:1 it follows that from constant volume considerations the velocities must also be in this ratio.

3.13 UPPER BOUND SOLUTIONS

Slip-line field solutions are usually laborious to obtain, so simplified approaches giving solutions which are greater than or equal to the actual load (upper bound) or less or equal to the actual load (lower bound) have been developed. The upper bound solution is of greater interest in metal forming. This method is particularly useful in plane strain problems where the solution can be obtained graphically. As might be expected with an upper bound solution there is some overestimation of load compared with the more exact slip-line field approach; this error need not, however, be serious and an overestimate is anyway preferable to an underestimate.

3.13.1 Use of upper bound solution for plane strain condition. Consider a piece of metal flowing plastically across a slip line of length *s* which has a velocity discontinuity of *u*. Work done/unit time in shearing along this slip line is the product of the force acting along the slip line and the velocity discontinuity. The force to produce shearing assuming unit depth, is the shear flow stress *k* times the length of the slip line.

$\therefore$ work done/unit time $= \mathrm{d}w/\mathrm{d}t = kus$ along a velocity discontinuity.

When applying upper bound solutions to plane strain problems the slip lines are approximated by straight lines. The plastic zone is divided into triangular areas and the magnitudes of velocity discontinuities are found by constructing a hodograph for the selected configuration. It is then possible to find the rate of working by summing the products of u and s at each velocity discontinuity and multiplying the sum by k,

i.e.
$$dw/dt = k\Sigma us \tag{3.13}$$

3.13.2 Calculation of indentation force using upper bound method. The indentation of a very thick block of metal by a smooth platen under plane strain conditions will be considered. Before the straight line velocity discontinuities are drawn, it is useful to look at the slip-line field (Fig. 3.24 (*a*)). The approximate field used for the upper bound solution is represented by six equilateral triangles shown in Fig. 3.24 (*b*). Due to symmetry about the centre line, the right-hand half of the deformation only need be examined.

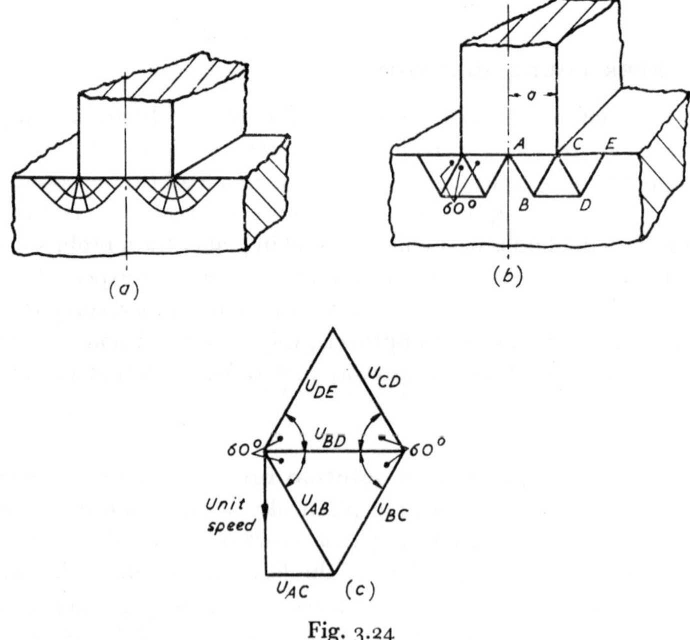

Fig. 3.24

The punch is assumed to move downwards with unit velocity. As the metal below ABDE remains rigid, shearing must occur along AB, BC, BD, CD and DE as the punch penetrates. The metal within triangle ABC moves downward in a direction parallel to AB and the second constraint applied by the base of the punch AC enables the first velocity triangle to be completed. On reaching discontinuity BC the metal is sheared and moves horizontally; shearing again occurs when CD is crossed. The direction of flow is now parallel to DE. Fig. 3.24 (*c*) shows the hodograph for the metal to the right of the centre line.

Applying the formula for rate of working, $dw/dt = k\Sigma us$ and assuming a punch of unit thickness and punch pressure p,

$$p \cdot a \cdot 1. = k(\mathrm{AB} \cdot u_{\mathrm{AB}} + \mathrm{BC} \cdot u_{\mathrm{BC}} + \mathrm{BD} \cdot u_{\mathrm{BD}} + \mathrm{CD} \cdot u_{\mathrm{CD}} + \mathrm{DE} \cdot u_{\mathrm{DE}})$$

The lengths of the velocity discontinuities s can be obtained from Fig. 3.24 (*b*) in terms of the half punch width a, and the sizes of the velocity changes u from the hodograph in terms of the punch velocity. In this example, all values of s are the same and equal to a, and the values of u are

$$p \cdot a = (10/\sqrt{3})a \cdot k$$

Hence $$p = (10/\sqrt{3})k$$

It is of interest to note that the slip-line field solution provides an answer 11% smaller than that obtained by the upper bound method. A closer approximation could be obtained by varying the proportions of the triangles.

3.13.3 Upper bound solutions for plane extrusion problem. Considering the plane extrusion problem previously discussed, it is easily shown that the extrusion pressure calculated from the slip-line field solution is approximately $3.8\,k$ N m^{-2} (lbf/in^2) of ram area.

Two upper bound solutions, both giving kinematically admissible velocity fields, are now discussed. Fig. 3.25 shows the simplest imaginable solution where the velocity discontinuities form a triangle bounded by the billet, the dead metal zone and the extruded section. The extrusion pressure

$$= k\Sigma u.s$$

$$= k\Sigma(\mathrm{AB} \cdot {}_1V_2 + \mathrm{AC} \cdot {}_3V_2 + \mathrm{BC} \cdot {}_2V_4$$

$$= k\Sigma(1.01 \times 1.50) + (0.95 \times 1.88) + (0.53 \times 3.23)$$

$$= 5.0k$$

Fig. 3.26 shows a more complicated configuration involving two triangles. In this case the extrusion pressure is found to be approximately $4.9k$. Although the upper bound solutions do not give very good agreement with the slip-line solution it would be possible to produce a configuration which agreed more closely. Bearing in mind the work hardening properties of metal, the upper bound solution may well be more accurate in many cases than the slip-line solution which is based on a rigid-plastic material concept.

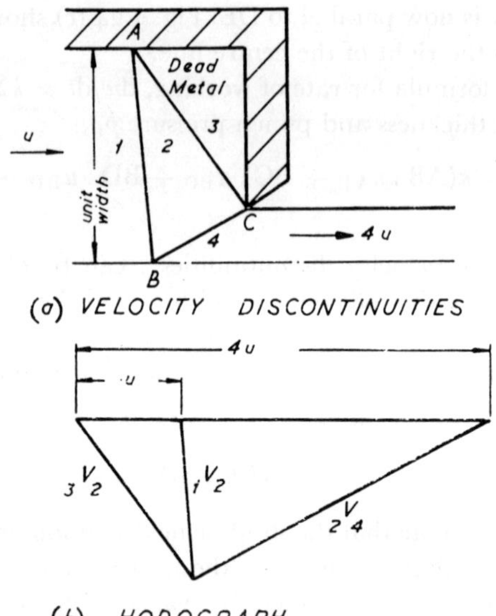

Fig. 3.25 Upper bound solution for plane extrusion

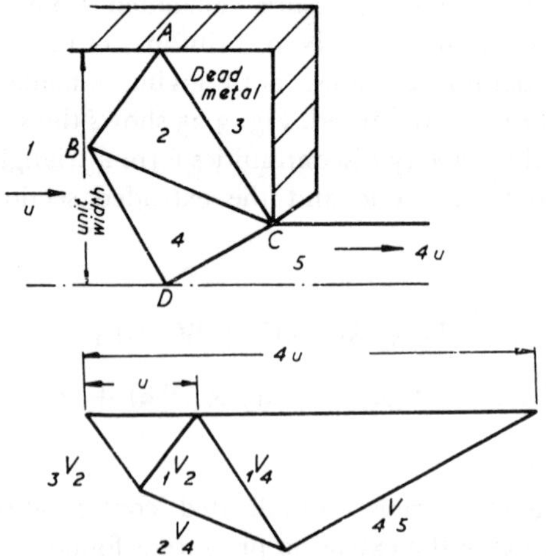

Fig. 3.26 Upper bound solution for plane extrusion

For the same problem, a solution based on the work formula $\delta w = Y\delta\bar{\varepsilon}$ gives an extrusion pressure of 2·4k. This ignores the redundant work performed which, with a large reduction of this type, is considerable.

3.14 RECENT DEVELOPMENTS IN METAL FORMING ANALYSIS

Most of the recent developments have hinged on the use of large computers for performing arithmetic and algebraic operations. Provided the general form of a slip-line field is known, it is possible to write a computer program which calculates the co-ordinates of the slip-line intersections using a step-by-step approach, and hence the nodal points of the hodograph can be computed. A particular advantage of this method is that it enables a range of slip-line fields to be plotted to depict changing deformation patterns, such as occur when metal is being forged. Similarly, variation in tool geometry or frictional conditions can be readily accommodated.

Finite element methods. These methods, which are already well established in the study of elasticity, are being extended into plasticity. It is claimed that they can predict accurately stress, strain, strain rate and temperature distributions during metal deformation, and that mechanical and metallurgical properties such as hardness, ductility and grain homogeneity can be deduced. Their application, however, requires a level of specialist knowledge which will probably limit their use in the foreseeable future.

Visioplasticity. This is the name given to a number of non-predictive techniques which rely on the analysis of displacements on a plane surface after deformation has occurred, to plot slip-line fields and distributions of strain, strain rate and stress. Visioplasticity has applications in both plane strain and axi-symmetrical deformation. Before processing, the workpiece is suitably prepared and an orthogonal grid orientated in the direction of flow, or an orthogonal pattern of touching circles, is etched or dyed on the surface, usually by photo-resist techniques. For most applications, the workpiece is cut into two plane parts which are reassembled after one part has had the grid applied to its inner face. However, for some applications, such as metal cutting, the grid can be on the outside of a large tube.

If a circular configuration is used, the circles deform after relatively small deformation to ellipses. The major axes of these ellipses provide a measure of the principal strains, while lines drawn at 45° to these axes give the slip-line directions at the centre of the ellipse. If a square grid approaches the deformation zone at a known velocity, the distance moved between adjacent grid intersections, in the x and y directions, is a measure of the respective velocity components u and v. When the deformation process is suddenly stopped the resulting grid provides a map from which the velocity field can be calculated.

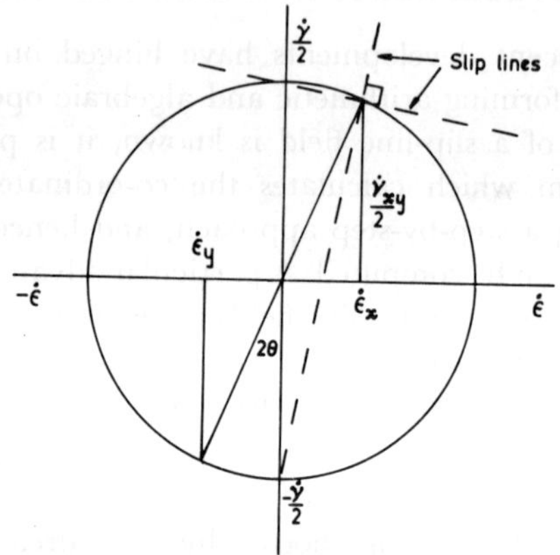

Fig. 3.27 Strain rate diagram for plane strain

For plane strain extrusion

$$\dot{\varepsilon}_x = \frac{\partial u}{\partial x} = -\dot{\varepsilon}_y = \frac{\partial v}{\partial y}$$

$$\dot{\gamma}_{xy} = \frac{\partial u}{\partial y} + \frac{\partial v}{\partial x}$$

Let θ be the inclination of the slip lines to the x and y axes at this point. Then from Fig. 3.27

$$\tan 2\theta = \frac{2\dot{\varepsilon}_x}{\dot{\gamma}_{xy}}$$

$$\dot{\bar{\varepsilon}} = \frac{2}{\sqrt{3}}\,\dot{\varepsilon}_1 = \frac{2}{\sqrt{3}}\left(\dot{\varepsilon}_x^{\,2} + \tfrac{1}{4}\,\dot{\gamma}_{xy}^{\,2}\right)^{\frac{1}{2}}$$

$$= \frac{2}{\sqrt{3}}\left[\left(\frac{\partial u}{\partial x}\right)^2 + \tfrac{1}{4}\left(\frac{\partial u}{\partial y} + \frac{\partial v}{\partial x}\right)^2\right]^{\frac{1}{2}}$$

If $\dot{\bar{\varepsilon}}$ is integrated along the streamlines it is then possible to construct a strain distribution.

By making suitable assumptions concerning the values of effective stress it is possible, using an involved mathematical procedure, to evaluate stress distributions. This approach can be adapted to axi-symmetrical operations. Subsequent comparison of forces calculated from the proposed stress distributions can be compared with values obtained from experiment to give a measure of the confidence which can be ascribed to the analysis.

4 Hot Forging and Rolling

4.1 FORGING PROCESSES

Forging is made up of a diverse group of hot and cold shaping processes and produces parts varying in weight from several tons to a few ounces. In this chapter hot forging has been grouped with hot and cold rolling, as both can be considered primary forming operations which are largely confined to the forge or steelworks. Cold forging is normally a factory forming operation and in consequence has been included with other factory forming processes in Chapter 6.

Hot forging can be split into three subdivisions: smith forging, closed die forging, and upset forging. Smith forging is the traditional art which uses simple tools, often for one-off or short batch work. The deforming force is either a squeeze from a press or the blow of a machine hammer. The second subdivision is closed die forgings, in which the part is produced between an upper and lower die, each of which has been machined with a half impression of the component shape. If the metal is deformed between closed dies by a series of hammer blows rather than by a press, the process is called drop forging. Thirdly, there is upset forging, in which the bar stock is thickened by pressure from a die to form a larger diameter, or is forced into another shape such as the hexagon head of a bolt.

Apart from shaping the metal, forging improves its structure and hence its mechanical properties. When ingots are used in heavy forging the force from the press penetrates to the centre of the material, breaking down and making more uniform the as-cast structure. For smaller forgings, billets and bars are used; these have been hot rolled and, due to the segregation of impurities, have acquired a fibre in the direction of rolling. This fibre may be considered similar to the grain in wood and gives the metal enhanced strength when stressed normal to the direction in which the fibre is running. Forging distorts the previously unidirectional fibre in such a way as to strengthen the component compared with a similar part manufactured by machining from the bar (Fig. 4.1).

4.2 SMITH FORGING

Smith forging is one of the oldest metal working arts. Although hand forging has been almost entirely replaced by machine forging, skill is still needed to decide when the work should be turned or when reheating is necessary. The metal is shaped by comparatively simple tools and larger forgings are handled mechanically by an overhead crane. The tools, which

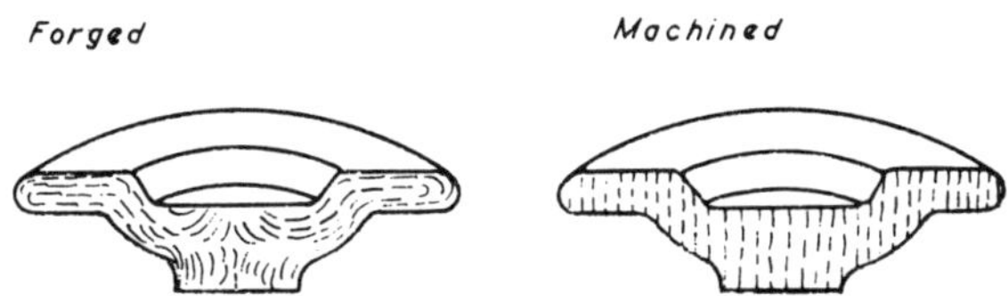

Fig. 4.1 Effect of forging on fibre structure

are often a pair of flat dies, are mounted either in a mechanical or hydraulic press or in a steam or mechanical hammer. Presses rather than hammers are used for heavy forgings as they provide deeper penetration into the metallic structure of the ingot.

Large forgings are produced for the heavy engineering and shipbuilding industry. For instance, flanged shafts for ships' propellors are made by forging first the flanges and then the shaft (Fig. 4.2 (*a*)). Open-ended cylinders, such as those used for pressure vessels, are also forged. Initially the cooled ingot is trepanned to remove a central core; after machining it is reheated, when the diameter can be increased and the

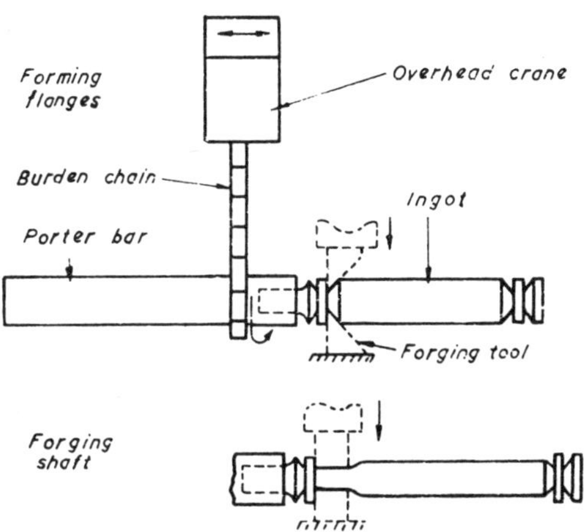

Fig. 4.2(*a*) Forging ship's propeller shaft

walls thinned by becking, as shown in Fig. 4.2 (*b*). After final shaping, forgings are heat treated to stress relieve them and to improve their mechanical properties.

In recent years automatic and semi-automatic, open-die, press forging machines have been introduced; the work is held in rail-mounted manipulators which feed it forward to match the high operating speeds of the forging press. An interesting new development is the swing forging machine Fig. 4.2 (*c*), which combines squeezing with axial feeding of the work. It can be used as an adjunct to bar and rod mills with two pairs of forging tools mounted at right angles to each other.

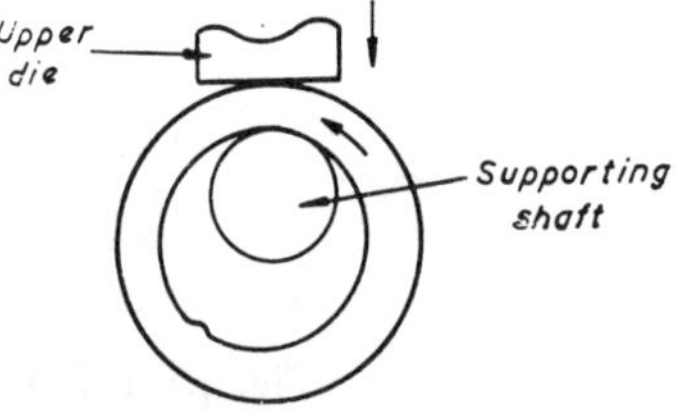

Fig. 4.2(*b*) Becking a cylinder

4.2.1 Hot bar forging. The object of this operation, shown diagrammatically in Fig. 4.3, is to reduce an ingot into a bar by squeezing it between two flat dies. If a bar of square cross section is required, the work should be turned 90° between passes. Bar forging, although a slower operation than rolling, is used when good penetration to the centre of the metal is required. For equivalent penetration by rolling, a massive and uneconomic rolling mill would be needed.

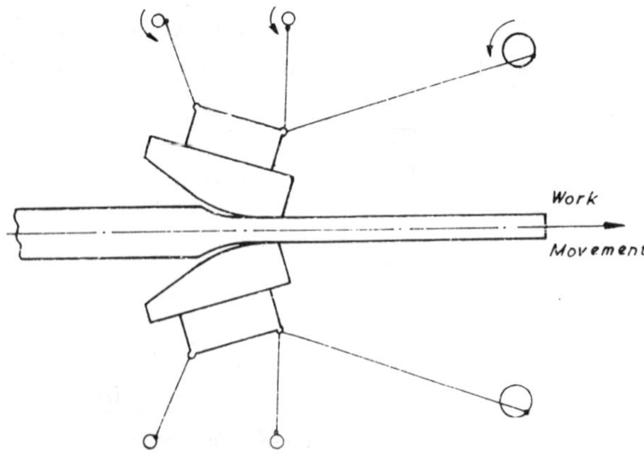

Fig. 4.2(*c*) Swing forging

Although the initial forging temperature of steel is around 1300°C the metal cools during shaping and must be returned to the furnace for reheating when its temperature has fallen to 750°C.

4.2.2 Product faults in bar forging. The size reduction obtainable at each pass is limited by the risk of forming laps, which introduce scale into the forging and produce serious weakness in the finished product (Fig. 4.4).

It is recommended that the squeeze ratio t_0/t_1 should not exceed 1·3 if laps are to be avoided; larger reductions are, however, possible if the edges of the dies are chamfered as shown in Fig. 4.3. If the material which is being forged is too thick relative to the platen breadth b, deformation will not penetrate sufficiently to improve the properties of the metal near the centre of the forging. To obtain satisfactory working at the centre, the unforged thickness t_0 should not exceed three times the platen breadth b.

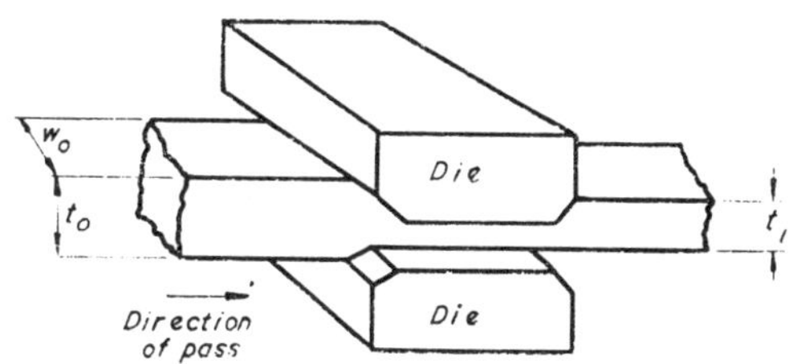

Fig. 4.3 Bar forging

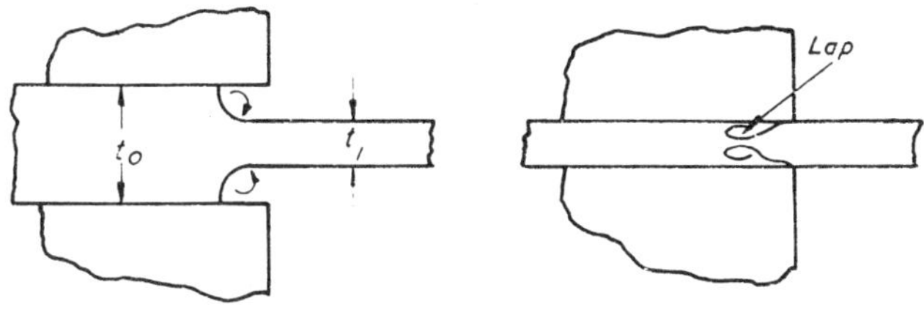

Fig. 4.4 Formation of laps

A fault liable to occur in forging presses with poor guide alignment is rhombic distortion (Fig. 4.5). If the ratio t_0/w_0 is kept below 1·5 this fault is unlikely to occur.

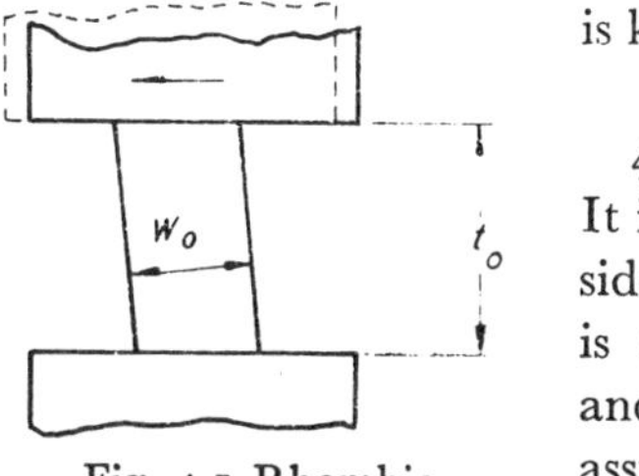

Fig. 4.5 Rhombic distortion

4.2.3 Analysis of stress in bar forging. It is assumed that, owing to the metal on each side of the plastic zone remaining elastic, there is negligible side spread under the platens and plane strain conditions apply. It is also assumed that no redundant work occurs, i.e. plane sections remain plane. Due to the impossibility of lubricating dies at forging temperatures, sticking friction occurs between work and platen, and the shear stress in the metal at the tool/metal interface reaches its shear flow stress k.

Considering the equilibrium of the small element of material being forged (Fig. 4.6 (a)),

$$tqw = t(q + \mathrm{d}q)w + 2k\,\mathrm{d}x\,w$$

$$0 = t\,\mathrm{d}q + 2k\,\mathrm{d}x$$

$$\therefore \qquad \frac{\mathrm{d}q}{\mathrm{d}x} = -\frac{2k}{t}$$

As p and q are principal stresses causing yielding (this assumption being reasonably true in the body of the material).

$$2k = p - q$$

Differentiating $0 = \mathrm{d}p - \mathrm{d}q$, the term $2k$ can be assumed constant in hot forging as there is no work hardening.

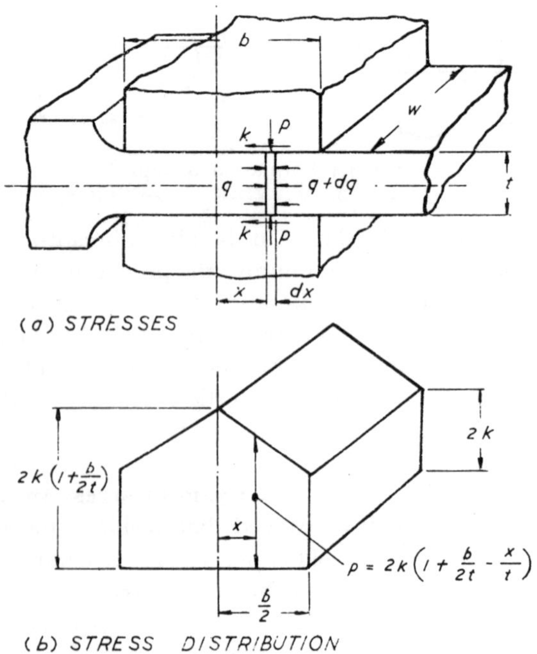

Fig. 4.6 Bar forging

It is now possible to substitute dp for dq and integrate the resulting expression dp/dx = $-(2k/t)$ to find the distribution of pressure under the platen.

$$p = -\left(\frac{2k}{t}\right)\int dx$$

$$p = -\left(\frac{2k}{t}\right)[x] + C$$

To find C, the boundary conditions at $x = b/2$ are considered; here $p = 2k$ as $q = 0$.

$$2k = -\left(\frac{2k}{t}\right)\left(\frac{b}{2}\right) + C$$

$$\therefore \qquad C = 2k\left(1 + \frac{b}{2t}\right)$$

Hence

$$p = 2k\left(1 + \frac{b}{2t} - \frac{x}{t}\right)$$

It will be seen from the above expression that the pressure distribution under the platen rises linearly from $2k$ at $x = b/2$ to a maximum of $2k[1 + (b/2t)]$ at $x = 0$ as shown in Fig. 4.6 (b). The triangular roof to the stress distribution is known as a friction hill and this type of stress distribution is common to forging and rolling operations. In this instance the average pressure under the platens $\bar{p} = 2k[1 + (b/4t)]$. The forging force is therefore

$$bw2k\left(1 + \frac{b}{4t}\right) \tag{4.1}$$

Readers seeking further information on large-scale hot forging are recommended to read an analysis of this process by Wistreich and Shutt.[3]

4.3 CLOSED DIE FORGING

By the use of closed dies, greater accuracy can be achieved and the rate of production increased beyond that obtainable with the open dies used in smith forging. A piece of heated metal is placed on the lower die block and then forced into the shape made by the upper and lower dies by blows from a machine hammer or by pressure from a mechanical or hydraulic press (Fig. 4.7). The amount of metal used is slightly in excess of that required by the component, the excess appearing as a flashing around the part. Usually a forging is brought to its final shape by taking it through a series of impressions, often within a single pair of die blocks.

Drop hammers are widely used for drop forging and are mechanical versions of the hammer and anvil of the blacksmith. The forging blow is obtained by the action of gravity on a heavy weight called a tup. The tup, to which the upper die is attached, is lifted to an appropriate height and then released (Fig. 4.8).

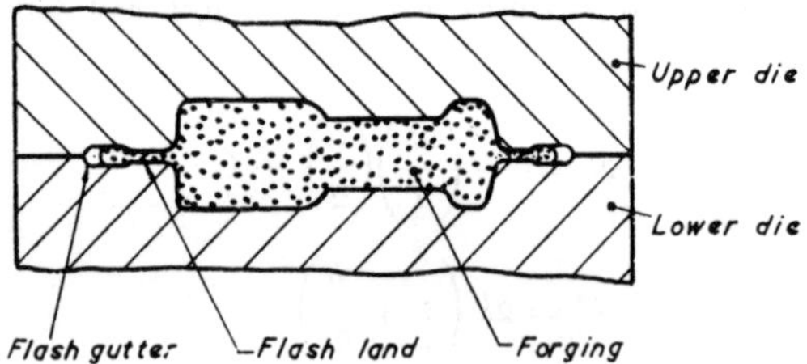

Fig. 4.7 Closed die forging

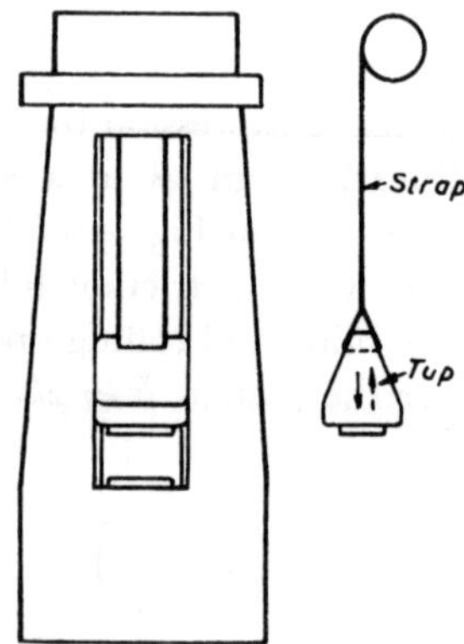

Fig. 4.8 Drop hammer

Drop hammers are relatively inexpensive and versatile machines but when large quantities of forgings are required press forging is preferable. Presses provide a faster rate of production as the die is filled in a single stroke. They can be readily automated and enable smaller draft angles to be used on the component due to mechanical ejection. In addition working conditions are improved as noise and vibration are reduced. Forgings can be pressed with cored holes in both vertical and horizontal axes.

High-velocity (high energy rate) machines can also be used for closed die forging. These machines are operated by compressed gases which complete the forging operation in a few milliseconds. The equipment is more compact and the cost of unit quantity of energy is less than with conventional forging equipment. A particular advantage is that the process is suitable for automation, provided the forging operation can be completed in a single blow. In addition, high energy processes can produce close tolerance forgings and thin webs. A major difficulty is short die life, which can result in high tool costs and poor machine utilization; it is however possible that, with improved die materials and lubricants, this disadvantage may be overcome. The operating principle of a high velocity forming machine is shown in Fig. 4.9: the piston is accelerated downwards when the high pressure acting on the top of the piston overbalances the low pressure acting on the underside.

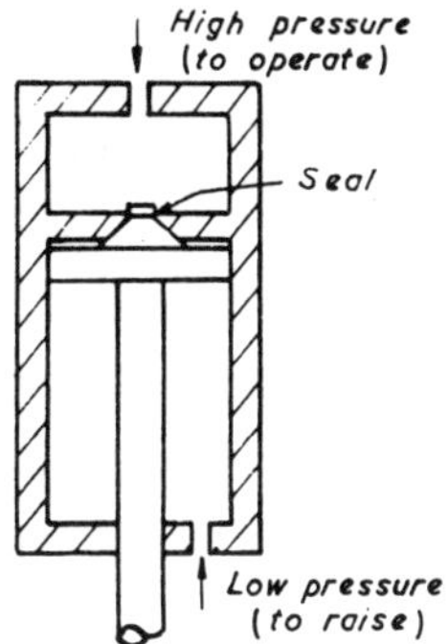

Fig. 4.9 High velocity forging machine (operating principle)

4.4 WOBBLE FORGING

This type of forging, which is also known as rotary forging, is illustrated in Fig. 4.10. The upper die rotates about an inclined axis, which itself rotates, while the die moves upwards. As a small proportion of the part only is being deformed at any particular time, the forces are reduced to between 5% to 10% of those required for the closed die forging of a similar part. The process was heralded in a patent filed in 1929 by H. F. Massey, but only recently has rotary forging equipment become commercially available.

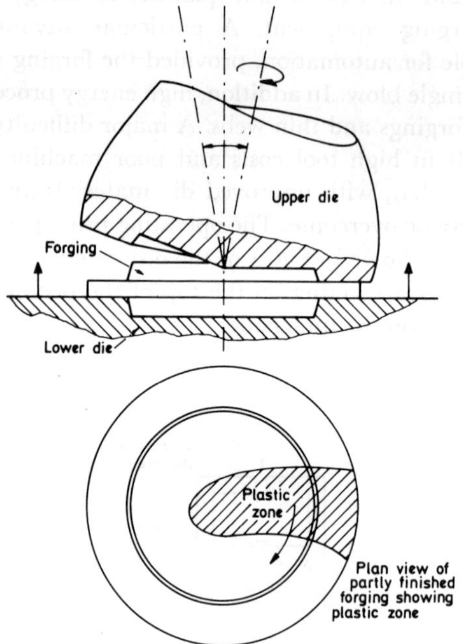

Fig. 4.10 Wobble forging

The following advantages are claimed over closed die forging:

(*a*) Lower capital costs resulting from the smaller forming forces.
(*b*) The possibility of scale-free forgings, as preheat temperatures need not be as high as for conventional forging processes.
(*c*) Very thin flanges may be forged.
(*d*) Closer thickness tolerances can be achieved.

Unlike closed die forging the products must be circular in form.

4.5 UPSET FORGING

This process, sometimes called machine forging, uses bar stock which is heated at the end to be forged. The bar is gripped in the fixed half of the die so that the requisite length projects. The forging blow is delivered by a moving die; simple shapes are produced in a single stage but more complicated shapes require several stages. An example of a simple upset forging operation and a part produced by multi-stage upset forging are shown in Fig. 4. 11.

4.6 HOT ROLLING

Rolling is one of the most important metal working processes and can be performed on either hot or cold metal. Material is passed between the cast or forged steel rolls of a rolling mill which compress it and move it forward. Rolling is a more economical method of deformation than forging if metal is required in long lengths of uniform cross section.

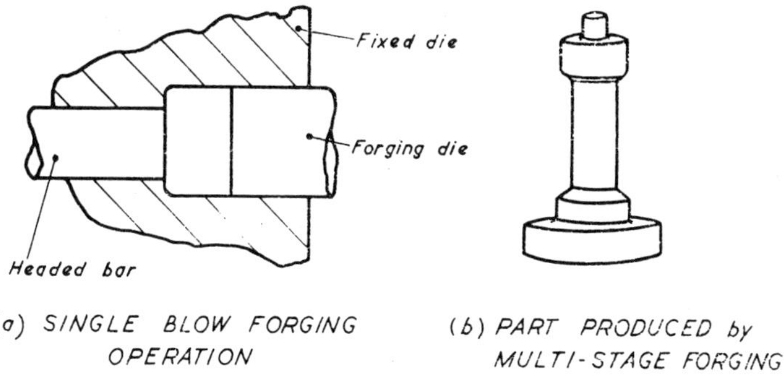

Fig. 4.11 Upset forging

4.6.1 Slabs and blooms. The ingot is first rolled into either slabs or blooms, which have no end use but are an intermediate stage in the rolling process. Slabs are rectangular and blooms square in cross section

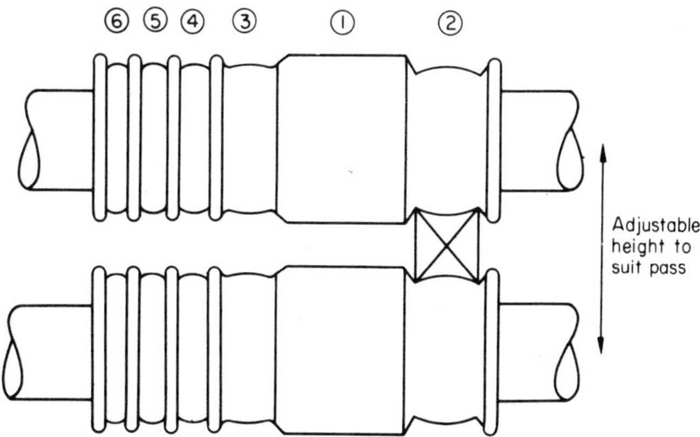

Fig. 4.12 Cogging mill rolls (the pass sequence is numbered)

and are produced by slabbing and cogging mills respectively. These large rolling machines have two rolls; slabbing mills have plain cylindrical rolls but those of cogging mills are more elaborate; the form of these is shown in Fig. 4.12. The direction of roll rotation is reversible, so that after the metal has been passed through the mill the rolls are reversed and the metal is passed back with the roll gap reduced. This passing backward and forward between the rolls continues until the desired change in shape has been achieved. If blooms are required, the work is turned mechanically by manipulators to maintain a square cross section.

4.6.2 Plates, strips and sections. Final products from hot rolling include plates, as used in shipbuilding, and a wide variety of bars and sections such as rolled steel joists. Plate and strip are normally rolled between two plain cylindrical rolls, but occasionally the rolls are arranged as a four-high mill (Fig. 4.13). This arrangement minimizes the tendency of the rolls to bow and produce stock which is thicker at the centre. The rolling of special section cannot be dealt with analytically and the design of the rolls is essentially empirical. It will be seen from Fig. 4.14 that the rolls are designed so that the cross section is brought, pass by pass, to its final shape.

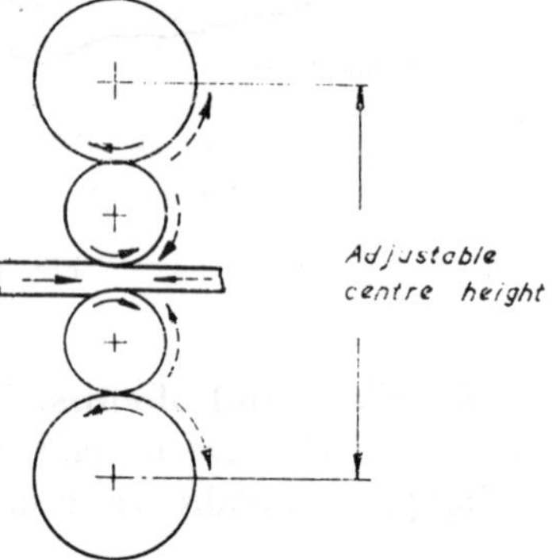

Fig. 4.13 Four high roll arrangement

When large quantities of similar sized strip and rod are required, several mills are often arranged for continuous rolling. Here the metal is passed from one mill to the next, until it emerges at the correct size. The speed of each rolling mill must be carefully adjusted so that the length of material between mills is kept constant.

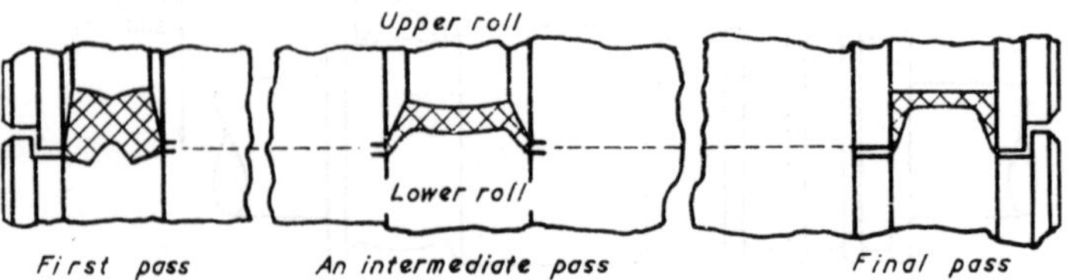

Fig. 4.14 Rolling of channel section from square section bar

4.6.3 Planetary rolling mills. Small diameter rolls are more effective than large ones in conveying rolling force to the deforming metal. This is due to the reduced area of contact which produces higher rolling pressures. Use is made of this principle in planetary mills, the roll arrangement being shown in Fig. 4.15. Serrated feed rolls force the strip forward to the planetary rolls, which consist of two heavy backing rolls surrounded by small diameter rolls mounted in cages. The two upper and lower cages are geared together so that corresponding small upper and lower rolls are brought into contact with the metal in synchronism. Each pair of rolls bites into the red hot metal along the arcs of contact and successively extends the strip. Maximum thickness reductions of 25:1 are possible with planetary mills compared with 2:1 ratios for conventional rolling. Planetary rolling mills have been used in conjunction with continuous casting, where their large reduction ratio enables one mill to produce continuous lengths of hot rolled strip.

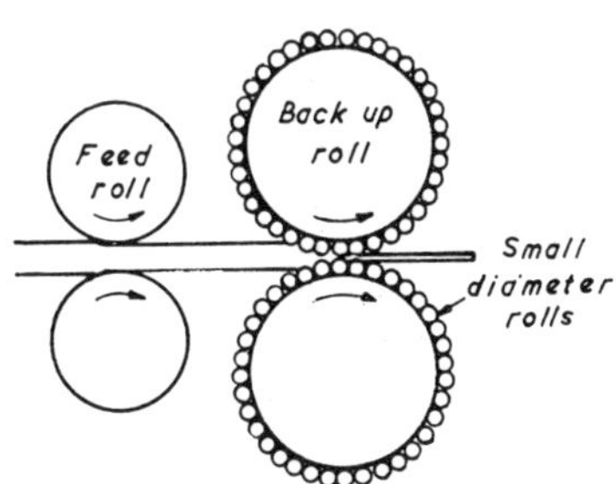

Fig. 4.15 Planetary rolling mill

4.6.4 Calculation of hot rolling forces. Approximate theories are available for calculating forces in rolling. Assuming that there is no elastic deformation of the rolls, the process can be represented schematically as in Fig. 4.16 (*a*).

Considering the forces acting on an elemental slice of material (Fig. 4.16 (*b*)),

$$(p + \mathrm{d}p)h + 2kR\mathrm{d}\phi \cos\phi = ph + 2sR\mathrm{d}\phi \sin\phi$$

$$\mathrm{d}(ph) = 2sR\mathrm{d}\phi \sin\phi - 2kR\mathrm{d}\phi \cos\phi \qquad (4.2)$$

The surface velocity of the rolls will vary between the velocity of the material at entry V_1 and its velocity at exit V_2. At some point along the deformation zone the velocity of the material will equal that of the rolls: this is called the neutral point. On the entry side of the neutral point the frictional force of the rolls on the material acts in the same direction as the material movement, but reverses on the exit side of the neutral point (Fig. 4.17). As with hot forging, the friction stress is k because the shear flow stress of the material is reached at the roll/work interface. To account for the change in direction of the frictional force, equation (4.2) for $\mathrm{d}(ph)$ becomes

$$\mathrm{d}(ph) = 2R(s \sin\phi \pm k \cos\phi)\mathrm{d}\phi \begin{array}{l} + \text{exit side} \\ - \text{entry side} \end{array}$$

$$\frac{\mathrm{d}(ph)}{\mathrm{d}\phi} = 2R(s \sin\phi \pm k \cos\phi) \begin{array}{l} + \text{exit side} \\ - \text{entry side} \end{array} \qquad (4.3)$$

To obtain an expression for the deforming force it is necessary that the above expression be integrated between the limits of ϕ_1 and o. A solution cannot be obtained analytically and a step-by-step method of integration

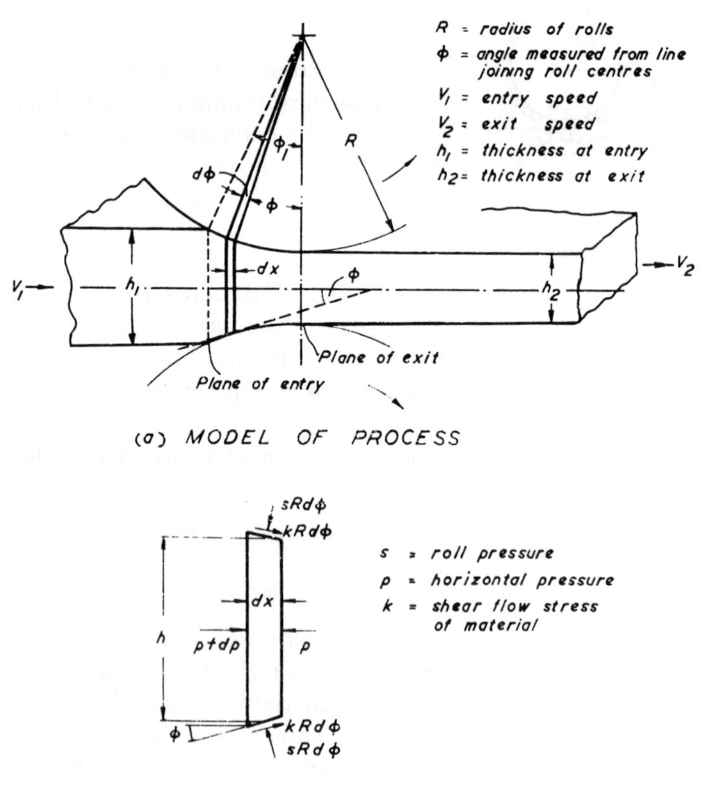

Fig. 4.16 Hot rolling

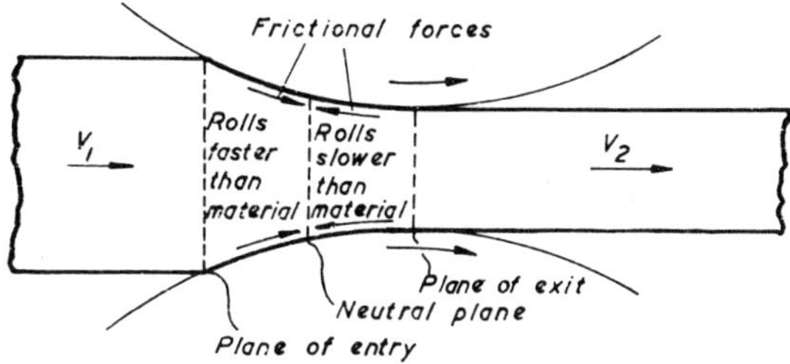

Fig. 4.17 Frictional forces of roll on material

has to be used. Some simplification can, however, be obtained by considering that the metal is deformed between flat platens (Fig. 4.18 (*a*)).

The angle ϕ now becomes zero and the thickness h is constant at h_2. The elemental slice $R\mathrm{d}\phi$ is replaced by $\mathrm{d}x$ and hence $\mathrm{d}\phi = (\mathrm{d}x/R)$.

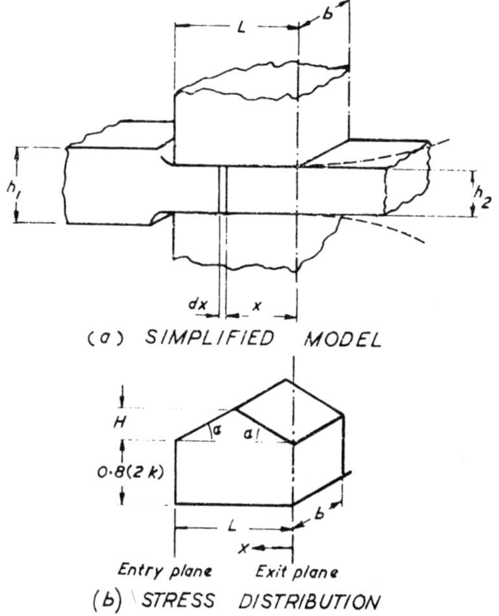

Fig. 4.18 Hot rolling

Equation (4.3) now simplifies to

$$Rh_2 \frac{dp}{dx} = 2R(\pm k) \begin{array}{l} + \text{ exit side} \\ - \text{ entry side} \end{array}$$

$$\frac{dp}{dx} = \pm \left(\frac{2k}{h_2}\right) \begin{array}{l} + \text{ exit side} \\ - \text{ entry side} \end{array}$$

Assuming that yielding is in plane strain

$$s - p = 2k$$

Differentiating and assuming k is constant

$$ds - dp = 0$$

$\therefore$

$$ds = dp$$

Substituting for dp

$$\frac{ds}{dx} = \pm \left(\frac{2k}{h_2}\right) \begin{array}{l} + \text{ exit side} \\ - \text{ entry side} \end{array}$$

$$s = \pm \frac{2k}{h_2} \int dx \begin{array}{l} + \text{ exit side} \\ - \text{ entry side} \end{array}$$

$$s = \pm \frac{2k}{h_2} [x] \begin{array}{l} + \text{ exit side} \\ - \text{ entry side} \end{array} + C$$

The constant C is evaluated by considering the exit plane and assuming that s equals s_2 at $x = 0$

$$C = s_2$$

$\therefore$

$$s = \frac{2k}{h_2}(x) + s_2$$

As there is no horizontal force p at the exit plane, the yield criterion in plane strain, $s - p = 2k$, becomes $s_2 = 2k$. Orowan and Pascoe[4] found a better value for s_2 to be 0·8 $(2k)$. Hence

$$s = \frac{2k}{h_2}(x) + 0{\cdot}8\,(2k)$$

$$s = 2k\left(\frac{x}{h_2} + 0{\cdot}8\right) \qquad (4{\cdot}4)$$

The gradient of the pressure distribution is given by

$$\frac{ds}{dx} = \frac{2k}{h_2}$$

It was also found that there was little loss of accuracy if the neutral point was assumed to be half-way along the arc of contact.

Using this assumption, the distribution of pressure under the rolls is shown in Fig. 4.18 (*b*). With the neutral point at $L/2$,

$$H = \frac{L}{2} \tan \alpha$$

where $$\tan \alpha = \frac{ds}{dx} = \frac{2k}{h_2}$$

Hence $$H = \frac{Lk}{h_2}$$

Roll force = average radial stress × area of contact under roll

$$F = \left[0{\cdot}8(2k) + \frac{H}{2}\right] Lb$$

$$F = Lb\left[0{\cdot}8(2k) + \frac{Lk}{2h_2}\right]$$

$$F = 2kLb\left[0{\cdot}8 + \frac{L}{4h_2}\right] \tag{4.5}$$

It will be seen from Fig. 4.19 that $h_1 = h_2 + 2R(1 - \cos \phi_1)$.

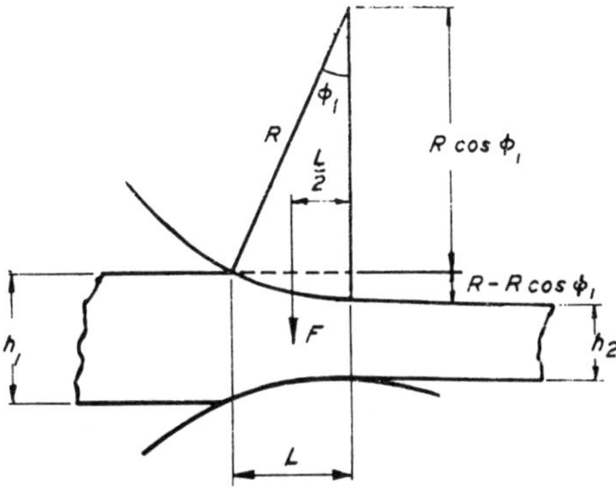

Fig. 4.19 Arc of contact in rolling

As $1 - \cos \phi_1 \simeq \phi_1^2/2$ for the small values of ϕ_1 found in rolling,

$$h_1 = h_2 + R\phi_1^2$$

and

$$\phi_1 = \sqrt{\left(\frac{h_1 - h_2}{R}\right)}$$

Arc of contact $= R\phi_1 \simeq L$

$$\therefore \qquad L = R\sqrt{\left(\frac{h_1 - h_2}{R}\right)} = \sqrt{[R(h_1 - h_2)]}$$

$h_1 - h_2$ is often referred to as the draft δ

Substituting δ for $h_1 - h_2$

$$L = \sqrt{(R\delta)} \qquad (4.6)$$

The torque per roll can be found by assuming that the roll force acts half-way along the arc of contact (Fig. 4.19).

i.e.

$$T = F \times \frac{L}{2}$$

4.6.5 Narrow stock rolling. The analytical work so far on hot rolling is only relevant to wide strip, where it is assumed that there is no side spread and plane strain conditions apply. Orowan and Pascoe have suggested that the stress distribution when rolling narrow strip is as shown in Fig. 4.20. It will be seen that there is now a transverse friction hill caused by the side spread of the material under the rolls.

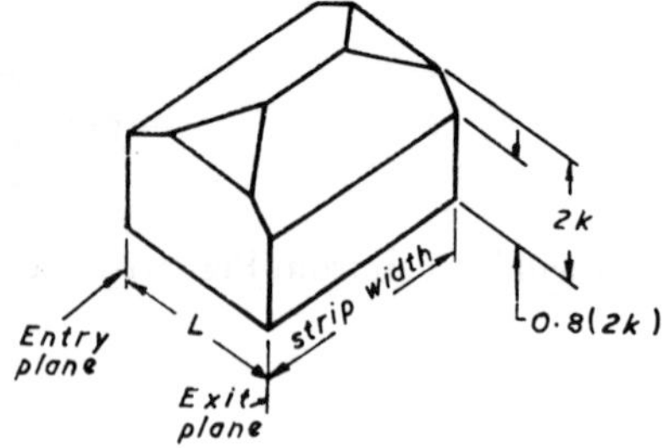

Fig. 4.20 Stress distribution in narrow strip rolling

4.7 COLD ROLLING

4.7.1 Cold rolling process. A large proportion of hot rolled steel is subsequently cold rolled into strip or sheet from which a multitude of pressed parts is manufactured. The hot rolled steel is first pickled to remove the scale from its surface; this is done by immersing in dilute sulphuric acid. After washing and drying it is rolled between ground rolls which give accurate control over thickness and impart a smooth surface finish to the metal. Two- and four-high roll arrangements, as used with hot rolling, can be employed. Two designs which make use of small diameter rolls are the Sendzimir mill, with its elaborate system of backing rolls (Fig. 4.21), and the recently introduced Saxl pendulum mill (Fig. 4.22). The pendulum mill is similar in principle to the planetary mill already described, but instead of a succession of peripheral rolls, a small freely rotating roll is mounted at the end of an arm which is forced to reciprocate at high speed over the deforming metal in the manner of a pendulum. In a particular operation the roll surface speed may be 2 m s^{-1} (400 ft/min) with an ingoing strip speed of 0·01 m s^{-1} (2 ft/min) and an outgoing speed of 0·1 m s^{-1} (20 ft/min).[5] Because of the very large reduction in thickness achieved by pendulum mills the temperature rise is significantly greater than that of conventional rolling and the tendency of the material to work harden is reduced.

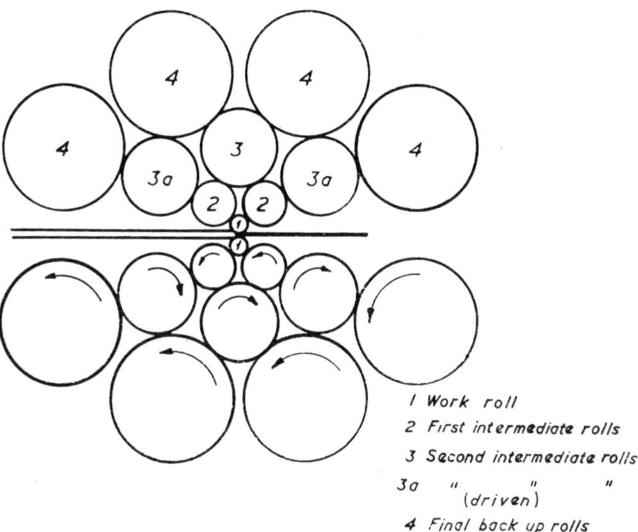

Fig. 4.21 Roll arrangement in Sendzimir mill

The ductility of work hardened material can be increased by annealing. Heating and cooling of the coils is carried out in a controlled atmosphere in order to protect the material from surface oxidization.

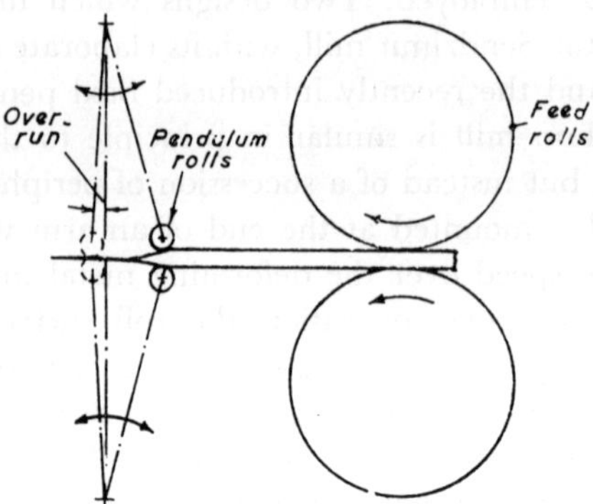

Fig. 4.22 Pendulum rolling mill

4.7.2 Calculation of cold rolling forces. A model similar to that used for hot rolling (Fig. 4.16 (*a*)) can be used for cold rolling. There are, however, some differences: the frictional force is μs not k, and the rolls are deformed in the contact zone from R to a larger radius R'. The yield stress of the metal is of course much higher than with hot rolling and work hardening occurs during deformation.

Equation (4.3) for horizontal equilibrium of the metal between the rolls now becomes

$$\frac{\mathrm{d}}{\mathrm{d}\phi}(ph) = 2R's(\sin\phi \pm \mu\cos\phi) \quad \begin{array}{l} +\text{ exit side} \\ -\text{ entry side} \end{array}$$

The cold rolling theory of Bland and Ford[6] makes the additional assumptions that the roll pressure s acts vertically and $\sin\phi = \phi$. These simplifying assumptions are reasonable for the small arcs of contact normal in cold rolling. Due to work hardening, the yield stress of the metal increases as it passes through the rolls, and when applying the yield criterion in plane strain, $s - p = 2k$, k must be treated as a variable. Hence

$$\frac{\mathrm{d}}{\mathrm{d}\phi}(s - 2k)h = 2R's(\phi \pm \mu) \quad \begin{array}{l} +\text{ exit side} \\ -\text{ entry side} \end{array}$$

$$\frac{\mathrm{d}}{\mathrm{d}\phi}\left(\frac{s}{2k} - 1\right)2kh = 2R's(\phi \pm \mu)$$

$$\left(\frac{s}{2k} - 1\right)\frac{\mathrm{d}}{\mathrm{d}\phi}(2kh) + 2kh\frac{\mathrm{d}}{\mathrm{d}\phi}\left(\frac{s}{2k}\right) = 2R's(\phi \pm \mu)$$

For most rolling conditions p is small compared with s, so that $s \simeq 2k$. Hence the first term in the above equation can be neglected, leaving

$$2kh \frac{\mathrm{d}(s/2k)}{\mathrm{d}\phi} = 2R's(\phi \pm \mu)$$

$$\frac{\mathrm{d}(s/2k)/\mathrm{d}\phi}{s/2k} = \frac{2R'(\phi \pm \mu)}{h}$$

As $h = h_2 + 2R'(1 - \cos\phi)$ and $1 - \cos\phi \simeq \phi^2/2$ when ϕ is small,

$$h = h_2 + R'\phi^2$$

$$\frac{\mathrm{d}(s/2k)\mathrm{d}\phi}{s/2k} = \frac{2R'(\phi \pm \mu)}{h_2 + R'\phi^2}$$

$$\frac{\mathrm{d}(s/2k)/\mathrm{d}\phi}{s/2k} = \frac{2\phi}{(h_2/R') + \phi^2} \pm \frac{2\mu}{(h_2/R') + \phi^2}$$

Integrating both sides of the equation

$$\ln\left(\frac{s}{2k}\right) = \ln\left(\frac{h_2}{R'} + \phi^2\right) \pm 2\mu \frac{1}{\sqrt{(h_2/R')}} \tan^{-1} \frac{\phi}{\sqrt{(h_2/R')}} + \text{constant}$$

If $H = 2\sqrt{(R'/h_2)} \tan^{-1} \sqrt{(R'/h_2)}\phi$ and since $h = h_2 + R'\phi^2$,

$$\ln\left(\frac{s}{2k}\right) = \ln\left(\frac{h}{R'}\right) \pm \mu H + \text{constant}$$

$$\frac{s}{2k} = C\left(\frac{h}{R'}\right) \mathrm{e}^{\pm\mu H} \quad \begin{matrix} + \text{ exit} \\ - \text{ entry} \end{matrix} \tag{4.7}$$

where C is the constant of integration.

No roll tension. The first condition to be considered will be that when there is no tension applied to the strip at entry or exit, i.e. p_1 and $p_2 = 0$. At entry

$$\frac{s_1}{2k_1} = C\left(\frac{h_1}{R'}\right) \mathrm{e}^{-\mu H_1}$$

but from the yield criterion in plane strain, $s_1 - p_1 = 2k_1$,

$$s_1 = 2k_1$$

$$\therefore \qquad \frac{2k_1}{2k_1} = C\left(\frac{h_1}{R'}\right) e^{-\mu H_1}$$

hence
$$C = \frac{R'}{h_1} e^{\mu H_1}$$

At exit

$$\frac{s_2}{2k_2} = C\left(\frac{h_2}{R'}\right) e^{\mu H_2}$$

As
$$s_2 - p_2 = 2k_2 \text{ and } p_2 = 0$$
$$s_2 = 2k_2$$

Also
$$\phi_2 = 0$$

hence
$$H_2 = 2\sqrt{\left(\frac{R'}{h_2}\right)}\tan^{-1}\sqrt{\left(\frac{R'}{h_2}\right)}\phi_2 = 0$$

Substituting for H_2 and s_2

$$1 = C\left(\frac{h_2}{R'}\right)$$

and
$$C = \frac{R'}{h_2}$$

Now that the values of C have been found for entry and exit conditions, the roll pressure equations can be rewritten and the friction hill plotted.

$$\text{On entry side of neutral plane } s^- = \frac{2kh}{h_1} e^{\mu(H_1 - H)} \qquad (4.8)$$

$$\text{On exit side of neutral plane } s^+ = \frac{2kh}{h_2} e^{\mu H} \qquad (4.9)$$

Where k is the value of shear flow stress at the reduction considered.

With roll tension. It is usual to apply tension to the strip in cold rolling as this reduces the amount of roll pressure and keeps thin strip flatter. If tensions of t_1 and t_2 are applied at entry and exit planes respectively the constants can again be evaluated by considering the yield criterion in plane strain to give roll pressure equations as follows:

On entry side of neutral plane $s^- = \left(1 - \frac{t_1}{2k_1}\right)\frac{2kh}{h_1}\,e^{\mu(H_1-H)}$ (4.10)

On exit side of neutral plane $s^+ = \left(1 - \frac{t_2}{2k_2}\right)\frac{2kh}{h_2}\,e^{\mu H}$ (4.11)

At the neutral point the roll pressures as given by the above equations are equal, and by equating them H_n, the value of H at the neutral point, can be found. From H_n the position of the neutral point ϕ_n can be determined

$$H_n = \frac{H_1}{2} - \frac{1}{2\mu}\ln\left\{\frac{h_1}{h_2}\left[\frac{1-(t_2/2k_2)}{1-(t_1/2k_1)}\right]\right\} \qquad (4.12)$$

and since

$$H = 2\sqrt{\left(\frac{R'}{h_2}\right)}\tan^{-1}\sqrt{\left(\frac{R'}{h_2}\right)}\phi$$

$$\phi_n = \sqrt{\left(\frac{h_2}{R'}\right)}\tan\frac{H_n}{2}\sqrt{\left(\frac{h_2}{R'}\right)} \qquad (4.13)$$

A typical distribution of stress under the rolls showing the effect of forward and backward tension in reducing roll pressure is shown in Fig. 4.23. The effect of other factors on the distribution of roll pressure can be seen in Fig. 4.24 (*a*), (*b*) and (*c*).

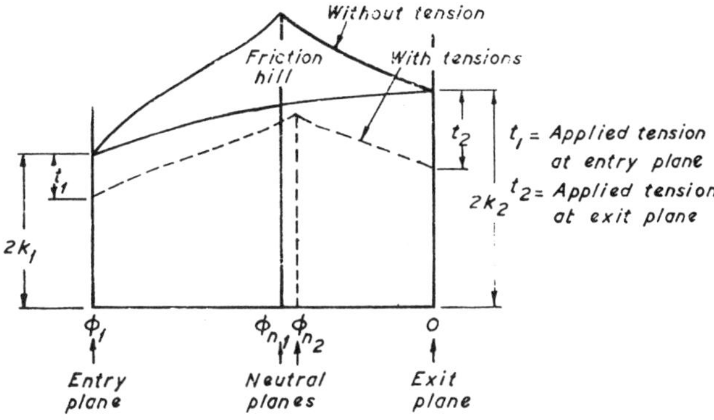

Fig. 4.23 Distribution of stress in cold rolling with and without strip tension

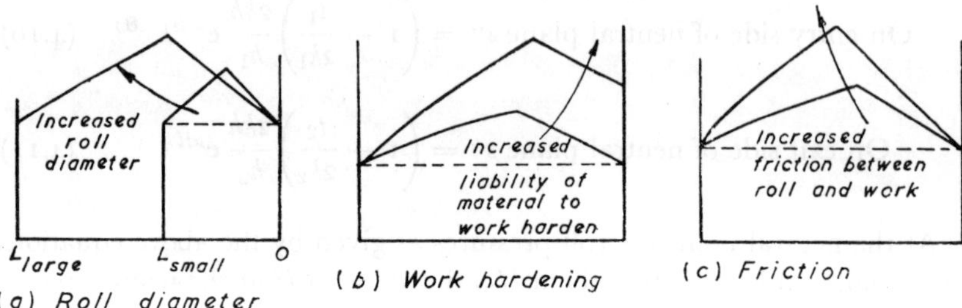

Fig. 4.24 Factors affecting stress distribution in cold rolling

Roll force. From Fig. 4.16 (*a*) it will be seen that

$$dx = R'd\phi \cos \phi \simeq R'd\phi[1 - (\phi^2/2)].$$

For cold rolling ϕ is small and $\phi^2/2$ can be neglected; therefore $dx \simeq R'\, d\phi$. The force per unit width of the roll F, can be found by integrating the roll pressure over the contact area

$$F = R' \left[\int_0^{\phi_n} s^+ \, d\phi + \int_{\phi_n}^{\phi 1} s^- \, d\phi \right] \tag{4.14}$$

$F = R' \times$ area under pressure distribution diagram.

Roll torque. The roll torque can be found by assuming that the roll force acts through the centre of the area of stress distribution. This centre may be assumed to be approximately $0{\cdot}45L$ from the roll centre line.

$\therefore$ torque per roll $T \simeq 0{\cdot}45FL$

Flattened roll Radius R'. Hitchcock's formula,[7] which assumes an elliptical distribution of roll pressure over the arc of contact, can be used to calculate the flattened roll radius. The formula is

$$R' = R\left(1 + \frac{2cF}{\delta}\right) \tag{4.15}$$

where R = unloaded roll radius, mm (in.)

$c = 1{\cdot}08 \times 10^{-5}\ \text{mm}^2\ \text{N}^{-1}$ ($1{\cdot}67 \times 10^{-4}\ \text{in}^2$ per tonf), for steel rolls

F = force per unit width of roll, N mm^{-1} (tonf per in)

δ = the draft $(h_1 - h_2)$, mm (in)

4.7.3 Simple assessment of roll load. By considering the metal to be homogeneously compressed between two well-lubricated platens the deformation model is greatly simplified. A similar simplification was assumed for hot rolling, Fig. 4.18 (*a*).

Projected area of contact $= Lb$

where L = projected length of arc of contact

b = width of strip being rolled

If homogeneous deformation is assumed, the load P required to compress the material between platens is given by the expression

$$P = Lb\,\bar{Y} \tag{4.16}$$

where $\bar{Y}$ = mean value of yield stress at entry and exit planes

It has been shown, towards the end of section 4.6.4, that the projected length of arc of roll contact L can be represented by

$$L = \sqrt{(R\delta)}$$

where R = the roll radius

δ = the draft, i.e. thickness reduction per pass

Hence substituting for L in equation (4.16)

$$P = b\bar{Y}\sqrt{(R\delta)}$$

Normally the strip thickness is very much less than the strip width, therefore plane strain deformation can be assumed and $\bar{Y} = 2\bar{k}$, where $\bar{k} = (k_{entry} + k_{exit})/2$. Friction can be taken into account by the use of a constant C; 1.2 is a frequently used value of C.

Hence

$$P = C\, b2\bar{k}\,\sqrt{(R\delta)} \tag{4.17}$$

Although this formula is simple it nevertheless can be used to calculate roll flattening, by means of Hitchcock's formula, and to determine the safe thickness reduction per pass for a given mill.

4.7.4 Effect of mill stiffness on roll gap. Although in most metal working operations the elastic deformation of the apparatus is negligible compared with the plastic deformation of the work, this is not so in cold rolling. Because of elastic deformation under load of the various mill parts, the thickness of metal rolled is often considerably in excess of the initial roll setting, h_0. The effect of load on roll gap is shown in Fig. 4.25; the stiffer the mill the steeper the 'elastic line.' Although an infinitely stiff mill is desirable, this cannot be achieved owing to the ever present elasticity of the rolls and roll necks.

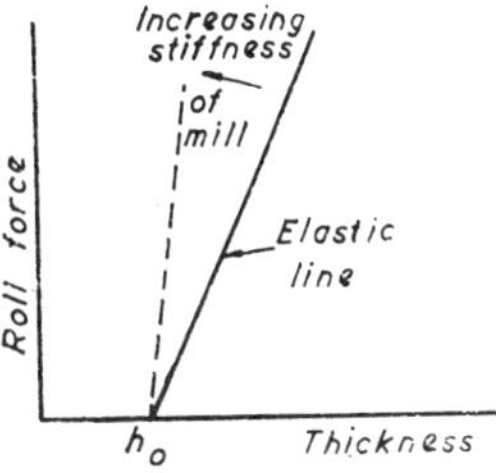

Fig. 4.25 Elastic deformation of mill

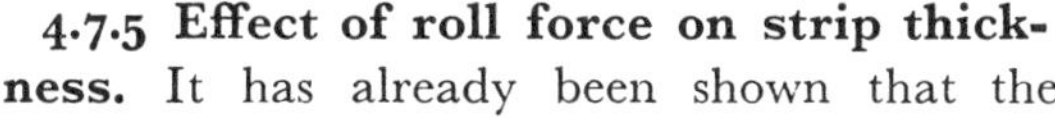

4.7.5 Effect of roll force on strip thickness. It has already been shown that the greater the draft $h_1 - h_2$ the larger the required roll force. The variation of thickness with roll force is called the 'plastic line.' Ideally this line should be as near horizontal as possible; factors reducing the slope of the 'plastic line' are low yield strength material, low friction between work and rolls, and increased applied tension. A typical 'plastic line' is shown in Fig. 4.26.

4.7.6 Gauge control. The intersection of the elastic and plastic lines determines the thickness of the rolled strip h_2 (Fig. 4.27). When rolling strip, the thickness must be controlled to within close limits. Many factors can cause the thickness of the strip to change. These include variation in thickness of the ingoing stock, and changes in material hardness, roll speed or lubrication. Rolling conditions must be rapidly changed to

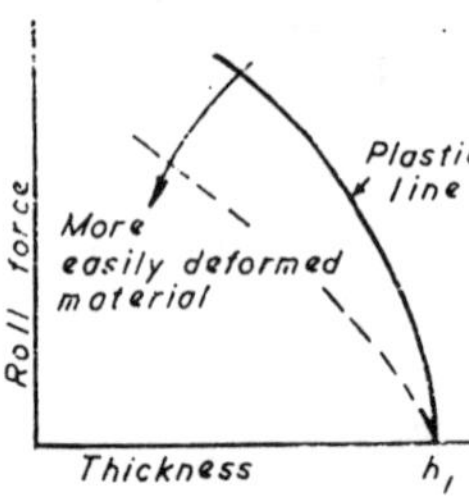

Fig. 4.26 Plastic deformation of work

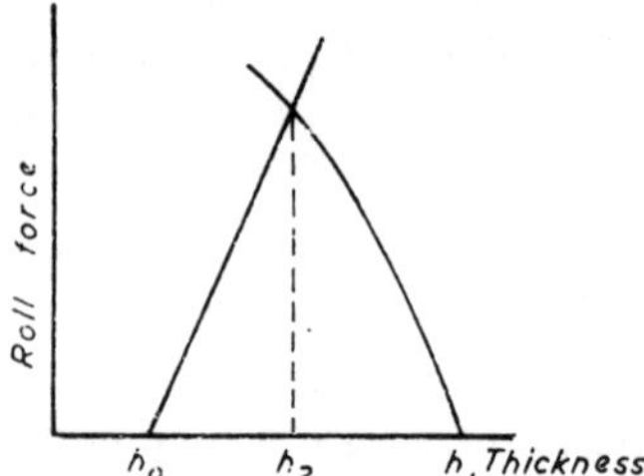

Fig. 4.27 Determination of strip thickness

maintain the correct gauge thickness. Perhaps the most obvious method is to adjust the roll gap, another is to change the tension applied to the strip. The effect of these two methods of gauge control is shown graphically in Figs. 4.28 (*a*) and (*b*).

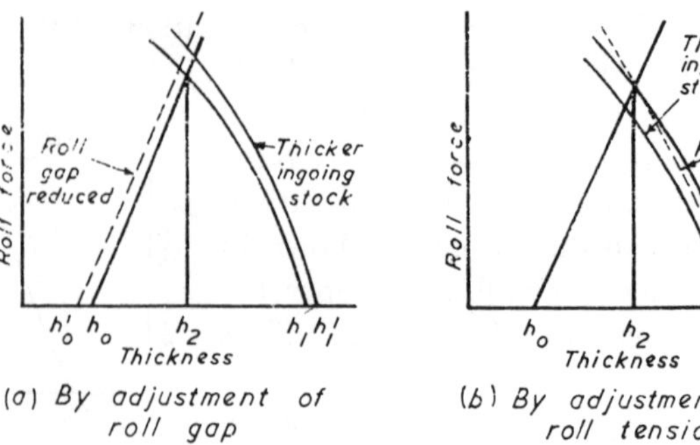

Fig. 4.28 Gauge control

Variation of applied tension is the more sensitive method of adjusting strip thickness and can be used with automatic control systems. The rolled strip passes a thickness sensing device, the signal from which is compared with another representing the desired thickness. Any resulting error signal is amplified and used to adjust tension and hence strip thickness.

4.8 TRANSVERSE WEDGE ROLL FORMING

This process hot rolls steel components having circular section from bar stock; Fig. 4.29 shows the principle of transverse wedge rolling. The rolls

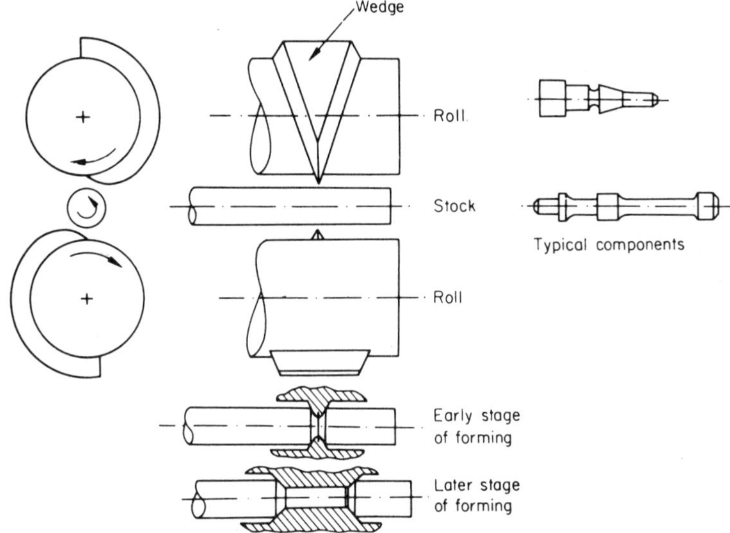

Fig. 4.29 Transverse wedge roll forming

progressively squeeze the component from its centre, elongating and shaping it into its final form; when completed it is parted off by the roll. The whole process lasts for only a few seconds and can be arranged for automatic operation. The standard of surface finish is comparable with that of hot rolling, with tolerances of around $\pm$ 1 mm (0·04 in). A diameter reduction ratio of 2 to 1 is possible and tool lives in excess of 50 000 parts are claimed.

4.9 COLD ROLLING OF ANNULAR PARTS

This process is also called flow forming and has been developed to produce annular parts, such as bearing races, in large quantities.

A blank, parted off from tube which has been machined on both its outside and inside diameters, is formed by rolling between appropriately shaped dies. The rolling of the inner races is shown in Figs. 4.30 (*a*) and (*b*). A method of rolling outer races is shown in Fig. 4.30 (*c*); this method is suitable only for light forming—heavier forming of outer races requires a mandrel and forming rolls, having a similar arrangement to that shown in Fig. 4.30 (*b*).

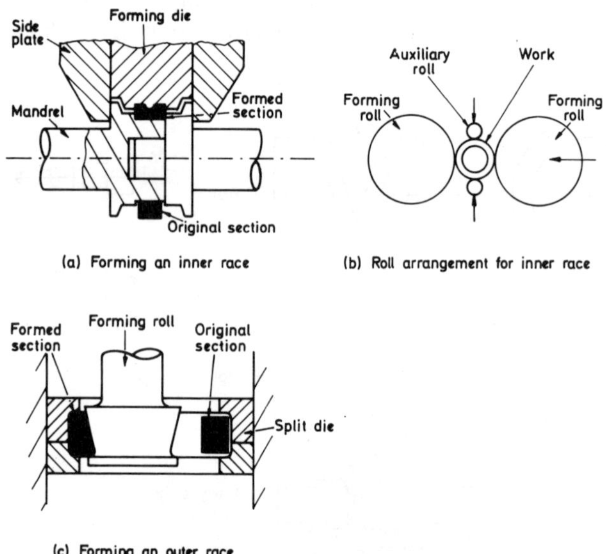

Fig. 4.30 Rolling of races

Compared with machining parts from tube, flow forming gives higher production rates with more consistent quality; there is also greater component strength and less distortion after heat treatment. In addition change-over times are less, thereby facilitating small batch production.

A considerable capital investment in rolling machines is, however, required, which can be justified only if machine utilization is sufficiently high.

5 Extrusion, Tube-making and Cold Drawing

5.1 EXTRUSION

The process of extrusion consists of forcing a billet of metal through a die to produce a continuous length of constant cross section corresponding to the shape of the die orifice. A simple analogy of the process is the squeezing of toothpaste from a tube. Hot extrusion only will be described in this chapter: cold extrusion will be dealt with in Chapter 6 with cold forging processes.

Although extrusion is commercially a comparatively recent metal forming process its conception can be traced back to a patent granted in 1797 to Joseph Bramah, a famous British engineer, for a machine to manufacture pipe from lead and other soft metals. There is no evidence that Bramah's machine ever produced pipes, but during the 19th century the extrusion of lead was successfully developed, first to make pipes and then to sheath electric cables. Copper alloys were first extruded in 1894 by A. G. Dick, a founder of the Delta Metal Co. The extrusion of copper alloys was the turning point in the development of the process and made extrusion one of the major methods of metal working.

The two basic methods are shown in Figs. 5.1 (*a*) and (*b*). Direct extrusion is the more popular method, the extrusion press being mechanically simpler. Indirect or inverted extrusion does, however, require less force and provides a better quality product, as it minimizes the amount of scale from the outside of the billet flowing into the extrusion. With both types the container can move or remain stationary. The more usual arrangement is that shown in Figs. 5.1 (*a*) and (*b*), where the container is stationary in direct extrusion, and moves in inverted extrusion.

5.1.1 Extruded products. Although copper and aluminium alloys are the most common metals to be extruded, the extrusion of steel and its alloys is now possible. Dick put forward the idea of extruding steel in 1893, but it was made commercially possible by the Ugine Séjournet process,[8] in which molten glass is used to lubricate the die and minimize the heat loss from the billet.

Extrusions can be either solid or hollow, the latter being produced from pierced or bored billets which are extruded around a mandrel (Fig. 5.2). Tapered holes can be produced by using long tapered mandrels which are fed forward at the same speed as the extruded metal.

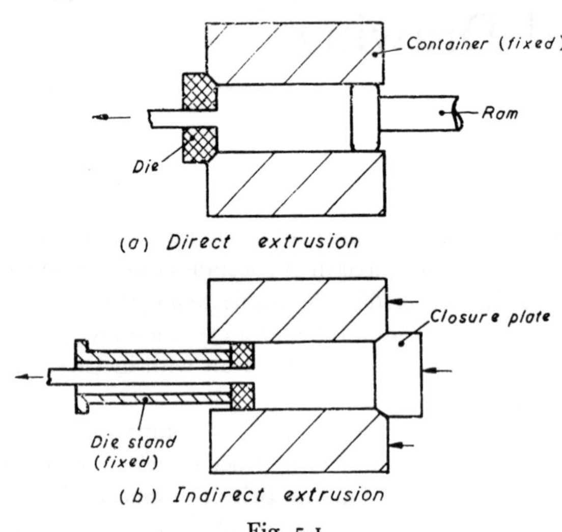

Fig. 5.1

Although large tonnages of simple shapes, such as tubes and hexagonal bars, are extruded, great economic advantages come from the production of complicated cross sections which otherwise could be produced only by expensive machining operations. Frequently extrusions are not required as bars, but are cut into pieces and machined to make components such as the bolt plates for doors shown in Fig. 5.3.

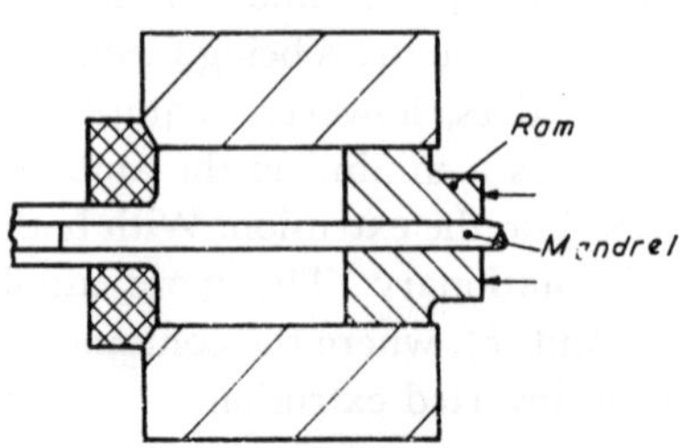

Fig. 5.2 Hollow extrusion

Stepped extrusions can be produced by using a two-piece die. The process is halted when a sufficient length of the smaller cross section has been extruded; the minor die is then removed and the extrusion completed using the larger aperture of the major die (Fig. 5.4).

A combination of forging and extrusion can also be used to manufacture some stepped parts. Poppet valves for internal combustion engines are produced by this method from heated steel slugs, as shown in Fig. 5.5.

5.1.2 Extrusion equipment. Extrusion presses are usually hydraulically operated, although mechanical presses are occasionally used. Capacities vary up to 200 MN (20 000 tonf), larger machines being used for heavier extrusions or stiffer metals such as steel or titanium. In most presses the container is placed horizontally, although some machines, particularly those of smaller capacity, have vertical cylinders.

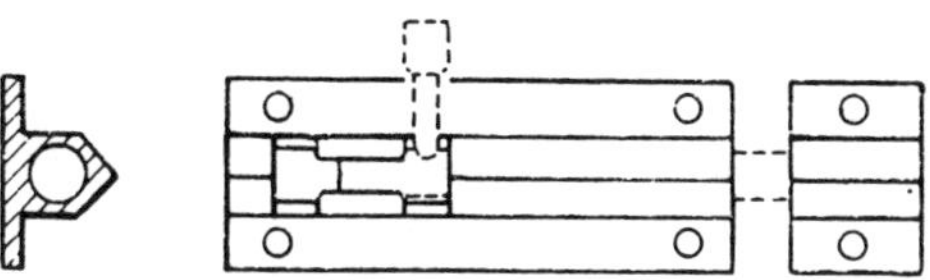

Fig. 5.3 Extruded bolt plates

As the type of extrusion described in this chapter is a hot working process, the inside of the container, the die and the mandrel have to withstand elevated temperatures as well as high pressures. The container is designed with a replaceable liner, which is shrunk into position. This arrangement reduces the bursting stress during extrusion and hence the chance of mechanical failure. When extrusion temperatures are below

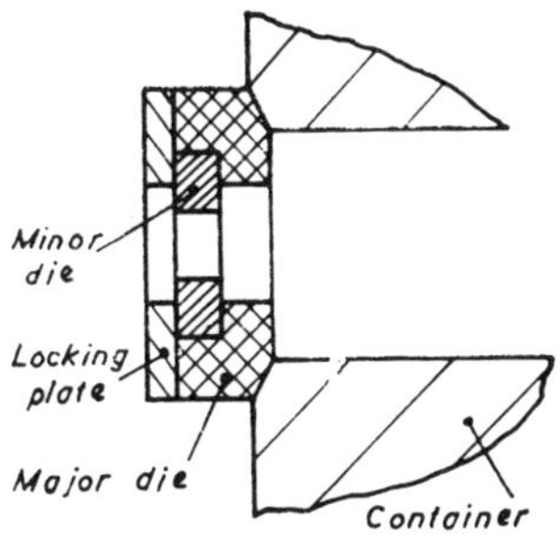

Fig. 5.4 Two piece die for stepped extrusions

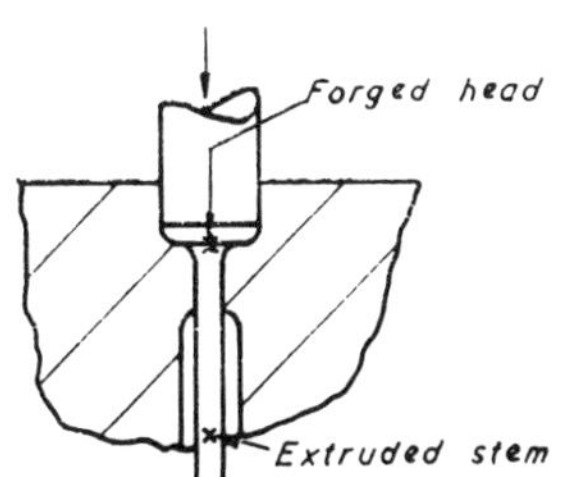

Fig. 5.5 Production of poppet valve

800°C nickel-chrome-molybdenum steels can be used for containers, but at higher temperatures an alloy steel which includes about 10% tungsten has to be employed. When heat treated the liner steels have an ultimate tensile strength of about 1400 N mm^{-2} (90 tonf/in^2).

Dies are produced either from heat resistant steels or tungsten carbide, the latter having a much longer die life.

Mandrels are subjected to particularly hard service, especially if they have to pierce the billet in the containers before extrusion. If small in diameter they heat up very quickly and may require cooling after each extrusion.

5.1.3 Heating of billets and containers. Billets should be uniformly preheated before extrusion. The preheat temperatures depend on the metal being extruded, reaching 1200°C for steel. Heating can be in specially designed furnaces or by low-frequency induction heaters. Induction heating is particularly useful when a temperature gradient is required to offset the rise in temperature which occurs at the ram end of a billet in direct extrusion. Temperature rises of 60°C during extrusion have been recorded with aluminium billets at high rates of deformation, and these can cause cracking of the extruded product.

Many extrusion presses are fitted with means of heating the container to prevent cooling during slow extrusion. Container heating is of particular importance when high-strength aluminium or magnesium alloys are being extruded, as these materials have to be extruded at low speeds.

5.1.4 Pressure variation during extrusion. Typical pressure variations during direct and indirect extrusion are shown in Fig. 5.6. Initially the pressure rises rapidly from zero as the billet is being expanded to fill the container completely. Extrusion then commences; a higher pressure is needed for direct extrusion, the additional force being required to overcome friction between the billet and the container wall. Pressure decreases in direct extrusion as the ram moves along the container and the total frictional force is reduced. Towards the end of the travel, the pressure for the direct method falls to the same value as that needed for inverted extrusion. The pressure rises in both graphs at the very end of the travel, due to the difficulty of making the thin plate of metal left in the container flow out of the die aperture.

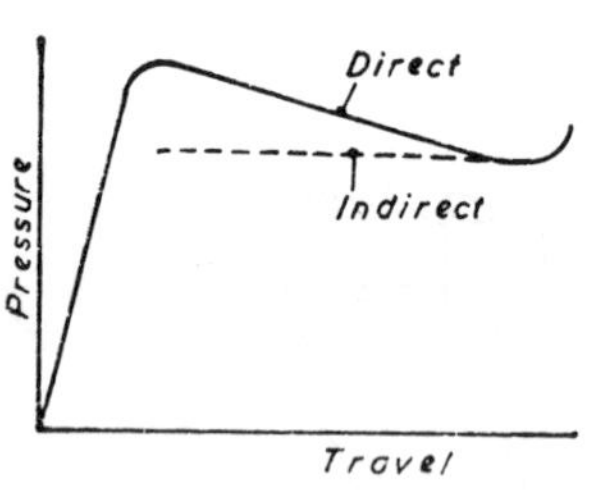

Fig. 5.6 Extrusion pressures

In practice, the whole of the billet is not extruded and a stub end, called the discard, is left in the container. If the discard were extruded the quality of the extrusion would be adversely affected due to oxide inclusions. When extruding brass billets it is usual to employ a ram sufficiently small in diameter to leave a thin tube of metal, called a skull, in the container. Most of the oxide layer and surface imperfections at the outside diameter of the billet are thereby left behind in the container. When aluminium alloys are extruded, turned billets are used to remove surface imperfections and preheating is performed in controlled atmosphere furnaces to minimize the formation of oxide.

5.1.5 Effect of extrusion speed and pressure. The faster the speed of extrusion, the closer the process comes to adiabatic conditions, due to lower heat loss, and the greater the temperature rise in the billet. Although extrusion pressure reduces as the billet temperature increases, a limit is set to the preheat temperature and extrusion speed by the onset of hot shortness. This appears as a cracking of the extrusion as it leaves the die, caused by the melting of the alloy constituent which has the lowest

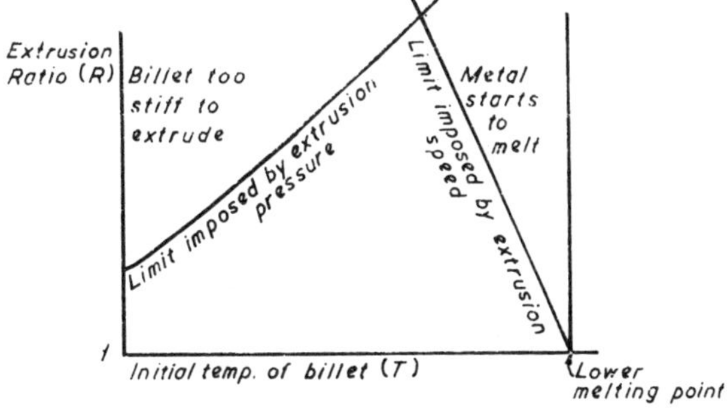

Fig. 5.7 Limitations imposed by given extrusion pressure and speed (*after Hirst and Ursell*)

melting temperature. At very slow extrusion speeds in an unheated container the billet becomes stiffer, requiring a higher extrusion pressure, and in extreme cases may become a 'sticker' which cannot be further extruded.

A diagram showing the interaction of billet temperature, extrusion speed and extrusion ratio from Hirst and Ursell[9] is shown in Fig. 5.7. The extrusion ratio R is used to indicate the degree of deformation and is A_1/A_2, where A_1 is the cross-sectional area of the billet and A_2 that of the extrusion. The effect of varying the extrusion pressure and extrusion speed is indicated in Fig. 5.8.

5.1.6 Determination of extrusion pressure. An early approach was based on the work done in homogeneous deformation with friction and redundant work being allowed for by an efficiency factor.

The work done per unit volume W/V in homogeneous deformation is given by

$$\frac{W}{V} = Y \int_{l_1}^{l_2} \frac{dl}{l} = Y \ln \frac{l_2}{l_1}$$

where Y is the yield strength of the material,

l_1 is the original length

and l_2 the extruded length.

Assuming constancy of volume i.e. $A_1 l_1 = A_2 l_2$ and substituting the extrusion pressure p for $W/V(W/V = pAl/Al)$

$$p = Y \ln \frac{A_1}{A_2}$$

An efficiency factor β is used to take account of the additional work needed to overcome friction and to allow for redundant work. The value

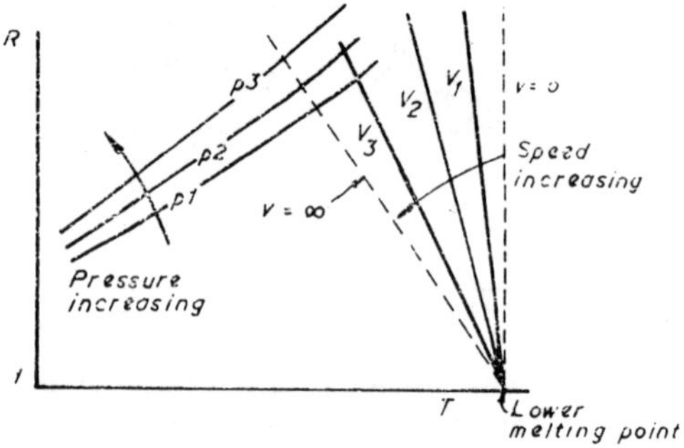

Fig. 5.8 Effect of varying extrusion pressure and speed
(after Hirst and Ursell)

of β can be obtained experimentally for a range of values of A_1/A_2 and in its final form the expression for extrusion pressure becomes

$$p = \beta Y \ln \frac{A_1}{A_2}$$

A second approach, applicable to lubricated extrusion, combines a stress evaluation method to find the effect of friction and a semi-empirical formula for frictionless extrusion pressure.

Consider a thin rigid slice of billet, width dx, being extruded by the direct method (Fig. 5.9). Since this part of the billet is rigid it can be assumed to be subject to a hydrostatic pressure p.

Resolving horizontally

$$\frac{\pi}{4} D^2 \, dp = \pi D \mu p \, dx$$

where D is the billet diameter

μ coefficient of friction between billet and container wall,
p horizontal pressure on the slice

$$\frac{dp}{p} = \frac{4}{D}\mu\, dx$$

Integrating between limits of O and L,

$$\frac{p}{p_0} = e^{(4\mu L/D)}$$

$$p = p_0\, e^{(4\mu L/D)} \qquad (5.1)$$

where p_0 is the frictionless extrusion pressure.

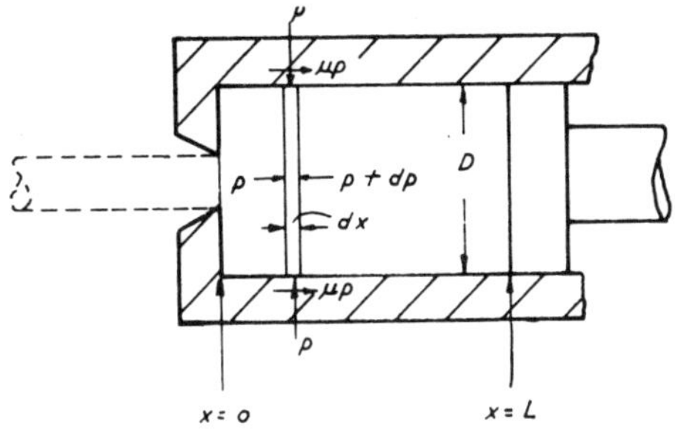

Fig. 5.9

A formula developed by Johnson[10] for the extrusion of short billets can be used to calculate the frictionless extrusion pressure p_0

$$p_0 = Y\left(0{\cdot}47 + 1{\cdot}2 \ln \frac{A_1}{A_2}\right)$$

where Y is the yield stress applicable to the mean extrusion temperature
A_1 is the cross-sectional area of the billet
A_2 is the cross-sectional area of the extrusion.

By substituting for p_0 in equation (5.1) the extrusion pressure p can be found.

$$p = Y\left(0{\cdot}47 + 1{\cdot}2 \ln \frac{A_1}{A_2}\right) e^{(4\mu L/D)} \qquad (5.2)$$

Hirst and Ursell suggested a method of finding the coefficient of friction at the container wall. This is done by extruding, under the same conditions, two billets of similar material but of different length and measuring the ram pressures, p_1 and p_2 at the commencement of each extrusion:

From equation (5.1) $$\frac{p_1}{p_2} = \frac{p_0\, e^{(4\mu L_1/D)}}{p_0\, e^{(4\mu L_2/D)}}$$

Hence $$\frac{p_1}{p_2} = e^{(4\mu/D)(L_1 - L_2)}$$

and $$\mu = \frac{D}{4(L_1 - L_2)} \ln\left(\frac{p_1}{p_2}\right)$$

5.2 TUBE MAKING

There are two basic types of tube, seamless and fabricated. Fabricated tube is formed from strip or plate and either welded along its joint or, for much electric conduit tubing, left unsecured. Non-ferrous tubes are usually extruded (Fig. 5.2), but are often finished by drawing.

5.2.1 Rotary forging. Most ferrous seamless tube is first rotary forged. This consists of two hot working processes of which the first is rotary piercing.

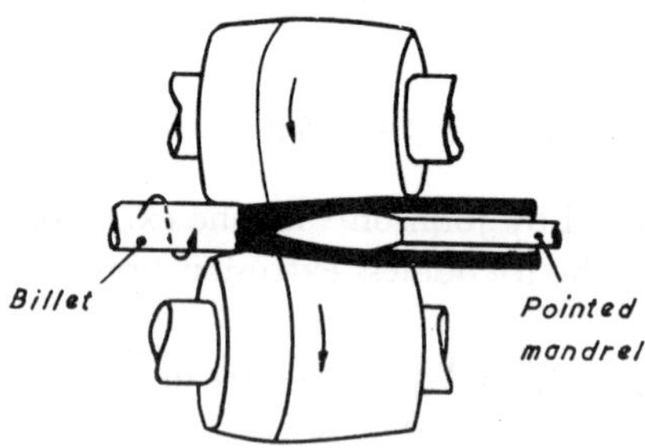

Fig. 5.10 Mannesmann mill

In rotary piercing a specially designed rolling mill is used. The two rolls of the mill are set at an angle to each other, so that the metal is not only deformed and fed forward, but is also rotated. Because of the small diameter of the rolls, the outside of the billet is deformed and a tensile stress produced at its centre. As a result, a cavity is induced at the centre of the billet; the formation of the cavity is assisted and controlled by a carefully profiled point, mounted on a mandrel. There are several mills used for rotary piercing; one of these, the Mannesmann mill, is illustrated in Fig. 5.10. The severe deformation produced by rotary piercing demands a steel free from faults and, due to the variation of temperature during piercing, one which remains ductile over a range of temperatures.

A three-roll method of tube piercing has been developed by Tube Investments Ltd. The billet is rotated between three shaped rolls which are orientated at 120° and have their axes inclined at a feed angle, so that the work is moved forward as well as rotated. In the three-roll method there is no tensile stress ahead of the piercer and the tendency of the billet

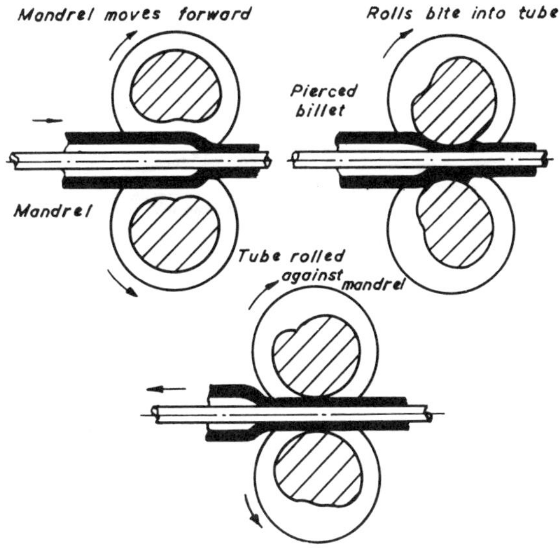

Fig. 5.11 Pilger mill

to open up and produce tears is avoided. This is particularly important if good quality hollow blooms are to be produced from continuously cast metal, as this process creates a central weakness which makes it unsuitable for piercing by the Mannesmann process. The use of three-roll piercing for continuously cast low carbon steel is described in an article by Metcalfe and Holden.[11]

The second stage in seamless tube production is to put the roughly formed tube on a mandrel and roll it to reduce its wall thickness. A Pilger mill is one which can be used for this operation, and the roll arrangement and operation sequence are shown in Fig. 5.11. The heated tube reciprocates under cam-shaped rolls, the profiles of which are shaped so that they first bite into the tube wall and then forge down the bite against the mandrel. During forging, the mandrel is pushed backwards against pneumatic pressure until the relieved portion of the roll profile is reached.

The tube is then moved forward so that the next bite can be taken. No theoretical analysis of rotary forging is available.

5.2.2 Tube drawing. If seamless tubes are required with either small diameters, thin walls or a smooth surface finish, they are completed by cold drawing. To remove the scale left by rotary forging the tube is first pickled in heated dilute sulphuric acid.

A draw bench is used to pull the tube through the drawing die. The die is mounted at one end of the bench and an end of the tube is collapsed so that it can be taken through the die and clamped to a carriage. The draw

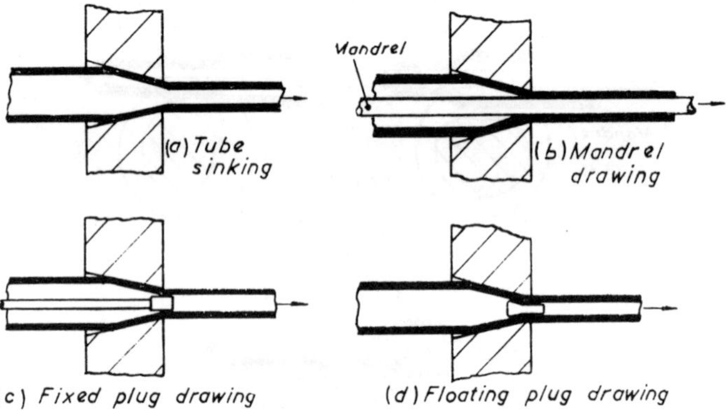

Fig. 5.12 Methods of tube drawing

bench can be operated either mechanically or hydraulically; often several tubes are drawn simultaneously through multiple dies to increase productivity.

As with wire and bar drawing (to be described later), adequate lubrication of the die surface must be maintained. Poor lubrication is undesirable as drawing forces are increased and, more seriously, metal transfer can occur in both directions between work and die. Metal transfer spoils the surface of the work and reduces die life, sometimes catastrophically. Lubrication is by means of soaps; the prior application of a phosphate coating is also of considerable benefit when drawing steel.

Dies are made from alloy steel or tungsten carbide. Tungsten carbide dies have a longer life and are not as liable to metal transfer as those made from alloy steel.

The simple drawing of tube through a die is called 'sinking', however, if an accurate internal diameter is required the tube is drawn on a mandrel or with a plug in position. The various methods of tube drawing are illustrated in Fig. 5.12.

5.2.3 Analysis of tube sinking. This analysis is similar to that first proposed by Sachs and Baldwin.[12] It is assumed that the wall thickness is small compared with the tube diameter, that the tube thickness remains constant during drawing, and that the normal stress acting on the transverse section is uniformly distributed.

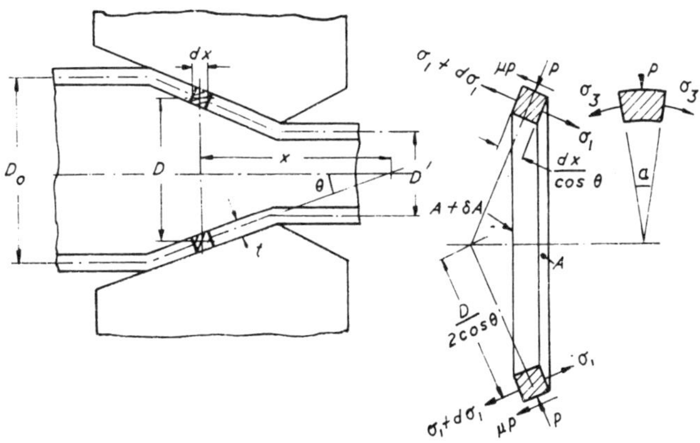

Fig. 5.13 Stresses in tube sinking

Under steady drawing conditions the forces acting on the elemental ring (Fig. 5.13) are in equilibrium.

Resolving horizontally

$$\cos\theta[(\sigma_1 + d\sigma_1)(A + dA) - \sigma_1 A] + p\sin\theta\,\frac{dx}{\cos\theta}\,\pi D + \mu p\cos\theta\,\frac{dx}{\cos\theta}\,\pi D = 0$$

Ignoring the products of infinitesimals

$$\cos\theta(A d\sigma_1 + \sigma_1 dA) + p\,dx\pi D(\tan\theta + \mu) = 0 \qquad (5.3)$$

Using the relationships

$$D = 2x\tan\theta \qquad dD = 2\,dx\tan\theta$$

$$A = \pi t D \qquad dA = \pi t dD = 2\pi t\,dx\tan\theta$$

Substituting in equation (5.3)

$$\pi t\cos\theta(D d\sigma_1 + \sigma_1 dD) + \tfrac{1}{2}\pi D dD p\left(\frac{\tan\theta + \mu}{\tan\theta}\right) = 0$$

Using B for $(\tan\theta + \mu)/\tan\theta$

$$\pi t \cos\theta(D\mathrm{d}\sigma_1 + \sigma_1 \mathrm{d}D) + \tfrac{1}{2}\pi D\mathrm{d}D\, p\, B = 0 \qquad (5.4)$$

Considering the forces perpendicular to the die surface acting on a small element of the tube wall

$$p \frac{D}{2\cos\theta}\alpha\frac{\mathrm{d}x}{\cos\theta} + t2\sigma_3\frac{\alpha}{2}\frac{\mathrm{d}x}{\cos\theta} = 0$$

$$p = -\frac{2\sigma_3 t\cos\theta}{D}$$

Substituting for p in equation (5.4) gives the general differential equation for tube sinking

$$D\mathrm{d}\sigma_1 + \sigma_1\mathrm{d}D = B\sigma_3\,\mathrm{d}D \qquad (5.5)$$

Using Tresca's modified criterion for yielding $\sigma_1 - \sigma_3 = Y_0$, where $Y \leqslant Y_0 \leqslant (2/\sqrt{3})Y$

Substituting in equation (5.5) for σ_3

$$D\mathrm{d}\sigma_1 + \sigma_1\mathrm{d}D = B\mathrm{d}D(\sigma_1 - Y_0)$$

Differentiating

$$\frac{\mathrm{d}\sigma_1}{\mathrm{d}D} + \frac{\sigma_1}{D}(1 - B) = -\frac{BY_0}{D}.$$

This is a linear differential equation of the form $(\mathrm{d}y/\mathrm{d}x) + Py = Q$ the solution to which is

$$y = \mathrm{e}^{-\int P\mathrm{d}x}\int Q\mathrm{e}^{\int P\mathrm{d}x}\,\mathrm{d}x$$

Hence

$$\sigma_1 = \frac{B}{B-1}Y_0\left[1 - \left(\frac{D}{D_0}\right)^{B-1}\right]$$

and

$$\sigma_1' = \frac{B}{B-1}Y_0\left[1 - \left(\frac{D'}{D_0}\right)^{B-1}\right] \qquad (5.6)$$

where σ_1' is the value of drawing stress at the throat of the die for a non-work-hardening metal.

5.3 WIRE DRAWING

Wire drawing is an old craft, the first drawing equipment having been introduced into this country from Germany in the 16th century. Today coiled hot-rolled rod, from which the scale has been removed, is drawn through a die and coiled on a motor driven block. Dies are normally manufactured from tungsten carbide, although diamonds are used for drawing very small diameter wire. Soaps are used as lubricants and are picked up by passing the undrawn wire through a container. Dry soap is used for slower drawing speeds and produces a dull finish; an aqueous solution of soap is used for faster drawing speeds and leaves a bright surface. The lubrication of steel wire can be greatly improved by phosphating.

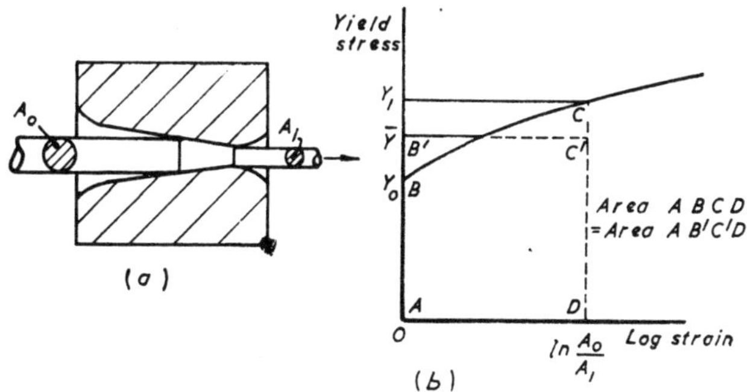

Fig. 5.14 Wire drawing

Several blocks and dies can be arranged one after the other so that progressive reductions are obtained within a single machine. Drawing speeds must be carefully matched so that the length of wire between blocks remains constant. Final speeds on some multiple die machines can be as high as several thousand feet per minute.

Ultrasonic vibrations have been applied to the die in the direction of drawing, the drawing stress is thereby reduced and in consequence a larger reduction is possible.

5.3.1 Forces in wire drawing. When wire is drawn, the work done can be considered to be the sum of the useful work, the work done against friction and the redundant work.

If it is assumed that the work material deforms homogeneously and there is no friction between the wire and the die, the work formula as described in Chapter 2.9 can be used:

$$\text{useful work/unit volume} = \int Y \frac{dA}{A}$$

The yield stress Y increases with strain due to work hardening but to simplify the equation an average value $\bar{Y}$ is assumed (Fig. 5.14 (*b*)).

$$\therefore \quad \text{useful work per unit volume} = \bar{Y}\int_{A_1}^{A_0}\frac{\mathrm{d}A}{A} = \bar{Y}\ln\frac{A_0}{A_1} \qquad (5\cdot7)$$

It has been shown in Section 2.10 that work done/unit volume is equal to the deforming pressure F/A which, in this instance, is the drawing stress. If the values of drawing stress for various strains are plotted, the curve will cross the stress/strain curve for the material. The point of intersection A (Fig. 5.15) indicates the maximum reduction possible without breaking the drawn wire. Both friction and redundant work increase the drawing stress and, as will be seen in Fig. 5.15, reduce the maximum permissible reduction.

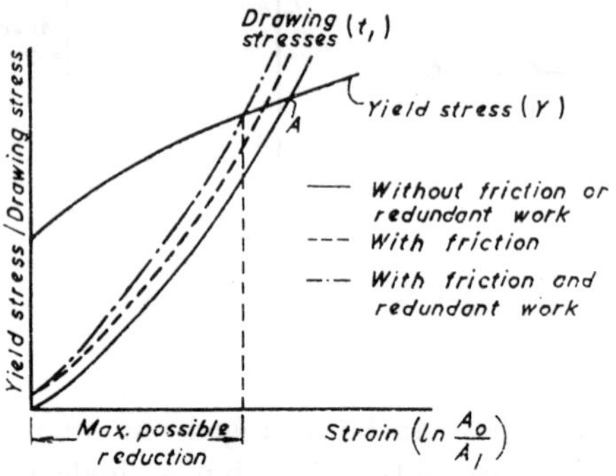

Fig. 5.15 Stresses in wire drawing

5.3.2 Analysis of wire drawing. The theory of wire drawing to be described is that proposed by Siebel.[13] A diagram of the process is shown in Fig. 5.16, where

q_m is mean die pressure on work,
μ is coefficient of friction between die and work,
F is the drawing force,
Q is the total force on die,
ρ is the angle of friction $= \tan^{-1}\mu$
θ is $\frac{1}{2}$ die angle,
l is length of contact between work and die.

Assuming equilibrium under steady drawing conditions and resolving forces horizontally

$$F = Q\sin(\theta + \rho) \qquad (5\cdot8)$$

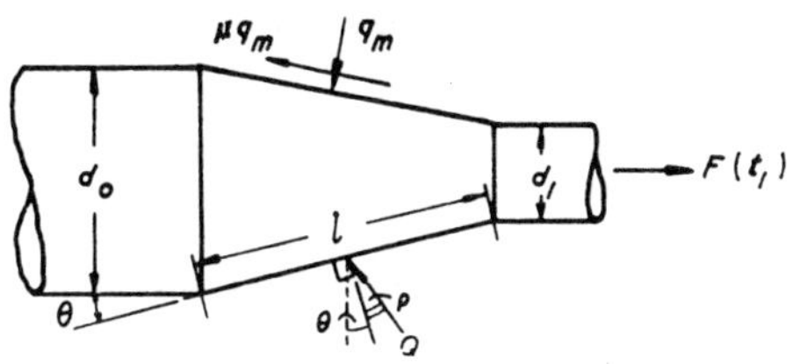

Fig. 5.16 Forces in wire drawing

As the die angle θ and the angle of friction ρ are small, it will be seen from equation (5.8) that Q is considerably greater than F and the process is basically one of compression.

The area of contact between work and die can be found by reference to Fig. 5.16,

$$l = \frac{\frac{1}{2}(d_0 - d_1)}{\sin \theta}$$

$$\text{Average circumference} = \frac{d_0 + d_1}{2} \pi$$

$$\text{Area of contact} \simeq \pi \frac{(d_0 + d_1)(d_0 - d_1)}{4 \sin \theta} \simeq \frac{A_0 - A_1}{\sin \theta}$$

Siebel assumed that the mean die pressure q_m reaches the mean yield stress of the material $\bar{Y}$, and as the area of contact between work and die is approximately $(A_0 - A_1)/\sin \theta$, a second expression for Q can be obtained

$$Q = \bar{Y} \frac{A_0 - A_1}{\sin \theta}$$

Substituting for Q in equation (5.8)

$$F = \bar{Y}(A_0 - A_1) \frac{\sin (\theta + \rho)}{\sin \theta}$$

By expanding $\sin (\theta + \rho)$ and assuming the following relationships (which are approximately true for small angles)

$$\tan \theta \simeq \theta \text{ and } \rho \simeq \sin \rho \simeq \mu$$

$$F \simeq \bar{Y}(A_0 - A_1) \left(1 + \frac{\mu}{\theta}\right)$$

As work done per unit volume equals drawing stress, t'

$$t' = \frac{F}{A_1} = \frac{Y}{A_1}(A_0 - A_1)\left(1 + \frac{\mu}{\theta}\right)$$

The useful work per unit volume has been found in equation (5.7) to be $\bar{Y} \ln (A_0/A_1)$ and by consideration of conventional and logarithmic strains

$$\bar{Y} \ln \frac{A_0}{A_1} \simeq \frac{\bar{Y}}{A_1}(A_0 - A_1)$$

$$\text{Hence } t' = \bar{Y}\left(1 + \frac{\mu}{\theta}\right) \ln \frac{A_0}{A_1} \tag{5.9}$$

By subtracting useful work, $\bar{Y} \ln (A_0/A_1)$ from (5.9) the remainder, $\bar{Y}\mu/\theta \ln A_0/A_1$ represents the frictional work per unit volume. If an expression for the redundant work could be found and added to equation (5.9) a complete expression for the drawing tension would be obtained.

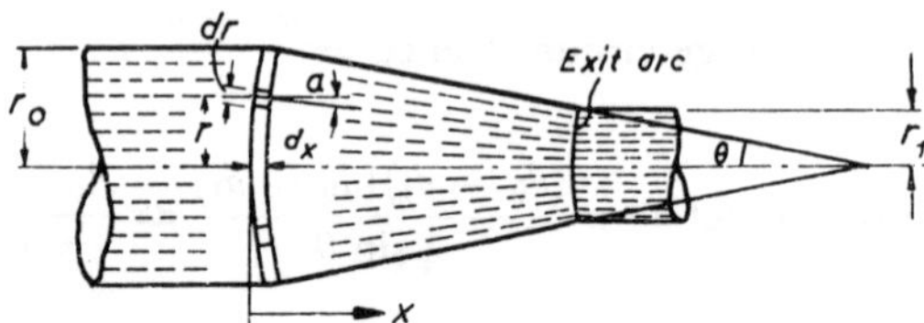

Fig. 5.17 Redundant work in wire drawing

Redundant work may be evaluated by considering Fig. 5.17. Almost all of it may be assumed to occur from shearing of the metal at the entry and exit arcs. When material crosses the entry arc at radius r, the work done per unit volume $= k\alpha = k(r/r_0)\theta$, where k is the shear flow stress of the material.

When an elemental ring of material radius r, thickness dr and length dx passes the entry arc:

$$\text{redundant work on elemental ring} = 2\pi r \, \mathrm{d}r \, \mathrm{d}x \, k\left(\frac{r}{r_0}\right)\theta$$

$$\text{redundant work at die entry} = 2\pi \, k \, \frac{\theta}{r_0} \, \mathrm{d}x \int_0^{r_0} r^2 \, \mathrm{d}r$$

$$\text{redundant work at die entry} = \tfrac{2}{3}\pi \, k\theta \, \mathrm{d}x \, r_0^2$$

volume of material at die entry $\simeq \pi r_0^2 \, \mathrm{d}x$

redundant work/unit volume at die entry $= \tfrac{2}{3}k\theta$.

Using Tresca's criterion of yielding $Y_0 = 2k$

$$\text{redundant work/unit volume at die entry} = \frac{Y_0\theta}{3}$$

The redundant work at exit from the die is given by a similar expression, $Y_1\theta/3$. It will be appreciated that $Y_1 > Y_0$ because of work hardening of the material.

Assuming an average value of yield stress $\bar{Y}$, where $\bar{Y} = (Y_0 + Y_1)/2$:

$$\text{Total redundant work/unit volume} = \tfrac{2}{3}\theta\,\bar{Y} \qquad (5.10)$$

The total drawing tension t can now be obtained by combining equations (5.9) and (5.10)

$$t = \bar{Y}\left[\left(1 + \frac{\mu}{\theta}\right)\ln\frac{A_0}{A_1} + \tfrac{2}{3}\theta\right] \qquad (5.11)$$

The maximum reduction in area which is possible without breaking the wire can be found by equating the total drawing stress (t) to the yield stress of the material ($\bar{Y}_1$).

5.4 BAR DRAWING

Large quantities of bars are finished by cold drawing. This process improves the strength of work hardening materials and imparts a smooth bright surface. The surface finish is frequently good enough to be incorporated unmachined in a component; for example, the hexagon heads of turned bolts. Drawing also accurately sizes the bar to within a few thousandths of an inch, an essential requirement for some types of machine tool collets.

As in tube drawing, a draw bench is normally used to pull the pickled hot rolled bars through rigidly mounted tungsten carbide or alloy steel dies. Sizes less than about 12 mm ($\frac{1}{2}$ in) in diameter are not produced on a draw bench but continuously round a drum. The coil is subsequently cut and straightened.

5.4.1 Analysis of flat strip drawing through wedge-shaped dies. The process is shown in Fig. 5.18 and it is assumed that the material is wide enough for plane strain to apply. In Fig. 5.18

t is drawing tension

p is normal pressure of die on material

μ is coefficient of friction between work and die
h is thickness of strip
θ is $\frac{1}{2}$ angle of the die.

Under steady drawing conditions, the forces on an elemental strip of material having unit width are in equilibrium.
Resolving horizontally

$$(t + \mathrm{d}t)(h + \mathrm{d}h) + 2\mu p \frac{\mathrm{d}x}{\cos\theta} \cos\theta + 2p \frac{\mathrm{d}x}{\cos\theta} \sin\theta = th$$

Ignoring products of differentials

$$t\,\mathrm{d}h + h\,\mathrm{d}t + 2\mu p\,\mathrm{d}x + 2p\,\mathrm{d}x \tan\theta = 0$$

Using the relationship $\mathrm{d}h = 2\,\mathrm{d}x \tan\theta$

$$t\,\mathrm{d}h + h\,\mathrm{d}t + \mu p \cot\theta\,\mathrm{d}h + p\,\mathrm{d}h = 0$$

$$h\,\mathrm{d}t + [t + p(1 + \mu\cot\theta)]\,\mathrm{d}h = 0 \qquad (5.12)$$

To integrate the above expression, a relationship between t and p must be obtained. This can be done by assuming that p acts vertically and

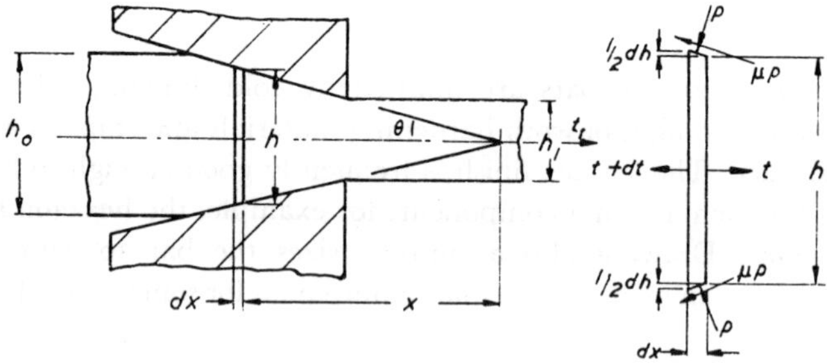

Fig. 5.18 Stresses in wide strip drawing

ignoring the component of μp. Stresses t and $-p$ now become principal stresses σ_1 and σ_3 respectively. As plane strain conditions apply,

$$\sigma_1 - \sigma_3 = 2k \text{ and } t + p = 2k$$

Substituting for p

$$\frac{\mathrm{d}h}{h} = -\frac{\mathrm{d}t}{t + (2k - t)(1 + \mu\cot\theta)}$$

$$= \frac{\mathrm{d}t}{t\mu\cot\theta - 2k(1 + \mu\cot\theta)} \qquad (5.13)$$

If the dies are straight, the coefficient of friction constant and the material non-work-hardening, θ, μ and k are constant.

Integrating equation (5.13)

$$\ln h + C = \frac{1}{\mu \cot \theta} \ln [t\mu \cot \theta - 2k(1 + \mu \cot \theta)]$$

or $$C_1 h^{\mu\cot\theta} = t\mu \cot \theta - 2k(1 + \mu \cot \theta)$$

where $$C_1 = e^{C\mu\cot\theta}$$

at entry, $t = 0$ and $h = h_0$. Hence evaluating C_1

$$C_1 = -\frac{2k(1 + \mu \cot \theta)}{h_0{}^{\mu\cot\theta}}$$

Substituting for C_1

$$-2k(1 + \mu \cot \theta)\left(\frac{h}{h_0}\right)^{\mu\cot\theta} = t\mu \cot \theta - 2k(1 + \mu \cot \theta)$$

$$\text{and } t = \frac{2k(1 + \mu \cot \theta)}{\mu \cot \theta}\left[1 - \left(\frac{h}{h_0}\right)^{\mu\cot\theta}\right] \tag{5.14}$$

The tension needed to draw the strip can be found by substituting h_1 for h in the above equation.

An allowance for work hardening can be made by using a value k where $\bar{k} = (k_0 + k_1)/2$.

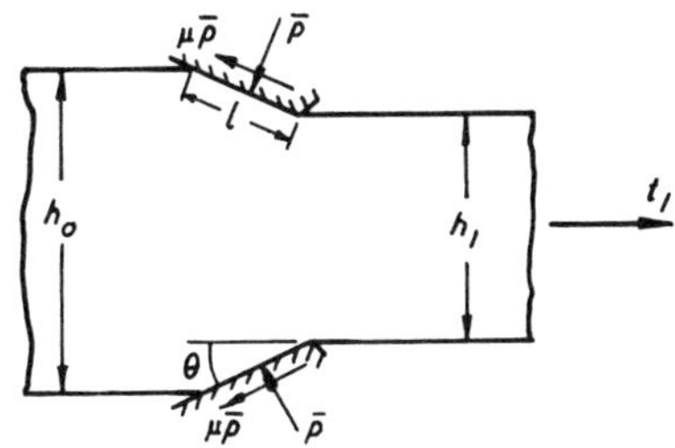

Fig. 5.19 Die pressure in wide strip drawing

The average die pressure $\bar{p}$ can be found by considering Fig. 5.19 and resolving forces horizontally over unit width of the strip.

$$t_1 h_1 = 2\mu l\bar{p} \cos \theta + 2l\bar{p} \sin \theta$$

but
$$l = \frac{h_0 - h_1}{2 \sin \theta}$$

$$t_1 = \left(\frac{h_0 - h_1}{h_1}\right) \bar{p}(1 + \mu \cot \theta)$$

$$\bar{p} = \frac{t_1}{[(h_0/h_1) - 1][1 + \mu \cot \theta]} \qquad (5.15)$$

5.4.2 Upper bound solutions for flat strip drawing. A slip-line field for frictionless drawing of flat wide strip with a thickness reduction of 1·25:1 appears in Fig. 5.20. Slip lines meet the die/material interface at 45° and are contained in a 45° isosceles triangle with its hypoteneuse along AB and A′B′. The field is completed by fans centred at A, B, A′

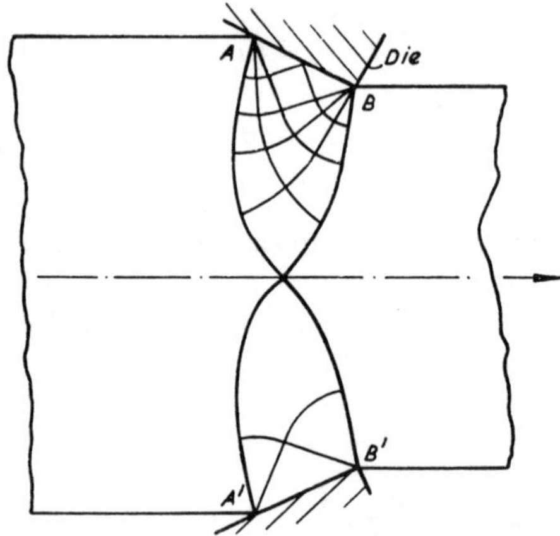

Fig. 5.20 Slip-line field for frictionless wide strip drawing

and B′ which are extended to meet at a point on the centre line. The graphical construction described in Chapter 3 can be used to extend the fans.

An estimate of the drawing tension can be obtained by the upper bound method. The simplified field and hodograph appear in Figs. 5.21 (*a*) and (*b*) respectively.

Taking unit strip width, and assuming h_4 and ${}_xV_1$ are unity

$$\frac{dw}{dt} = k\Sigma us$$

$$t_4 h_{4x} V_4 = k[{}_1V_2\,\mathrm{AC} + {}_2V_3\,\mathrm{CB} + {}_1V_3\,\mathrm{CD} + {}_3V_4\,\mathrm{BD}]$$

$$t_4\, 1 \times 1{\cdot}25 = k[1{\cdot}02]$$

$$t_4 \simeq 0{\cdot}82k$$

If there is friction along the die/metal interface, it is necessary to know both the value of μ and of the normal stress on the interface before the

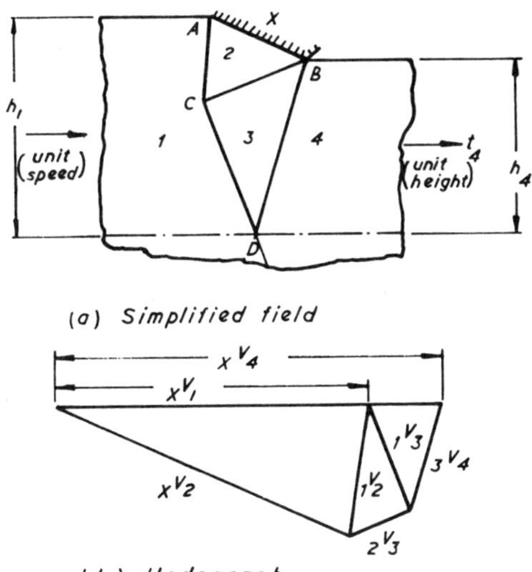

Fig. 5.21 Wide strip drawing

drawing tension can be found. The value of the normal stress cannot exceed $2k$, and if μ is $0{\cdot}1$ the frictional stress is equal to $0{\cdot}1 \times 2k$.

Hence,
$$t_4 h_{4x} V_4 = k[2\mu\, {}_xV_2\,\mathrm{AB} + 1{\cdot}02]$$

$$t_4 = 0{\cdot}92k$$

If t_4 is calculated from equation 5.14 for similar drawing conditions

$$t_4 = \frac{2k(1 + 0{\cdot}1 \cot 25)}{0{\cdot}1 \cot 25}\left[1 - \left(\frac{1}{1{\cdot}25}\right)^{0{\cdot}1 \cot 25}\right]$$

$$t_4 = 0{\cdot}54k$$

The work formula gives

$$t_4 = \sqrt{3}k \ln \frac{h_1}{h_4}$$

$$t_4 = 0{\cdot}39k$$

It will be seen that the second and third methods produce a lower drawing tension than the upper bound solution. Equation 5.14 does not account for redundant work and the work formula neither accounts for redundant work nor friction.

6 Sheet Metal Forming and Cold Forging

6.1 This chapter is concerned with the factory forming processes which produce components that are subsequently assembled into finished products. It is distinct from Chapters 4 and 5, which dealt with primary forming operations for the manufacture of bars, sheets, tubes and forgings —the raw material for factory machining and forming.

Sheet metal blanking and conventional forming methods are discussed, followed by a description of the newer techniques of sheet metal forming. The last section is concerned with cold forging.

6.2 BLANKING

Before a component is formed from sheet metal, it usually has to be cut from a roll, strip or sheet. Sometimes, as in certain car body pressings, the cutting is done after forming. The term blanking is used when the outside profile of a part is cut, and piercing denotes the cutting of a shape from within a profile. Some parts, such as washers, can be manufactured in two stages within a single tool by piercing and blanking.

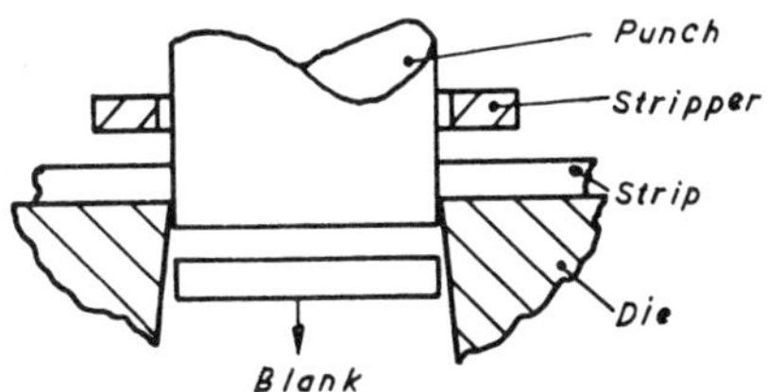

Fig. 6.1 Blanking of circular shape

The process of blanking consists of shearing between a hardened steel or tungsten carbide punch and die (Fig. 6.1). The punch, which is fitted to the ram of the press, first penetrates the metal causing plastic deformation; it then shears it and pushes the cut piece from the sheet. A stripper plate, mounted above the material, forces it from the punch on the return stroke of the press ram.

The depth of penetration before failure occurs is within the range 20–40% of the thickness for most engineering materials. For a given material the punch penetration at failure will also be affected by the clearance between punch and die. When failure occurs, cracks are usually propagated from both sides of the material.

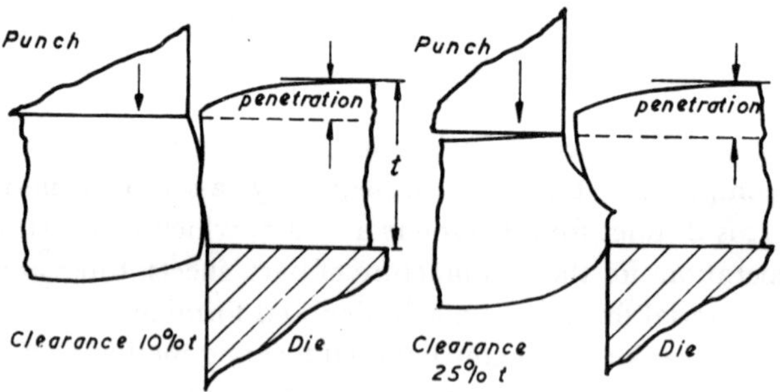

Fig. 6.2 Effect of punch and die clearance on blanked edge

The clearance between each side of the punch and the die is normally stated as a percentage of the material thickness. Very small clearances should be used for most non-ferrous materials but a clearance of 10% is usually suitable for mild steel. The effect of clearance in blanking mild steel is shown in Fig. 6.2. In service the sharp corners of punch and die will wear to a curved profile; this has the effect of increasing clearance and the surfaces of both have to be periodically reground to maintain the quality of the work.

6.2.1 Improvement of the blanked edge. The somewhat rough edge left by blanking will be unacceptable for certain applications. It can be improved by a subsequent shaving operation in which the periphery is sized by a sharp die about 0·4 mm (0·015 in) smaller than the original part. If a pierced hole is too rough it can be improved by a reaming operation. Two processes however, fine blanking and finish blanking, will provide a smooth square edge at the original blanking or piercing operation.

Fine blanking. This process was originally used in the manufacture of watch parts but it is now widely used for office machinery and automobile components. Surface finishes of 12 μ in CLA on the shear edge are possible with dimensional consistency of 0·005 mm (0·0002 in) between batches. The material being blanked is held firmly between a grip ring and the top of the die (Fig. 6.3), the knife-edge on the grip ring prevents metal flow at the top of the blank. Fracture of metal under the punch is inhibited and separation is achieved under conditions of pure shear. An ejector operates from beneath the press to push the blank out of the die and to support it during blanking. The total clearance between punch and die is less than 0·012 mm (0·0005 in) and accurate tool setting is essential to prevent the tool entering the die. A triple action hydraulic press is usually used for fine blanking, these presses are considerably more expensive than conventional mechanical presses and have significantly slower production rates.

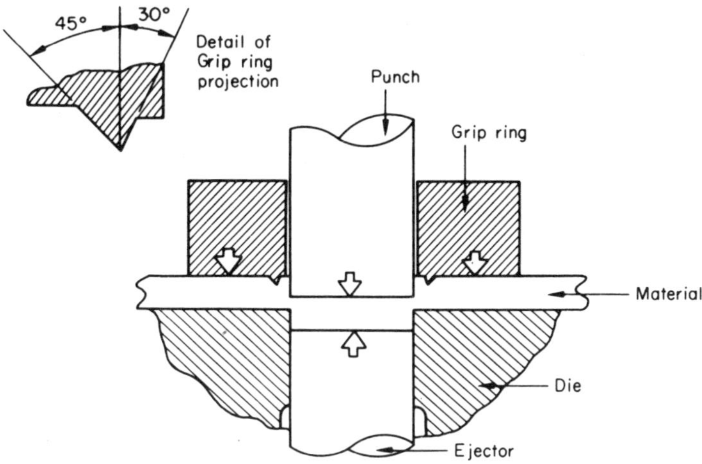

Fig. 6.3 Fine blanking (component partly blanked)

Finish blanking. In this process the clearance between punch and die is also low. The workpiece is not constrained, as in fine blanking, but the corner of the punch is radiused; if a clean pierced hole is required the die rather than the punch is radiused.

High velocity blanking. When punch speeds are increased to 10 m s^{-1} (30 ft/s) from those generally used in blanking, 0·6 m s^{-1} (2 ft/s), a cleaner fracture surface and less doming of the blank is achieved. In fine and finish blanking early fracture of the workpiece is inhibited; in high velocity blanking, however, fracture should occur as a straight crack between corners of the punch and die at low punch penetration. High velocity blanking is therefore effective with materials, such as mild steel, which are sensitive to strain-rate. Large punch/die clearances, in the order of 0·225 mm (0·009 in), are possible, thus simplifying tool making and tool setting. Peak blanking loads are, however, 3 times greater than in conventional blanking and in consequence of the high shock loading it is difficult to design tools with commercially viable lives.

6.2.2 Blanking forces. There is poor correlation between the blanking and yield stresses in tensile and shearing tests. However some idea of blanking forces may be obtained by multiplying the area being sheared by 1·5 × the ultimate shear stress of the material.

i.e. $$F = 1{\cdot}5\tau_{ult}lt \qquad (6.1)$$

where τ_{ult} is the ultimate shear stress of the material,
l is the perimeter of the punch,
t is the thickness of the metal.

Due to the complex compressive stress system existing in blanking, it is probable that fracture occurs at shear strains of the order of 300–400%, i.e. at a stress considerably greater than the ultimate tensile stress found

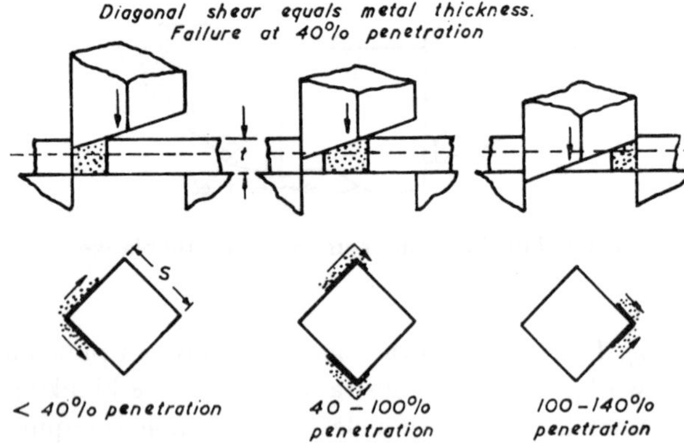

Fig. 6.4 (*a*) Stages in the production of a square blank

by a uniaxial tensile test. Hence, although it is a fair approximation to assume pure shear, the failure stress should be taken as approximately 1·5 times the ultimate shear stress, $(1{\cdot}5\sigma_{ult}/\sqrt{3})$.

To reduce blanking force, the bottom of the punch or die may be ground at a small angle. This application of 'shear' to the punch means that the metal fails progressively rather than almost instantaneously. 'Shear' is sometimes applied to the die rather than to the punch, to minimize blank distortion which occurs with a 'sheared' punch. The stages in producing a square blank using a punch with diagonal 'shear' equal to the metal thickness are shown in Fig. 6.4 (*a*), it being assumed that the blank fractures at 40% penetration. In this example the blanking force increases until the punch has penetrated to 40% of the metal thickness and shearing begins. The force remains at its maximum value as shearing proceeds across the blank; at 100% penetration the area being sheared starts to decrease and with it the shearing force. The shearing force has fallen to zero at 140% penetration. The approximate magnitude of the blanking force at any instant can be found from the formula $F = 1{\cdot}5\tau_{\text{ult}}lt$ by substituting the appropriate value of l. It is of interest to note that, in this example, the maximum blanking force with 'shear' is well under half that which would be required for conventional blanking. The variation in punch force with penetration for a rigid-plastic and for

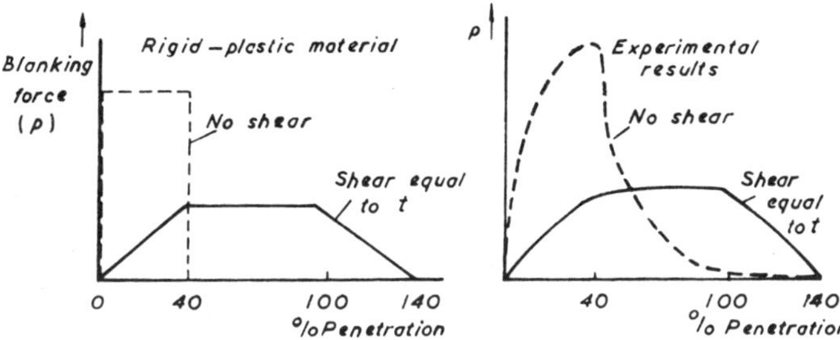

Fig. 6.4 (*b*) Variation of force in blanking—failure at 40 per cent penetration

an actual material are shown in Fig. 6.4 (*b*). The shearing of metal bars and circular blanks has been investigated by Chang and Swift,[14] and by Chang[15] respectively.

6.2.3 Bar cropping. The preparation of billets from bar for hot and cold forming is an important industrial operation. It can be done by sawing, abrasive cut-off or parting-off in a lathe. All these processes involve a material loss which can be economically significant with high value materials. Cropping, however, is a rapid method of billet production which avoids any material loss; open blade cropping is shown in Fig. 6.5. This process has the disadvantage of producing distorted billet ends which may be unsatisfactory for cold forging without a preforming operation.

Two recently introduced methods of bar cropping produce billets

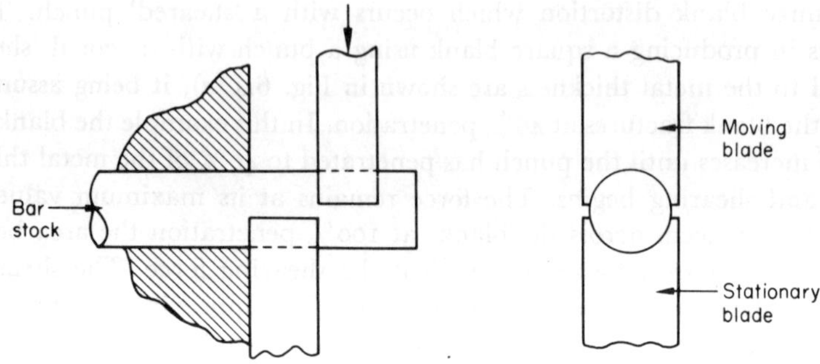

Fig. 6.5 Open blade cropping

having ends of greatly improved squareness. High speed cropping with a blade speed of about 10 m s^{-1} (30 ft/s) greatly reduces distortion. Satisfactory cropping is only possible if the offcut is supported or has sufficient inertia to prevent it from bending during shearing. With a blade speed of 10 m s^{-1} a length/diameter ratio of 5:1 is needed if inertial support is to be used, compared with a 1:1 ratio for cropping at conventional speeds.

The second method of bar cropping is the Hungarian Veres process in which an axial thrust greater than the yield stress of the material is applied before and during cropping (Fig. 6.6). The result is a square end and a clean cut, even on billets having a length/diameter ratio of around $\frac{1}{3}$. The Veres cropping machine is necessarily more expensive than equipment used in competitive methods of billet preparation, it is, however, likely to show considerable savings when used to crop high cost (non-ferrous) materials, particularly with low length/diameter ratios.

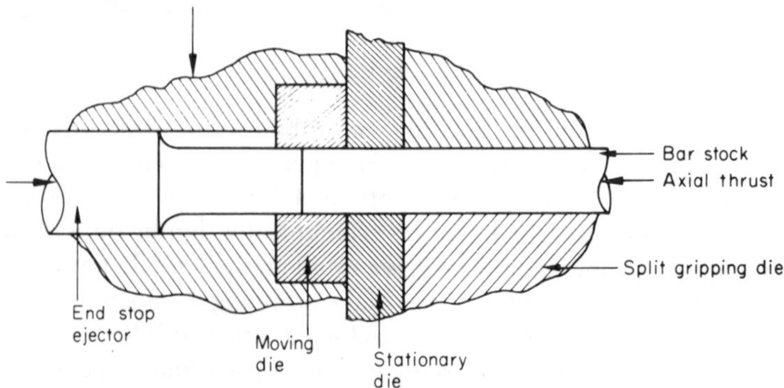

Fig. 6.6 Cropping under high axial thrust

6.3 PRESSING, DEEP DRAWING, BENDING AND STRETCH FORMING

This is an important group of manufacturing operations in which sheet metal is shaped by plastic deformation. The distinction between pressing and drawing is not clear-cut, although there is little inward flow of the metal in pressing because of constraint at the blank periphery, whereas with deep drawing the inward flow is considerable. Simple bending is the most straightforward of the operations and consists of forming in one plane only. Stretch forming is more complex and produces a shape by extending and thinning the metal over a profile. Pressing and drawing operations are a combination of stretching and bending.

Punches and dies may be single or multi-stage. Multi-stage tooling can be designed so that various piercing, forming and blanking operations are performed by a single tool. This reduces production time and is more economical where the quantity of parts justifies the higher cost of multi-stage tooling. Tools used for the above operations are mounted in machine tools called presses, which are either mechanically or hydraulically operated.

6.3.1 Bending. Bending is a comparatively simple operation involving plastic deformation. Two typical bending operations are shown in Fig. 6.7. The material being bent is raised to its yield point Y on both sides of the neutral axis: on one side the deformation is in tension while on the other it is in compression (Fig. 6.8). The magnitude of the bending

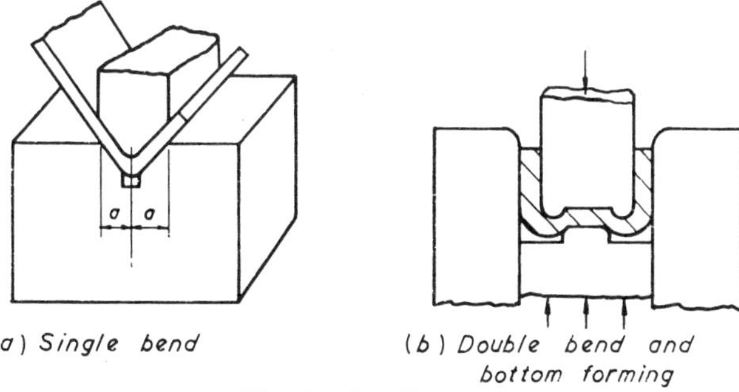

Fig. 6.7 Bending tools

moment can be found by equating it to the moment of resistance offered by the metal about its neutral axis

i.e.
$$M = Y\left[\left(b\frac{t}{2}\right)\frac{t}{4} + \left(b\frac{t}{2}\right)\frac{t}{4}\right]$$

$$M = Y\frac{bt^2}{4} \tag{6.2}$$

The force to initiate bending can be found in simple cases by considering the geometry of the tool and equating the bending moment produced by the tool force to that required to bend the material. For the single bend shown in Fig. 6.7 (*a*), tool force $F = Y(bt^2/2a)$ where a is the distance between the nose of the tool and the point where the material is supported.

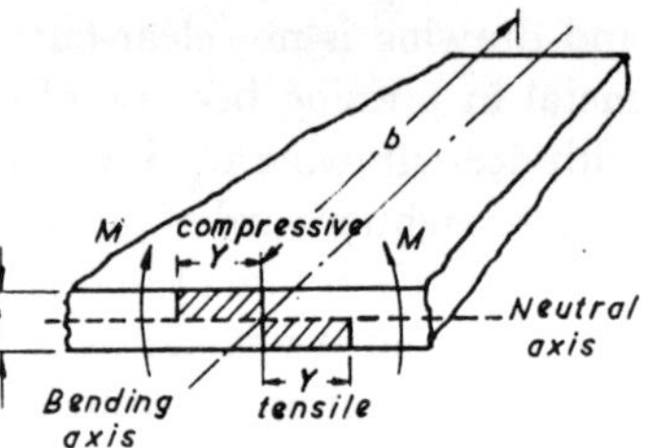

Fig. 6.8 Plastic bending

Bent material will recover elastically when the deforming force is removed. This spring-back can be overcome by bending the metal a few degrees more than the desired angle or by setting the punch travel so that it compresses the bent metal beyond its yield stress. This latter operation is known as planishing and is analysed in Section 6.5.1.

In bending it is desirable that the metal be bent at right angles to its 'fibre'. Fibre results from the redistribution and elongation of impurities, grains and second phases during previous working operations. The sheet is more liable to fail with the fibre, i.e. along the length of impurities and elongated grains, than across them.

6.3.2 Stretch forming. A piece of sheet metal can be formed by stretching it over a profile. This process is best performed on a stretch-forming machine; presses may however be used, providing there is sufficient clamping force at the edge of the blank to prevent inward flow of the metal. Two examples of stretch forming are shown in Fig. 6.9.

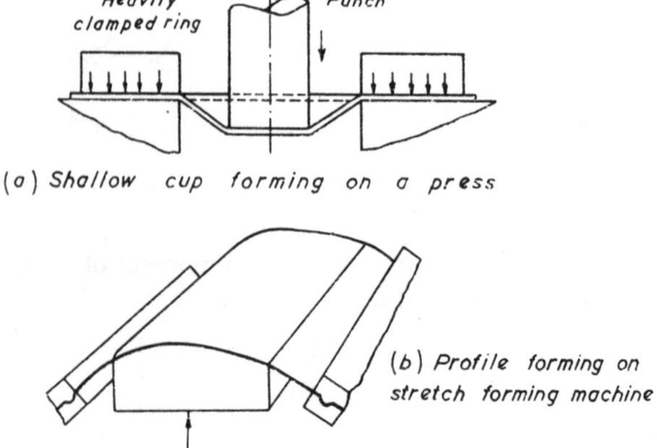

Fig. 6.9 Stretch forming operations

Good quality steel should be used for stretch forming. Stresses must be kept below those which cause failure by plastic instability, a type of failure which occurs in standard tensile tests. At first extension is uniform, and the load has to be gradually increased to maintain plastic extension as the metal work hardens. Eventually, further extension can be obtained without increasing the load: this is because the increasing level of stress in the shrinking cross-sectional area can no longer be contained by the work hardening of the metal. Instability is said to occur at this point and is indicated by necking of the material.

If A is the cross-sectional area at any instant and F is the uniaxial stretching force, then σ, the direct stress in the material, is given by $\sigma = F/A$.

The load sustained by the metal reaches a maximum value at the point of instability

i.e.
$$\mathrm{d}F = 0$$

As
$$F = A\sigma,$$

$$\mathrm{d}F = \sigma\,\mathrm{d}A + A\,\mathrm{d}\sigma = 0$$

and
$$\frac{\mathrm{d}\sigma}{\sigma} = -\frac{\mathrm{d}A}{A}$$

Assuming constancy of volume

$$\frac{\mathrm{d}\sigma}{\sigma} = -\frac{\mathrm{d}A}{A} = \frac{\mathrm{d}l}{l} = \mathrm{d}\varepsilon$$

$$\frac{\mathrm{d}\sigma}{\mathrm{d}\varepsilon} = \sigma$$

Hence instability in uniaxial stressing occurs when the subtangent Z to the stress/strain curve (Fig. 6.10) equals unity.

i.e.
$$\mathrm{d}\sigma/\mathrm{d}\varepsilon = \sigma/1$$

It is possible to represent the stress/strain curve by the equation $\sigma = A(B + \varepsilon)^n$ where $B = 0$ for an annealed material and n represents the liability of the material to strain harden.

$$\frac{\mathrm{d}\sigma}{\mathrm{d}\varepsilon} = nA(B + \varepsilon)^{n-1} = nA\,\frac{(B + \varepsilon)^n}{B + \varepsilon} = \frac{n\sigma}{B + \varepsilon}$$

but
$$\frac{\mathrm{d}\sigma}{\mathrm{d}\varepsilon} = \frac{\sigma}{Z} = \frac{n\sigma}{B + \varepsilon}$$

$\therefore$
$$Z = \frac{B + \varepsilon}{n}$$

and
$$\varepsilon = Zn - B$$

Hence the larger the value of n and the smaller the value of B, the greater the strain at which instability occurs. Therefore an annealed material which has a high strain hardening rate is best suited to stretch forming.

If the metal is subjected to equal biaxial tension, the stress at which instability occurs will be greater and the subtangent under these conditions is 2 (Fig. 6.10).

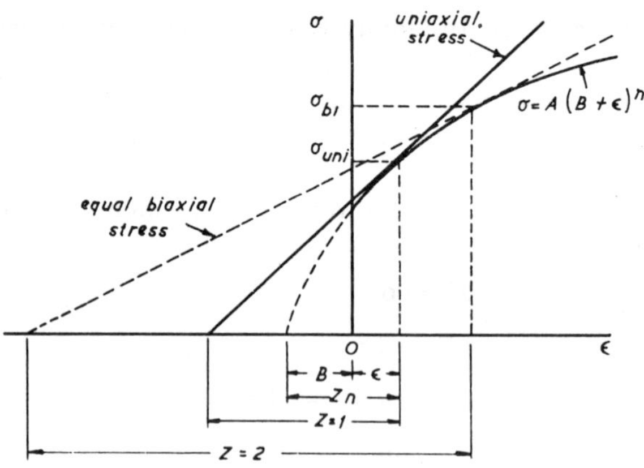

Fig. 6.10 Plastic instability

6.3.3 Deep drawing. The essential features of deep drawing are illustrated in Fig. 6.11 which shows a partly drawn flat-bottomed cup being produced from a flat blank. The pressure ring which bears on the upper surface of the blank prevents wrinkling of the metal while it is being drawn radially over the surface of the die. Pressure may also be applied to the base of the cup by means of a pressure pad.

The drawing of cylindrical cups has been analysed by Chung and Swift,[16] but despite the comparative simplicity of the final shape, the

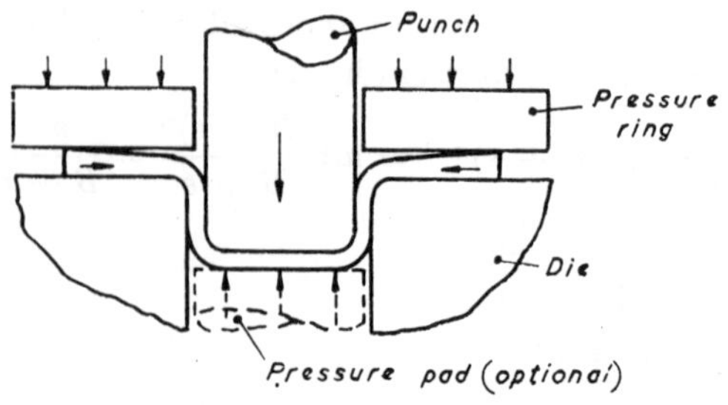

Fig. 6.11 Deep drawing of flat bottomed cylindrical cup

analysis is very complex and here it will be treated descriptively. Unlike other metal working operations analysed elsewhere in this book, in which consistent deformation has been assumed, the type of deformation undergone by various portions of the blank in deep drawing varies considerably. Stages in the drawing of a flat-bottomed cylindrical cup are shown in Fig. 6.12. It will be seen that zone 0–1, the central circular portion of the

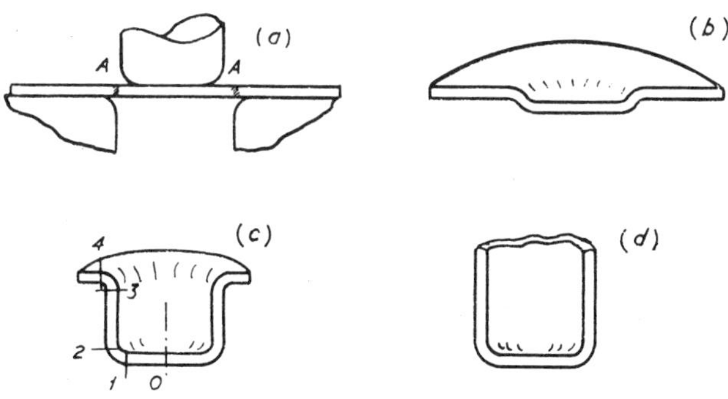

Fig. 6.12 Stages in cup drawing

blank, is stretched over the base of the punch. The rest of the blank is drawn radially inwards across the top of the die and, owing to the continually shrinking diameter, will thicken towards its outer edge. As the blank is pulled over the die radius, it is bent and unbent in tension and then drawn vertically downwards under tension to form the wall of the cup.

The distribution of tensile and compressive strains caused by bending and unbending in tension over the die radius is shown in Fig. 6.13. It will be seen that the strains, and hence work hardening, are greatest at the outer fibres. It should also be noted that the neutral plane is shifted from the central surface because of the presence of the tensile stress. When the blank undergoes plastic bending or unbending under tension there is an instantaneous thinning of the metal. This thinning can be serious at the start of drawing, since the blank is still at its original thickness: later it progressively thickens as the blank diameter is reduced by radial drawing. The variation in thickness of a partly drawn blank is shown in Fig. 6.14. Near the base of the cup wall, thinning caused by bending and unbending under tension produces two necks, separated by a narrow band of thicker material which has escaped bending and has been subjected only to pure tension (zone AA, Fig. 6.12 (*a*)). If the cup ruptures, it is likely to fail in tension at one of the necks, for here not only is the metal thin but the level of stress is highest.

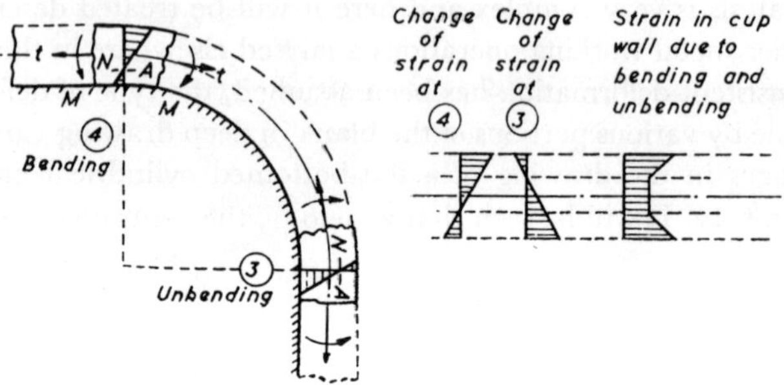

Fig. 6.13 Bending and unbending under tension

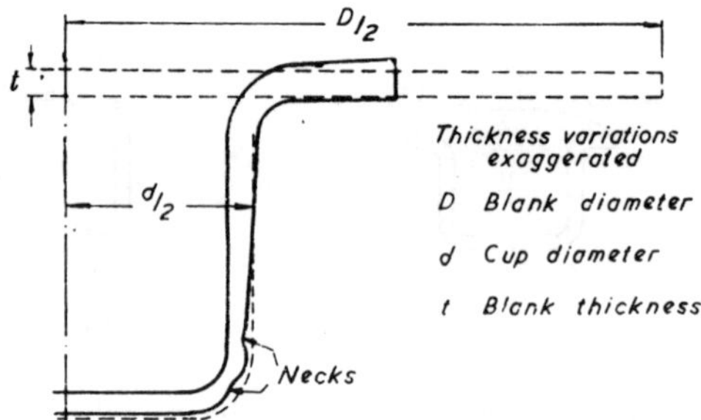

Fig. 6.14 Variations in thickness of partly drawn blank

A simple formula which can be used to find the punch force needed to perform a given drawing operation is

$$F_p = K\sigma_{ult}\pi dt \tag{6.3}$$

where F_p is the drawing load
K is a constant ($\simeq$ 0·8)
σ_{ult} is the ultimate tensile stress of material
πdt is the cross-sectional area of the cup wall.

The press chosen must be at least able to maintain the drawing load throughout its working stroke.

Various methods are used to increase the depth of cup which can be drawn. One is to roughen the end of the punch head to reduce thinning of the metal at the base of the cup and to lubricate the die radius to reduce drawing friction. Another is to use tractrix curves instead of radii on the die to provide a more gradual unbending of the material. Apart from modifying the tool, differential annealing can be employed to soften the metal towards the edge of the blank. With a harder central zone, thinning of metal around the bottom of the punch is reduced and a cup of more uniform thickness is produced.

Other factors which affect the magnitude of the drawing ratio are the ductility of the blank material, its thickness, and the magnitude of the punch and die radii.

6.3.4 Redrawing. If a sufficiently deep draw cannot be obtained in a single operation, one or more redrawing operations will be necessary: work hardening will also occur, and interstage bright annealing may be

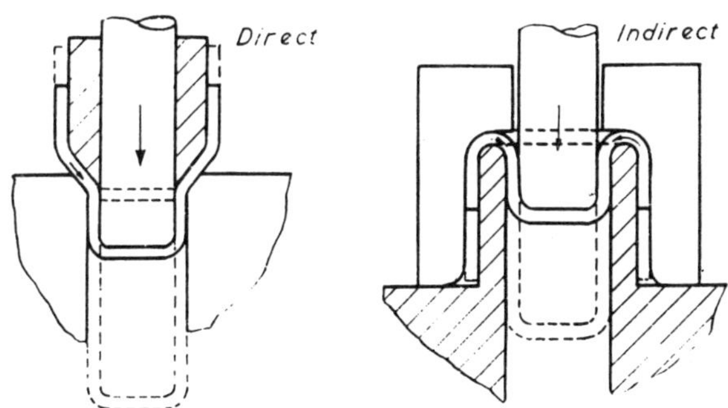

Fig. 6.15 Redrawing of cups

needed to restore ductility. Two types of redrawing are possible and these are shown in Fig. 6.15. The inverted method has an advantage over the direct as a second bending is avoided, although with the inverted method the radius of the first draw is limited to what can be accommodated by the redrawing tool.

6.3.5 Ironing. It can be seen from Fig. 6.14 that the thicknesses of drawn cups vary along their walls. The ironing process consists of pushing a cup through a die, which reduces the wall thickness to a constant value as well as increasing the depth of the cup. The KMTE process, which has been developed from the Keller process can produce a very deep cylindrical cup in a single punch movement by combining drawing with a series of ironing operations (Fig. 6.16).

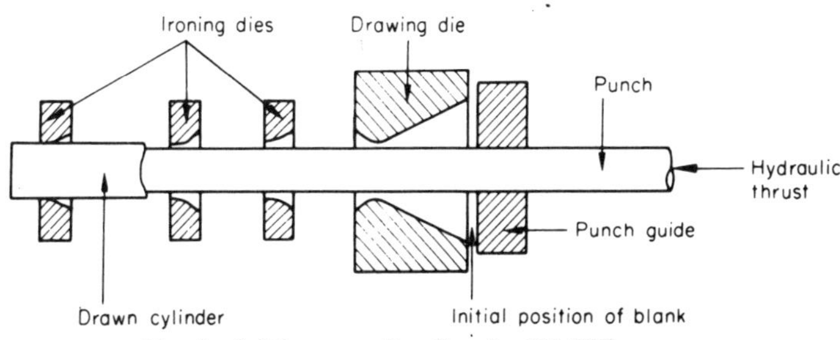

Fig. 6.16 Diagram of tooling for KMTE process

6.3.6 Forming limits in sheet metal. The amount of deformation which sheet metal will undergo before it fractures depends chiefly on the drawing quality of the material, on the design of the tooling and on the frictional conditions between the tool and workpiece. Although fracture and usually necking must be avoided it is economically desirable to use a metal having a drawing potential just sufficient for the job, as material costs tend to increase with drawability. Tooling design is dictated by the component shape, although this shape may have to be reached in a number of stages, often with interstage bright annealing. Manufacturing cost increases with the number of stages, hence if maximum deformation is achieved at each stage of shaping cost will be minimized.

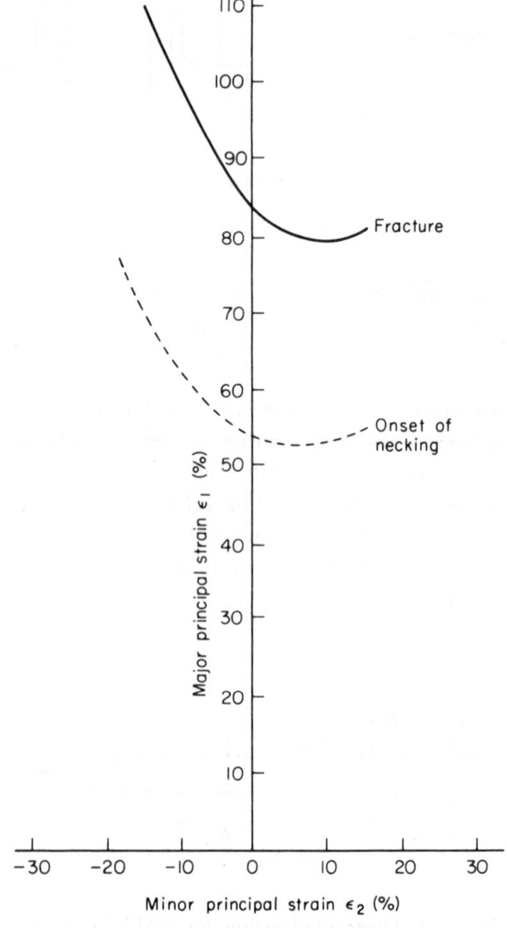

Fig. 6.17 A typical forming limit diagram

Forming limit diagrams are useful in determining how much deformation can be incurred without causing (a) necking, or (b) failure. They are produced in the laboratory, from similar material to that used for the pressing, by analysing the principal strains which occur at the onset of necking and at fracture. These strains are then plotted on a forming limit diagram, an example of which is illustrated in Fig. 6.17. In most pressings necking is to be avoided and the "onset of necking" line on the forming limit diagram is of greater interest than the "fracture" line. The construction of forming limit diagrams is discussed by Veerman, Hartman, Peels and Neve.[17] Principal strains are detected by marking the material to be pressed in a pattern of small etched circles which distort into ellipses as the metal is formed. A suitable pattern and the effects of forming on the grid circles are shown in Fig. 6.18. After the component has been formed the grid pattern deformation can be measured in the critical areas, thus enabling actual strains to be compared with those on the forming limit diagram.

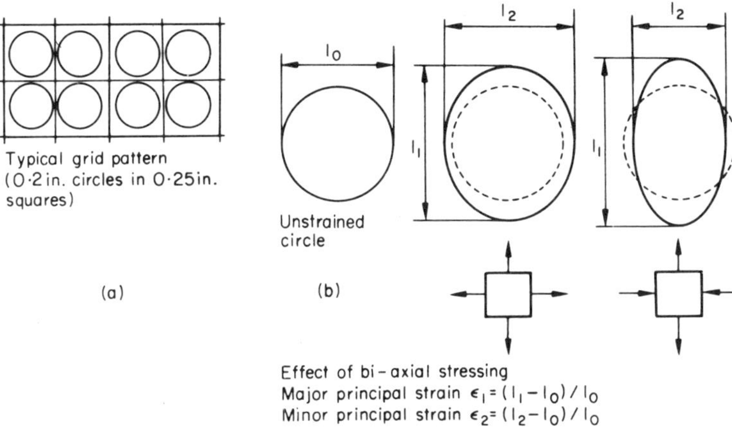

Fig. 6.18 Grid patterns in sheet metal forming

The effects of lubrication on the formability of pressings are difficult to predict and are usually evaluated by trial and error. Newsprint can be placed between the tool and the sheet metal to test the effect of reduced friction over a large area, whereas localized friction can be introduced by sticking emery paper at appropriate places on the tool. The effect of lubrication may or may not be advantageous. In the case of large radii, e.g. a hemispherical punch, good lubrication is advantageous, allowing a more even distribution of strain because the material slips easily over the punch surface. With small radii however, strain becomes highly localized and fracture is likely to occur if the metal is allowed to slip away from highly strained areas due to good lubrication. Formability can also be affected by the surface roughness of the work materia!, smooth material surfaces act as if they had been lubricated but too smooth a surface can cause seizure of the tool and work. Forming speed and pressure, material properties and die design can all affect lubrication and a satisfactory lubricant for one component design may be quite unsuitable for another. Sheet metal forming is therefore still largely intuitive, although a more quantified approach is being evolved through techniques such as the forming limit diagram.

6.3.7 Hydroforming. This is a less common method of deep drawing (Fig. 6.19), in which the die is replaced by a rubber diaphragm backed by hydraulic pressure of about 75 N mm^{-2} (5 tonf/in^2). As the punch moves upwards, a hydrostatic pressure is produced which forces the blank to wrap round the punch. At the same time the displaced oil is pushed out of a relief valve.

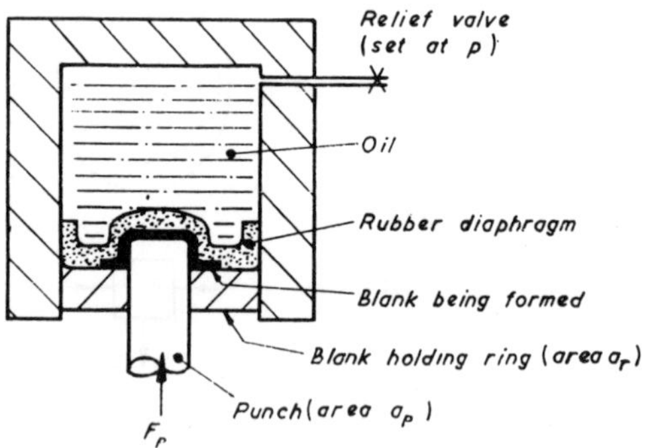

Fig. 6.19 Hydroforming

The closing force F_c is about 5 MN (500 tonf) for machines able to draw 0·3 m (1 ft) diameter blanks, and is given by adding the closing force on the punch to that on the pressure ring

$$F_c = p(a_p + a_r) \tag{6.4}$$

where p is the backing pressure,
a_p is the area of the punch,
a_r is the area of the pressure ring.

The punch drawing force F_p can be found by adding the force necessary to draw the cup to that required to displace the oil through the relief valve,

$$F_p = K\sigma_{ult}(\pi dt) + pa_p \tag{6.5}$$

where K is a constant (frequently $\simeq 0{\cdot}8$)
σ_{ult} is the ultimate tensile strength of the metal,
πdt is the cross-sectional area of the cup wall.

The punch power is the product of F_p and the punch penetration rate V_p.

It will be seen by reference to equations (6.3) and (6.5) that the power needed for hydroforming is greater than that for conventional drawing by an amount $p\, a_p V_p$. This term represents the power needed to force the displaced oil out of the relief valve and is usually considerably greater than that needed for the drawing operation itself. Therefore for a given installed power, faster penetration speeds and consequently higher rates of production are possible with conventional drawing. Another disadvantage of hydraulic forming is that the capital cost of equipment is considerably greater than that of conventional presses capable of deforming similar blank diameters. The life of the rubber diaphragms is low but on one type of press the diaphragm is eliminated and pressurized oil is brought into direct contact with the workpiece.

On the other hand, hydraulic forming has some distinct advantages. Tooling is less expensive and can be quickly manufactured. More severe deformation and greater complexity of deformed shape are possible than with conventional forming because of the hydrostatic pressure; in consequence a part which might require two or three conventional drawing operations can often be hydroformed in one. The absence of a pressure ring means that the component is not scored as in press drawing. Hydraulic forming is therefore usually limited to small batches of parts with complex drawn shapes.

6.3.8 Liquid bulge forming. This process has been developed in Japan,[18] where it is used to produce components for the cycle and car industries. A tube is held in a split die, a hydrostatic pressure is applied internally and as the tube bulges it is axially compressed (Fig. 6.20 (*a*)). The oil pressure in the tube must be maintained below bursting level but high enough to prevent wrinkling and neck formation at the mouth of the bulge, the relationship between hydrostatic pressure and axial compression for successful bulging of a steel tube 40 mm diameter, 2 mm wall thickness is shown in Fig. 6.20 (*b*). Liquid bulge forming has been used to produce up to four bulges in a tube as well as to manufacture stepped hollow shafts. A wide range of tube diameters (12 mm to 300 mm) have been employed.

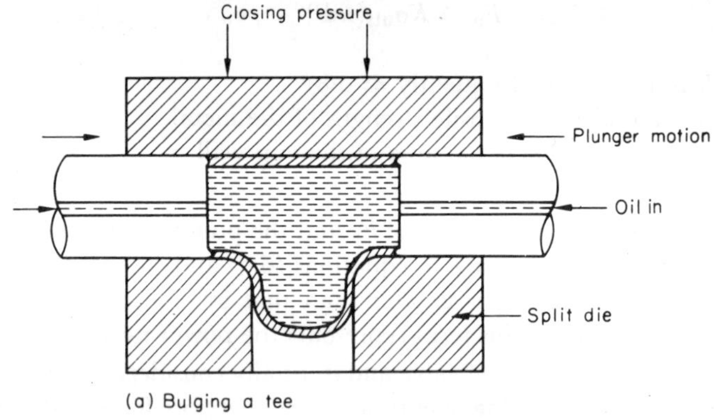

(a) Bulging a tee

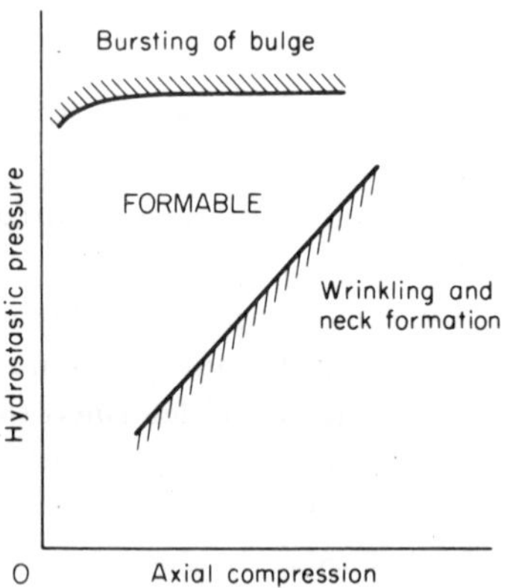

(b) Formable region for a steel tube

Fig. 6.20 Liquid bulge forming

6.4 HIGH-VELOCITY FORMING

Some unconventional methods of sheet metal forming such as explosive, electrohydraulic and electromagnetic forming, have created considerable interest in recent years. An alternative name for this group of processes is high-energy-rate forming, due to the high rate at which energy is transferred to the workpiece. The major application of high-velocity forming lies in the use of chemical explosives to manufacture large sheet metal components.

Machine tools which produce a high-velocity blow have also been developed. As their application is chiefly in the field of forging they were described in Section 4.3.

6.4.1 Use of chemical explosives. In this process a chemical explosive is detonated near to the blank while both are submerged in water. The blank is either formed in a female die or free formed over the edge of a ring-shaped die (Figs. 6.21 (*a*) and (*b*)). No punch is required, and if a shaped die is used it can often be cheaply produced from concrete lined with epoxy resin. Apart from the low tooling cost, the cost of the explosive is negligible and the capital expenditure on firing facilities is much less than that on comparable presses. Explosive forming can therefore be economic when large components are needed in small quantities.

The mechanics of the process cannot be treated analytically but the following sequence is thought to occur after the explosion has been detonated. First, a high-pressure gas bubble is formed almost instantaneously. The pressure of this bubble is estimated to be 7000 N mm^{-2} (10^6 lbf/in^2) and causes a pressure wave to move out spherically through

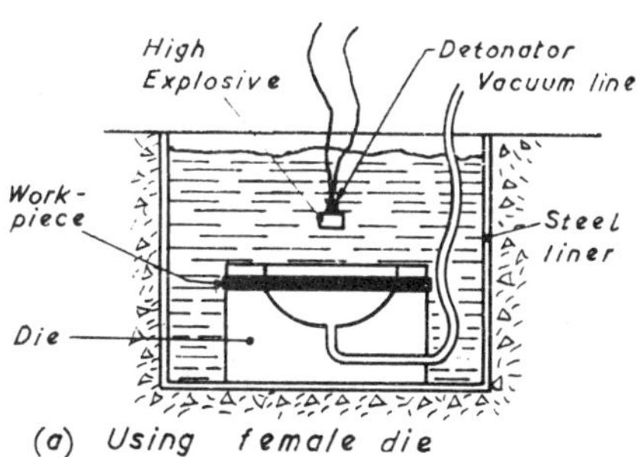

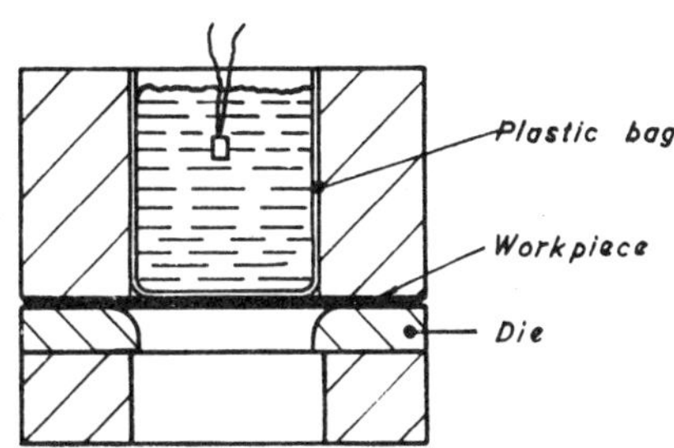

Fig. 6.21 Explosive forming

the water. Energy from the explosion is transferred to the workpiece by the pressure wave as well as by cavitation, water hammer and diffraction.

The gas bubble continues to expand but may oscillate in size and send out secondary pressure waves with further transfer of energy to the workpiece.

6.4.2 Electrohydraulic forming. This is a method of converting electrical energy into energy of deformation. Capacitors are employed as the energy store, and water is used as the medium through which energy is transferred to the workpiece. There are two methods of presenting the electrical energy: (*a*) as a spark across two permanent electrodes, (*b*) along a thin wire which is melted by the discharge. A simplified representation of the process is shown in Fig. 6.22. The discharge of energy through a wire gives a lower production rate than the sparking plug method as the wire has to be renewed after each firing, but by shaping the wire a more efficient utilization of the energy is obtained. The electrical discharge causes pressure waves to be sent through the water and their impact will stretch a clamped blank or produce more complicated shapes in a female die.

Electrohydraulic forming is similar in many ways to explosive forming, but the release of energy can be more closely controlled and, unlike explosive forming, it is suitable for use within the factory. Due to the high capital cost of the requisite banks of capacitors the process is limited to the production of smaller components than those possible with explosive forming. The cycling time using permanent electrodes is, however, faster than that for explosive forming, although the handling time is similar.

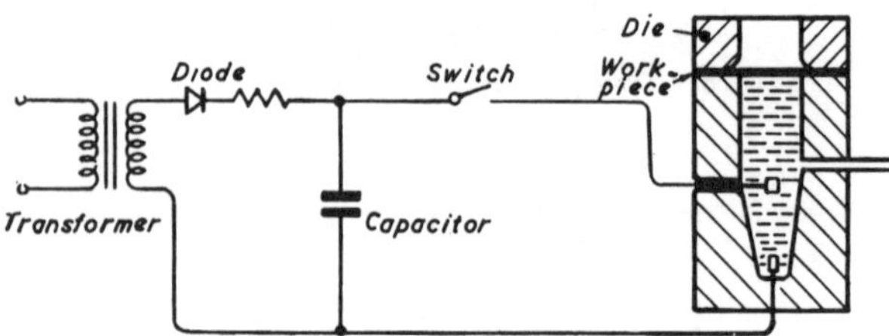

Fig. 6.22 Electrohydraulic forming

6.4.3 Electromagnetic forming. If two conductors carrying current are in close proximity, they are mutually either attracted or repelled. When a sheet metal component is one of these conductors, the force produced can be used to deform it. The other conductor, which is not electrically connected to the component, is in the form of a coil which closely conforms to the shape of the component. As in electrohydraulic forming, energy is stored in high-voltage high-capacity capacitors which can be charged from the mains *via* a transformer and a rectifier. When energy is released to the coil, the high-frequency oscillatory current induces eddy currents in the workpiece, creating an opposing magnetic field. This produces stresses at the surface of the work of about 70 N mm^{-2} (10 000 lbf/in^2), and causes it to accelerate away from the coil and to be formed against the die.

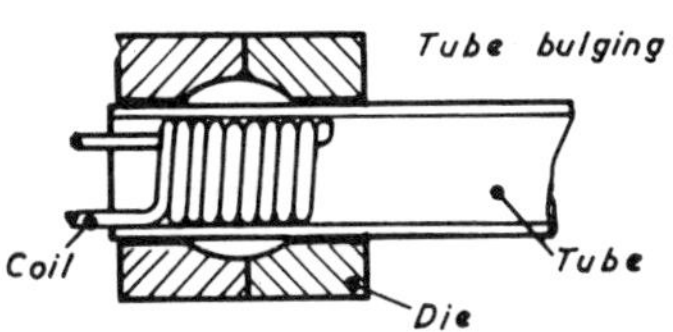

Fig. 6.23 Electromagnetic forming

High conductivity materials such as copper and aluminium are normally used for magnetic forming, although non-conducting materials can be formed if they are first rendered conducting by coating them with copper foil. At present the amount of work done is limited to low levels of deformation because of the difficulty of producing work coils which are strong enough to withstand the large magnetic forces and the high cost of storing large quantities of electricity in capacitors. Typical work includes the bulging of aluminium tube and the shrinking of copper bands on insulators; 'pancake' type spirally wound coils can be used to dimple or bulge flat sheets. The bulging of tube is shown in Fig. 6.23.

Despite the restricted range of operations performed by this process it has the advantages of not needing a transfer medium, a low noise level and a high rate of output. Electromagnetic forming is employed in mass production, particularly for swaging of aluminium tube.

6.5 COLD FORGING PROCESSES

Cold forging is the name given to a group of forming operations in which the component shape is obtained by causing a solid piece of metal to flow under the action of a deforming force. The term 'cold' is used to imply that shaping occurs at temperatures below that of recrystallization.

Considerable savings can be obtained in material cost compared with machining, as with cold forging almost all of the material is utilized in the finished component. There is no production of swarf as in metal cutting, and this group of processes is sometimes referred to as chipless machining. An additional advantage of cold forging is that most metals work harden when cold worked, giving them enhanced strength and allowing, for instance, plain carbon steel to be used in place of more expensive alloy steels. As the metal is being deformed cold rather than hot, considerably higher deforming forces are needed; these frequently involve the use of heavy and expensive capital equipment. Tooling costs are also high, and in consequence cold forging is normally used when large quantities of similar components are required. Cold forging is limited to ductile metals and the range of shapes produced is not as wide as can be obtained by machining.

Cold forging operations include cold extrusion, a process of great potential; cold heading, a well-established process used chiefly to produce small parts from bar and rod; and form rolling which, in the main, produces thread forms on existing components. Other cold forging processes include planishing, embossing and coining. Planishing is a simple operation which uses smooth dies, frequently to flatten blanks; analytical treatment of cold forging operations of the planishing type appears in Section 6.5.1. Embossing raises a pattern on thin sheet metal and is used to produce some types of military badge and buttons. Coining also produces a raised pattern but on thicker blanks; different designs on each side can be obtained if required, as on coins and medals. Greater force is required for coining than for a comparable embossing operation but, due to the complicated shapes usually produced, analysis of these operations is not normally attempted.

6.5.1 Analysis of cold forging between smooth platens. A wide strip of thickness t is forged with platens of breadth b and plane strain conditions are assumed to apply.

When $b/t = 1$. If $b = t$ and there is no friction between the work and the platen, the slip-line field is as shown in Fig. 6.24 (*a*). As σ_1 is zero (there

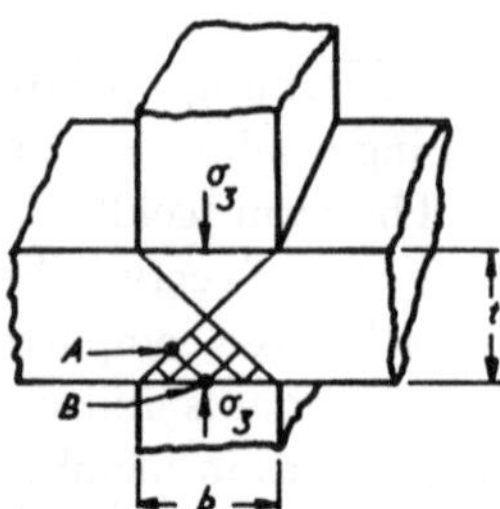

Fig. 6.24(*a*) Slip-line field when $b/t = 1$

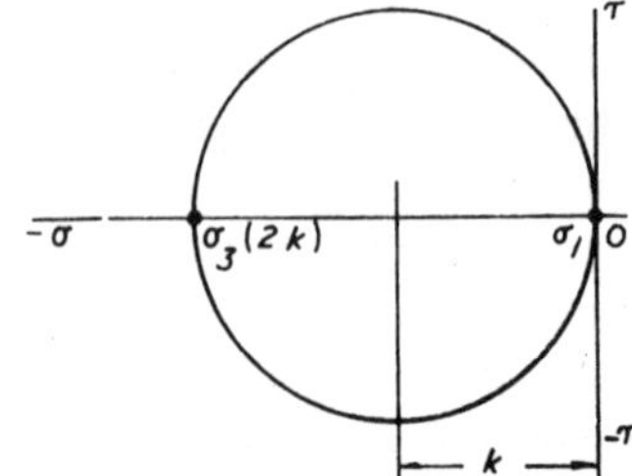

Fig. 6.24(*b*) Mohr stress circle when $b/t = 1$

being no externally applied horizontal stress) the hydrostatic stress p will equal k at the boundary of the plastic zone.

Using Hencky's equation on the α slip line AB

$$p_B - 2k\phi = p = k$$

but $\phi = 0$ as AB is a straight line

$$\therefore \qquad p_B = k$$

$$\sigma_3 = p_B + k$$

$$\therefore \qquad \sigma_3 = 2k$$

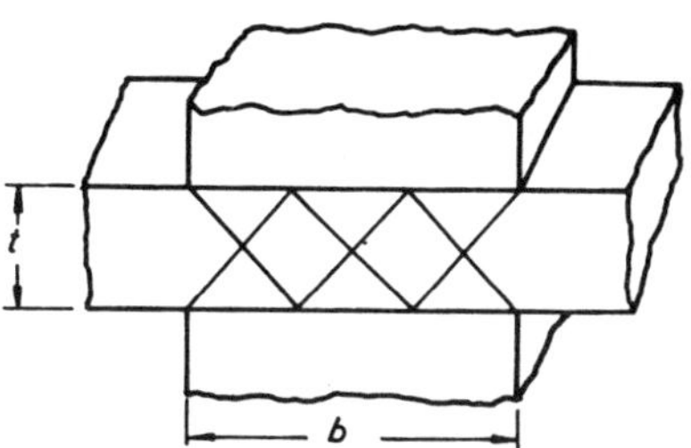

Fig. 6.25 Slip-line field when b/t is integral

In simple cases such as this, σ_3 could be found without the use of a slip-line field. It can in fact be obtained by considering the Mohr stress circle (Fig. 6.2 4(b)).

When b/t = whole number > 1. The field shown in Fig. 6.25 is for a broad platen where $b/t = 3$. Again $\sigma_3 = 2k$ and the deforming force will be proportional to the platen breadth. Where the ratio of platen breadth to work thickness is not integral, a slightly higher pressure is needed because of a more complicated pattern of metal deformation.

Upper bound solution (b/t not integral). In this instance a b/t ratio of 1.4 has been assumed, and the slip-line field shown in Fig. 6.26 (a) has been approximated to by the straight line velocity discontinuities shown in Fig. 6.26 (b). From Fig. 6.26 (b) a hodograph can be constructed by assuming that each platen moves inwards with unit velocity. Metal in zone ACB moves downward with the platen but is constrained by the rigid metal to slide parallel to AB; this produces a velocity discontinuity along AB of magnitude ${}_1V_2$. The horizontal constraint AC imposed by the upper platen enables the first velocity triangle A'B'C' to be drawn. The metal movement in the other three zones is then considered and found to follow a similar pattern. The complete hodograph is shown in Fig. 6.26 (c). The values of u and s can now be found from Figs. 6.26 (c) and (b) respectively and substituted into the formula for rate of work done. A discussion of the upper bound method appears in Chapter 3.13.

$$\frac{dw}{dt} = k\Sigma us$$

Considering unit width of the platen

$$4pa = k[(AB \times {}_1V_2) + (BD \times {}_1V_3) + (EB \times {}_4V_2) + (BG \times {}_4V_3)]$$

$$4pa = 4k(a \sec\theta \operatorname{cosec}\theta)$$

$$p = k \sec\theta \operatorname{cosec}\theta$$

In this instance $\theta = 36°$ and $p = 2{\cdot}1k$.

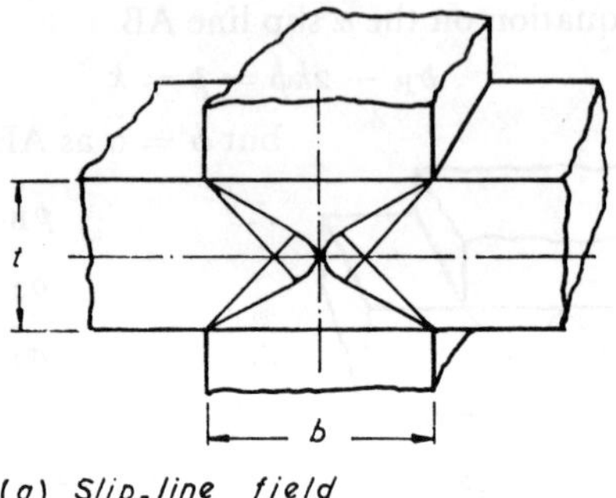

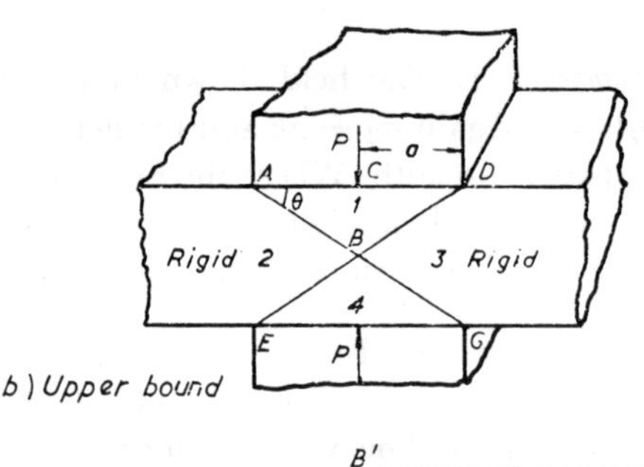

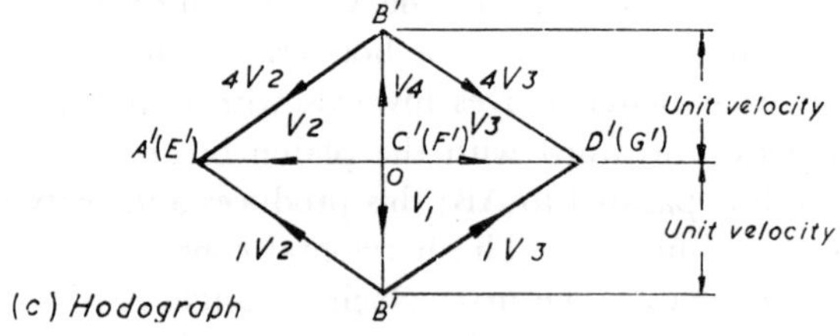

Fig. 6.26 When b/t not integral

It will be recalled that when b/t is a whole number, $p = 2k$. The higher value of p obtained in this example is due to the overestimation of pressure inherent in the upper bound method and, to a smaller extent, to the more complicated pattern of metal deformation if b/t is not integral.

6.5.2 Cold forging between platens with Coulomb friction. Lubrication is more effective when metal is forged cold and, unlike hot forging, the yield stress at the surface of the material is not usually reached.

Referring to Fig. 6.27 (*a*) and adopting a similar approach to that used for hot bar forging, it is easily shown that

$$\frac{\mathrm{d}p}{\mathrm{d}x} = -\frac{2\mu p}{t}$$

$$\frac{\mathrm{d}p}{p} = -\frac{2\mu\,\mathrm{d}x}{t}$$

Integrating, $\ln p = -\dfrac{2\mu x}{t} + C$

$$p = A\mathrm{e}^{(-2\mu x/t)}$$

At $x = b/2$ the horizontal stress is zero,

$\therefore$
$$p = 2k$$

$$2k = A\mathrm{e}^{(-\mu b/t)}$$

and
$$A = 2k\mathrm{e}^{(\mu b/t)}$$

the pressure distribution under the platen is given by

$$p = 2k\mathrm{e}^{(\mu/t)(b-2x)} \tag{6.6}$$

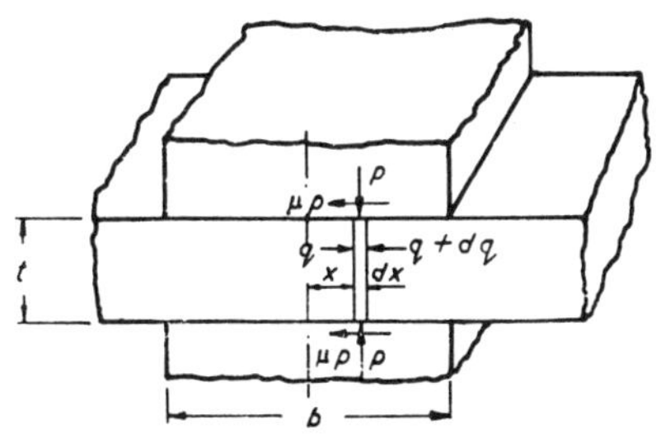

(a) Stresses on narrow strip

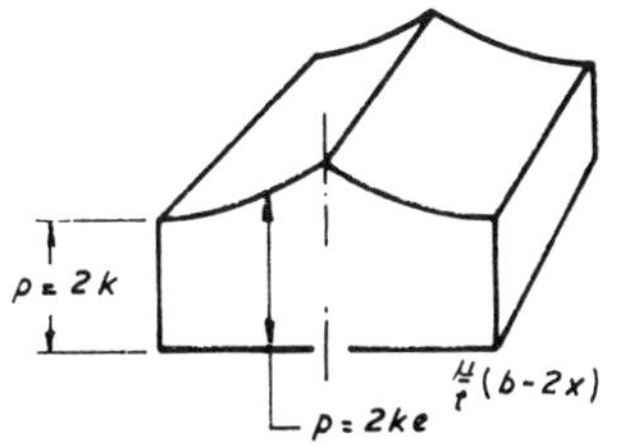

(b) Stress distribution with Coulomb friction only

Fig. 6.27 Cold forging

Mean pressure

$$\bar{p} = \frac{2k}{b}\left[2\int_0^{\frac{b}{2}} e^{(\mu/t)(b-2x)}\, dx\right]$$

$$\bar{p} = \frac{4k}{b} e^{(\mu b/t)} \int_0^{\frac{b}{2}} e^{(-2\mu x/t)}\, dx$$

$$\bar{p} = \frac{2kt}{\mu b}\left(e^{(\mu b/t)} - 1\right)$$

$$\bar{p} = 2k\frac{t}{\mu b}\left(1 + \frac{\mu b}{t} + \frac{\mu^2 b^2}{2t^2} + \ldots - 1\right)$$

If μ is small $\bar{p} \simeq 2k\left(1 + \frac{\mu b}{2t}\right)$ (6.7)

The slope of the sides of the friction hill is an exponential curve, as shown in Fig. 6.27 (*b*).

6.5.3 Cold forging between platens with sticking and Coulomb friction. Mixed friction conditions can occur when forging with broad platens or where the coefficient of friction is high. This produces a friction hill, as shown in Fig. 6.28. At the centre, in the sticking zone, the slope of the hill is a straight line, which merges into an exponential curve towards the edges of the platen, where there is Coulomb friction.

The distance x_t, from the centre line of the platens to the point along them at which the friction conditions change can be found, because at

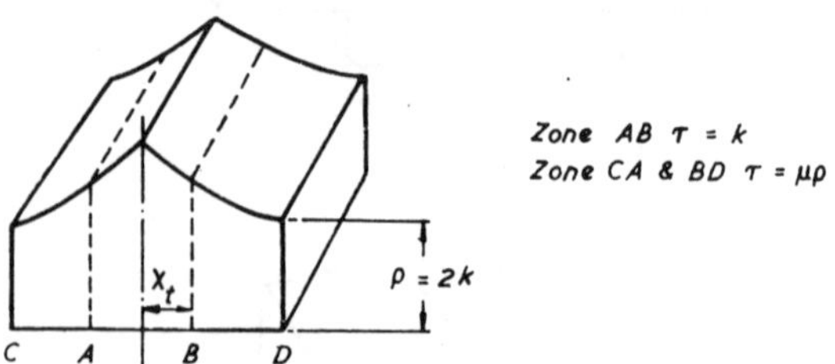

Fig. 6.28 Stress distribution with Coulomb and sticking friction

this point $\mu p = k$. It has already been shown (equation (6.6)) with Coulomb friction

$$p = 2ke^{(\mu/t)(b-2x)}$$

$\therefore$ at x_t, $$p = \frac{k}{\mu} = 2ke^{(\mu/t)(b-2x_t)} \tag{6.8}$$

$$e^{(\mu/t)(b-2x_t)} = \frac{1}{2\mu}$$

Hence $$x_t = \frac{b}{2} - \frac{t}{2\mu}\ln\frac{1}{2\mu} \tag{6.9}$$

It will be seen that for $x_t > 0$

$$\ln\frac{1}{2\mu} < \frac{\mu b}{t}$$

i.e. sticking friction will not occur for values of $\ln 1/2\mu \geqslant \mu b/t$.

Also if $\mu \geqslant 0{\cdot}5$, $\ln 1/2\mu \leqslant 0$.

Hence for coefficients of friction equal to or greater than 0·5, sticking friction occurs over the whole of the platen area and the friction hill is made up of straight lines.

The average pressure required to deform the metal can be found by adding the vertical forces in the slipping and sticking zones and dividing by the total area over which they act.

Considering unit platen width and measuring from the centre to one extremity.

Slipping zone. From equation (6.6)

$$p = 2ke^{(\mu/t)(b-2x)}$$

vertical force in slipping zone

$$F_c = 2k\int_{x_t}^{\frac{b}{2}} e^{(u/t)(b-2x)}\,dx$$

$$F_c = \frac{t}{\mu}k\{e^{(u/t)(b-2x_t)} - 1\}$$

But from equation (6.8), at $x = x_t$

$$p = \frac{k}{\mu} = 2ke^{(u/t)(b-2x_t)}$$

hence $$\frac{1}{2\mu} = e^{(u/t)(b-2x_t)}$$

and $$F_c = \frac{tk}{\mu}\left(\frac{1}{2\mu} - 1\right)$$

Sticking zone. From Section 4.2.3

$$\frac{\mathrm{d}p}{\mathrm{d}x} = -\frac{2k}{t}$$

$$p = \frac{-2k}{t}x + C$$

At $$x = x_t,\ p = \frac{k}{\mu} = \frac{-2kx_t}{t} + C$$

hence $$C = k\left(\frac{1}{\mu} + \frac{2x_t}{t}\right)$$

Vertical force in sticking zone

$$F_s = 2k\int_0^{x_t}\left(\frac{1}{2\mu} + \frac{x_t}{t} - \frac{x}{t}\right)\mathrm{d}x$$

$$F_s = 2k\left(\frac{x_t}{2\mu} + \frac{{x_t}^2}{2t}\right)$$

The average pressure on the platen, $\bar{p}$ is given by

$$\bar{p} = \frac{F_s + F_c}{b/2}$$

$$\bar{p} = \frac{2}{b}\left\{2k\left(\frac{x_t}{2\mu} + \frac{{x_t}^2}{2t}\right) + \frac{tk}{\mu}\left(\frac{1}{2\mu} - 1\right)\right\}$$

$$\bar{p} = \frac{2k}{b}\left\{\frac{x_t}{\mu} + \frac{{x_t}^2}{t} + \frac{t}{2\mu^2} - \frac{t}{\mu}\right\}$$

From equation (6.9), substituting $(b/2) - (t/2\mu)\ln(1/2\mu)$ for x_t and rearranging terms

$$\bar{p} = 2k\left\{\frac{1}{2\mu}\left(1 - \ln\frac{1}{2\mu}\right) - \frac{t}{2b\mu^2}\left[\ln\frac{1}{2\mu} - \frac{1}{2}\left(\ln\frac{1}{2\mu}\right)^2 - 1\right] + \frac{b}{4t} - \frac{t}{b\mu}\right\}$$

This rather cumbersome formula reduces to something much simpler if a value of μ is assumed, for instance if $\mu = 0{\cdot}2$; then

$$\bar{p} = 2k\left(0{\cdot}21 + 1{\cdot}29\frac{t}{b} + 0{\cdot}25\frac{b}{t}\right)$$

6.5.4 Cold extrusion. The production of components by the cold extrusion of steel was first achieved in Germany in 1934. It was kept a military secret because of its value in munitions production, and elsewhere little was known of the process until 1945.

The two basic processes are forward and backward extrusion, as in the hot extrusion process. In forward extrusion the punch moves vertically downwards and the metal flows in the same direction as the punch (Fig. 6.29 (*a*)). Although this figure shows a shallow cup being extruded, forward extrusion is equally suitable for extruding solid billets. Backward extrusion is illustrated in Fig. 6.29 (*b*), where it will be seen that the metal flows backwards up the punch in the opposite direction to the punch movement.

More complicated shapes can be produced in a single operation by a combination of forward and backward extrusion (Fig. 6.29 (*c*)). If, as is frequently the case, the required shape cannot be obtained in a single operation, a series of extrusion operations is performed until the desired shape is obtained.

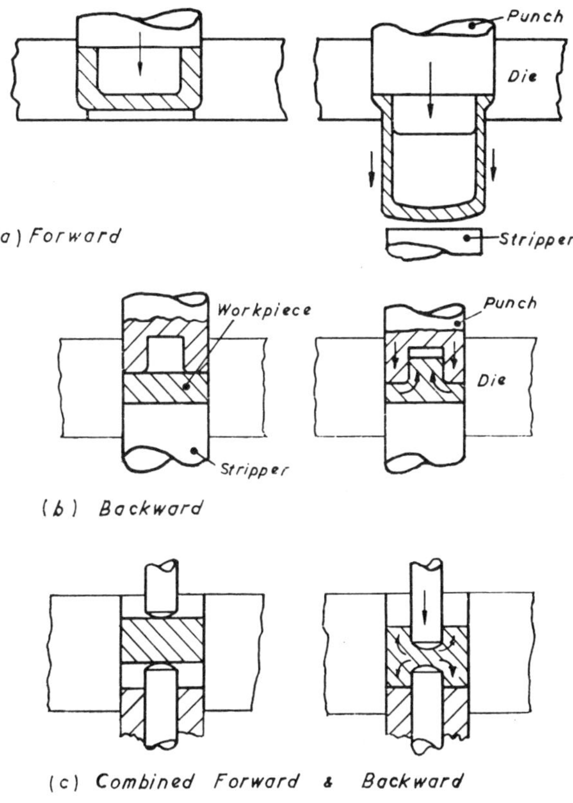

Fig. 6.29 Cold extrusion

Steel for cold extrusion. The choice of steel for use in cold extrusion is important as this will determine the possible amount of deformation. As a rough guide, a change of cross-sectional area of at least 25% should be possible with punch stress <2500 N mm^{-2} (160 tonf/in^2). In a uniaxial tensile test the steel should have a low yield stress, a slow rate of work hardening, and a considerable extension before fracture (Fig. 6.30). At present cold extrusion is limited to low and medium carbon steels, although no doubt harder steels will be used as advances are made in tool materials.

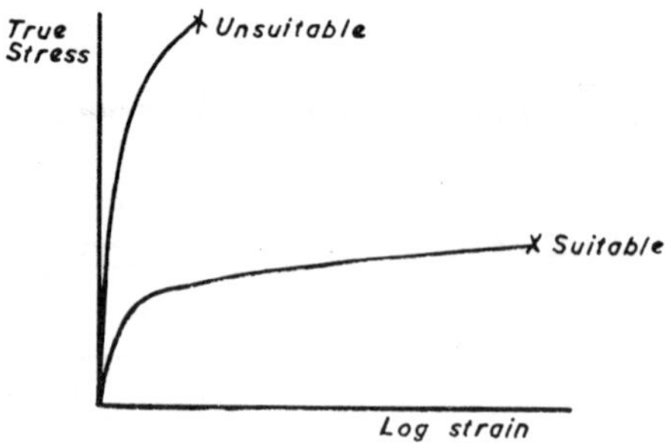

Fig. 6.30 Stress strain curve in tensile test

Effect on mechanical properties. After cold extrusion, the grain structure of the steel becomes severely distorted and subject to internal stresses. Although ductility is reduced, the yield strength is sometimes doubled thus enabling plain carbon steels to be used in place of more costly alternatives such as nickel-chrome steels. The residual stress in the component depends on both the degree of deformation and the temperature attained during deformation. When severe deformation occurs, temperature rises of 300°C are possible and this will produce some stress relief. If more than one extrusion is necessary on the same component, the part is annealed between operations. Recrystallization produced by annealing will give a fine grain structure in heavily deformed parts but a coarser structure in parts subjected to light deformation.

Treatment prior to extrusion. The steel slugs, which are subsequently extruded, undergo a number of treatments before deformation. The most important are annealing, cleaning, phosphating and lubricating.

Annealing ensures that the material is in a soft state before extrusion. After annealing, the metal is cleaned and the surface oxide removed by pickling in heated dilute sulphuric acid. A coating of zinc phosphate is then provided by a bonderizing treatment, and finally the part is immersed in a suitable lubricant, usually sodium stearate.

The provision of an effective barrier between work and tool during extrusion to prevent metal-to-metal contact is essential. This is achieved by the porous phosphate coating, which not only acts as a vehicle for the lubricant but remains tenaciously attached to the surface of the part at the extremely high pressures associated with cold extrusion.

Tooling. Tools have to withstand very severe operating conditions when extruding steel. A common cause of failure is fatigue caused by the rise and fall of direct and bending stresses during extrusion. To avoid fatigue failure, notch stresses must be minimized by rounding corners and avoiding machining marks perpendicular to the metal flow. Eventually tools will wear as a result of surface grooving caused by the flow of the extruding metal. When this occurs, friction increases and the deforming force rises rapidly.

Tools are expensive, and in consequence tool life has an important bearing on the economics of the process. Life varies considerably and an upper limit of 50 000 parts is not unusual, although with smaller deformations and more stages a tool life of 500 000 pieces is possible. Tool parts such as the ends of punches, which are subjected to particularly severe wear, can be designed so that they are easily replaceable. Dies are frequently made in the form of liners shrunk into bolster rings; this compressive prestressing of the die enables it to withstand higher bursting stresses. Punches and dies are usually manufactured from high-speed steel.

Calculation of deforming forces. A relationship, proposed by Johnson,[19] for extrusion pressure p has given good agreement with experimental results when extruding mild steel in the lower range of commercial extrusion ratios.

This is $p = Y_m \varepsilon_m$

where $\varepsilon_m = 1{\cdot}5 \ln (A_0/A_1) + 0{\cdot}8$

Y_m is the mean yield stress

A_0 is the original cross-sectional area

A_1 is the extruded cross-sectional area.

The value of Y_m can be found by a plane strain compression test on the material (see Appendix 2) and by plotting true stress against natural strain. From this curve Y_m, the mean value of stress over the range of strain from zero to ε_m is determined (Fig. 6.31). Difficulty is found in obtaining true stress/strain curves at high values of strain because of the

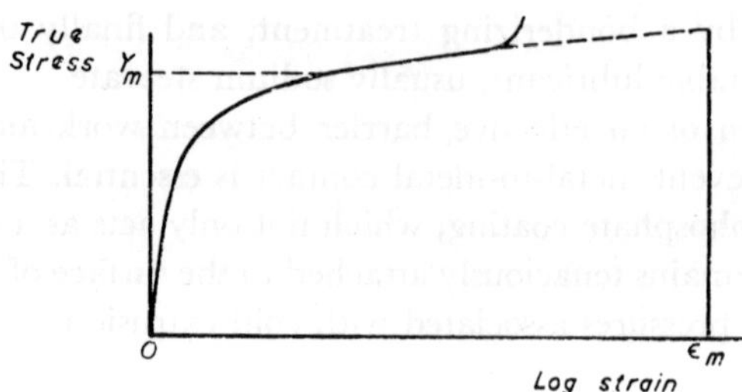

Fig. 6.31 Compressive stress strain curve

inadequate lubrication of the interface between the test piece and the platens. The curve is, however, fairly flat at high strains and can if necessary be extrapolated with little loss of accuracy.

6.5.5 Warm forming. This process is sometimes referred to as warm forging or warm extrusion. The billet is pre-heated to below its recrystallisation temperature, thus enabling deforming forces to be reduced to less than those needed for comparable cold working, but still permitting some strain hardening to occur. The pre-heat temperatures for steels are normally around 500°C to 600°C. Warm forming is particularly useful for hard heat resisting steels which cannot be cold worked and may contain phases which would melt if shaped at high temperature.

The component shapes are limited to those which do not require a complex metal flow. If a special shape feature such as an undercut is required, or if the dimensional tolerances are close, the part will have to be machined. Mechanical properties of warm formed parts are good and their surface finish is much better than parts which are hot worked.

It is critically important that there should be adequate lubrication between tool and workpiece during forming. This is due to the high stresses at the tool/workpiece interface and to the large increases which occur in the workpiece area. Lubrication is, however, more difficult than with cold extrusion because of the elevated working temperatures. Molybdenum disulphide can be used up to 300°C, but above this temperature a variety of lubricants based on colloidal graphite are employed.

6.5.6 Hydrostatic extrusion. This interesting method of cold working, illustrated in Fig. 6.32, is one of potential value but as yet little used in industry. Compared with conventional extrusion it has the advantage that there is no direct contact between the billet and the container wall. Some of the oil which is used to apply pressure to the billet leaves with the extrusion and greatly improves die lubrication. By using differential extrusion, i.e. extruding into a second pressurized container, brittle materials can be extruded. Differential extrusion requires that primary container pressures

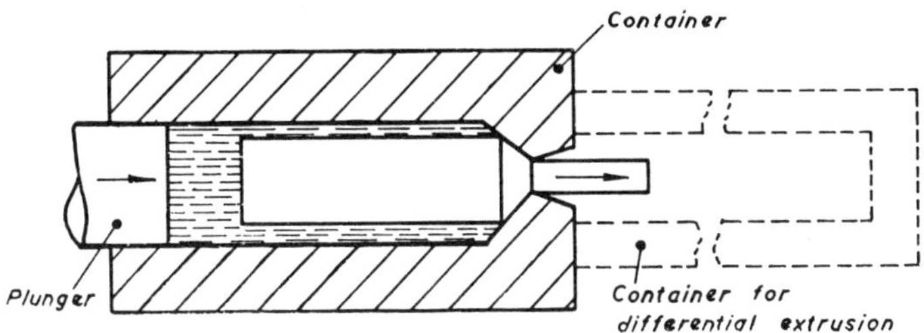

Fig. 6.32 Hydrostatic extrusion

are increased by about 70% to around 4000 $N\,mm^{-2}$ (250 tonf/in^2), which although technically possible, is commercially unattractive. In simple hydrostatic extrusion there is little control over the rate of extrusion. Extrusion rate can be controlled if the container pressure is maintained below expulsion level and the process is assisted by a plunger making contact with the billet, as with conventional extrusion, or by applying a drawing force to the extrudate.

Two problems militating against the wider application of hydrostatic extrusion are the relatively short fatigue life of containers and the difficulty of ensuring the reliability of high pressure seals under production conditions.

6.5.7 Cold heading. Cold heading consists of forming a head on the shank of a work piece: the material normally used is low carbon steel wire. A typical cold heading operation, the manufacture of a rivet, is shown in Fig. 6.33. Parts with heads too large to produce in a single blow are formed in two or more stages. Special-purpose machine tools with fast production rates are used for cold heading. Millions of fasteners such as rivets and bolts are manufactured by this process, which is usually more economical than machining when large quantities are required.

6.5.8 Form rolling. Form rolling produces comparatively complex forms, such as screw threads on machined or formed blanks. Almost all standard external thread forms can be rolled, with the exception of square threads; splines, worms, gear teeth and knurled surfaces can also be form rolled. Rolling dies are made from hardened steel and have an appropriate form ground on their surface. The work is rolled between the dies, which move radially inwards until the full form is obtained on the component.

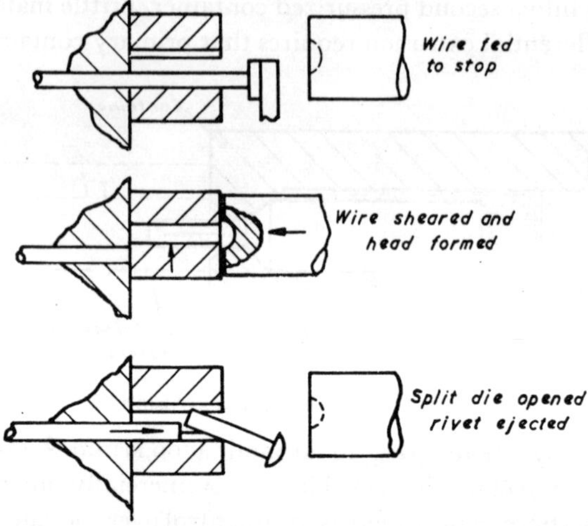

Fig. 6.33 Production of rivet in single blow split-die header

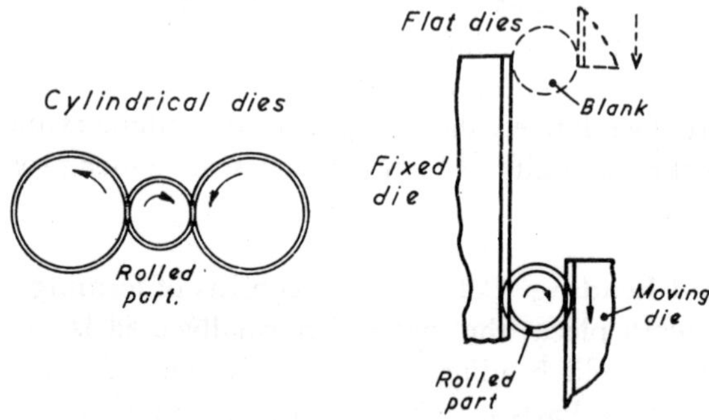

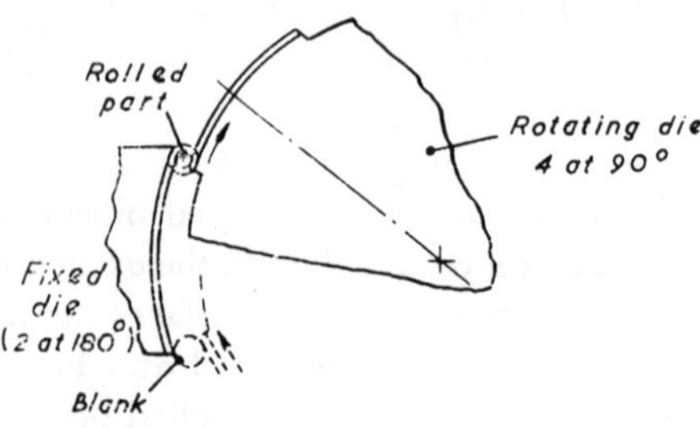

Fig. 6.34 Thread rolling

In thread rolling, blanks slightly in excess of the effective diameter of the thread are used, the profile being achieved by forcing the metal to flow from the thread roots into the unfilled die form at the crests. The component is work hardened by rolling, and the burnishing effect of the dies leaves an excellent surface finish.

Standard machine tools, such as automatic and capstan lathes, can be fitted with thread rolling attachments; these have one or more roller dies. Flat and curved dies are used on special purpose thread rolling machines which are frequently fed from heading machines. Production rates on special purpose machines can be in excess of 500 parts/min. Three different die arrangements are shown in Fig. 6.34.

6.5.9 Flow turning. Basically there are two flow turning processes, one which converts discs of metal into hollow shaped components of approximately conical form, and one which elongates a preformed tube by reducing its wall thickness.

When flow turning a cone the process resembles spinning, except that when spinning the section thickness remains substantially constant,

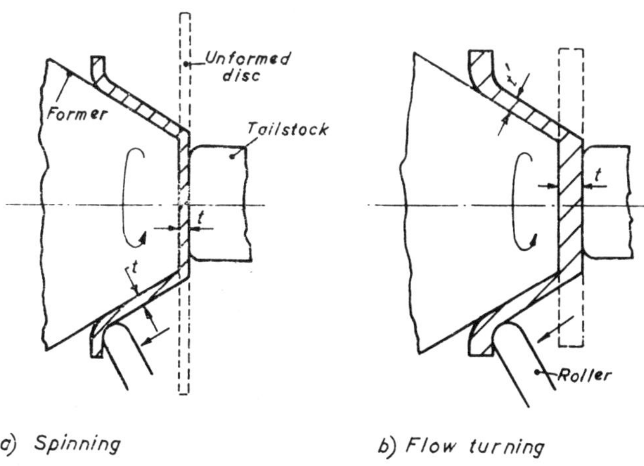

Fig. 6.35

whereas when flow turning the section is significantly reduced. The essentials of the two processes are shown in Fig. 6.35. It will be seen that spinning is essentially a stretching process, the diameter of the disc being progressively reduced as the cone is formed. A high radial tensile stress is induced in the disc causing plastic flow, and the conical deformation is due mainly to bending and unbending under tension.

Flow turning does not involve a significant decrease in the diameter of the disc, and the disc is relatively stress free. The process is predominantly compressive and as an approximation was considered by Kalpakcioglu[20] as a case of simple shearing. Fig. 6.36 illustrates the shearing of an element of the cone, where shear strain $\gamma = (\delta x/\delta y) = \cot \alpha$, and $t' = t \sin \alpha$.

Work done in shearing/rev

$$= 2\pi \,.\, F \,.\, r = \text{volume sheared/rev} \times (\gamma \,.\, k)$$

$$= 2\pi \,.\, r \,.\, t' \,.\, f \,.\, \gamma \,.\, k$$

where F = tangential force on roller,

r = instantaneous radius of cone,

f = feed/rev,

k = shear flow stress = $Y/\sqrt{3}$,

$$F = t \,.\, \sin \alpha \,.\, f \,.\, \cot \alpha \,.\, (Y/\sqrt{3}),$$

$$F = t \,.\, f \,.\, (Y/\sqrt{3}) \cos \alpha.$$

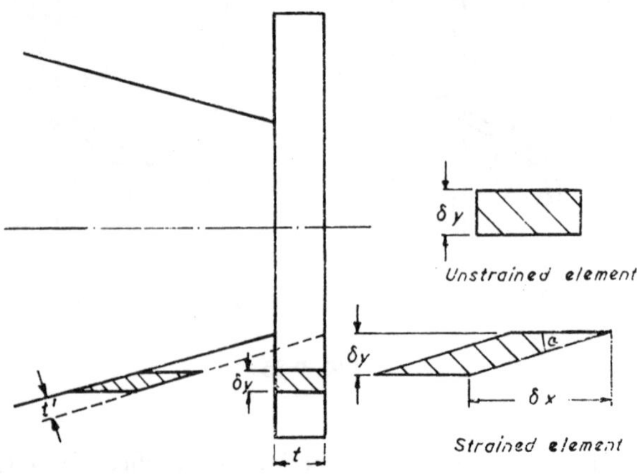

Fig. 6.36 Shearing action when flow turning a cone

For a strain hardening material, an average yield stress Y can be taken, giving

$$F = t \,.\, f \,.\, (Y/\sqrt{3}) \cos \alpha \tag{6.10}$$

This analysis does not allow for redundant work, and for small instantaneous cone diameters it underestimates the force by a factor of about 2. However, for cone diameters over about 0·5 m (20 in), the agreement with experiment is reasonable.

Cones can be flow turned to $\pm$0·05 mm ($\pm$0·002 in), with surface finishes of 0·15–0·2 μm (6–8 μin). The process can be used for plate thicknesses up to 19 mm ($\frac{3}{4}$ in) with stainless steels or nimonics, and up to 38 mm ($1\frac{1}{2}$ in) with some non-ferrous materials. Although most ductile metals can be flow turned, cold titanium usually requires pre-heating. The apex angle of the cone is limited by the metal used, but given favourable conditions angles as small as 30° can be achieved without preforming.

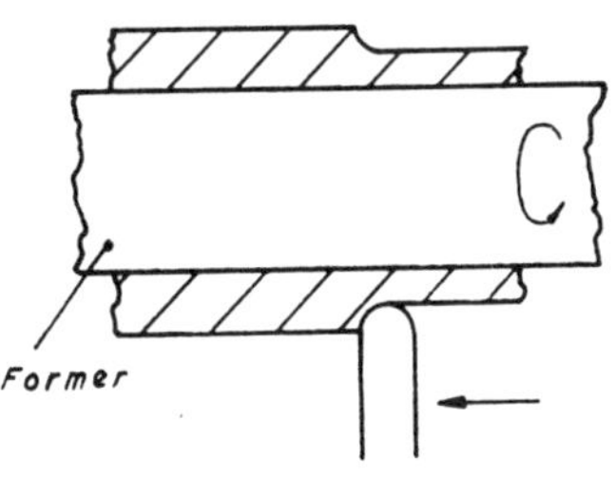

Fig. 6.37 Flow turning a tube

The application of flow turning to the reduction of tubes is shown in Fig. 6.37. There does not appear to be a reliable analysis of this process, although analogies have been drawn between flow turning and plane extrusion. A variant of this process which has considerable commercial application is thread rolling, described in section 6.5.8.

Flow turning does not involve a high investment in tooling and can be performed on rigid lathes with powerful motors and copying attachments to guide the roller. It is used to produce small quantities of simple hollow components, often of considerable size, where alternative tooling and equipment would be costly and deformation forces prohibitively large.

6.5.10 Impact extrusion. This process is popular for the manufacture of large quantities of components from soft, ductile materials such as aluminium or lead. It is a cold extrusion process, and forward or backward extrusion is possible; conventional crank presses giving ram speeds of about 0·6–0·9 m s^{-1} (2–3 ft/s) are used.

Complex forms can be produced, the particular advantage being where thin-walled cylinders and tubes, possibly with a flange or heavy base are required. The resulting mechanical properties are good.

Recent work[21] using high-velocity machines with ram speeds of 15–90 m s^{-1} (50–300 ft/s), has resulted in products from impact extrusion with better finish and improved straightness. Also, a wider range of materials, including steel and titanium, has been successfully extruded.

Maximum extrusion pressures increase with velocity, but where the pressure reaches very high values the effect can be offset by preheating the blanks. In practice, ram speed is generally limited to 25 m s^{-1} (80 ft/s), particularly in forward extrusion, due to break-off caused by thermal or inertial effects. Rapid deceleration causes high tensile stresses in extrusions, and failure may occur due to necking. Thermal break-off occurs particularly in metals of low thermal conductivity, and appears to be due to the localized near-adiabatic temperature rise in the deformation zone, which may cause thermoplastic instability and result in what appears to be a brittle fracture.

6.6 SUPERPLASTIC ALLOYS

These materials, which are expensive, specially prepared fine grain alloys, can be shaped from pre-heated sheet at very low loads using inexpensive tooling. Although many metals can be produced in superplastic form, aluminium and zinc based alloys have shown the best commercial promise.

The two basic methods of shaping are female and male forming. In female forming a pre-heated sheet is placed over a die cavity and air pressure or a vacuum is used to induce the sheet to assume the die shape. This process is suitable only for shallow parts, due to thinning at the corners of the formed parts. Male forming uses a shaped punch which is pushed into the clamped pre-heated sheet. In addition compressed air is normally used to ensure that the sheet conforms to the tool profile. Non-uniform thicknesses also result from this process, unless the sheet is blown first into a bubble of correct size and then collapsed on to the tool.

As only one die has to be manufacturered, from relatively inexpensive and easily modified cast iron or cast aluminium, tooling costs are very much lower than for a pair of conventional drawing dies. The most favourable production range quoted[22] for superplastic aluminium is for a total order quantity of between 100 to 5000 parts. Forming times are much longer than for conventional drawing: exceptionally up to 30 minutes are required for some deep drawn parts.

After forming, the parts are heat treated to strengthen them; superplastic aluminium alloy can be converted to a strength approximating to that of mild steel. Typical parts manufactured from superplastic alloys are covers, panels and housings. Older superplastic alloys had poor mechanical properties and little commercial application. However, considerable growth in demand is now expected, particularly for parts made from superplastic aluminium alloys.

7 Cutting Tool Geometry and Tool Materials

7.1 In this chapter some loosely connected aspects of metal cutting are discussed. Tool nomenclature has been dealt with first so that the various tool angles can be defined for use in the other sections. The direction of chip flow across the tool is next considered, to allow a concept of 'effective' rake to be established.

In practice the steady state conditions on which cutting theories are based are often disrupted by the formation of discontinuous chips, and tool geometry is altered by the presence of a built-up edge. The last section considers the influence of different cutting tool materials on cutting speeds and tool geometry.

7.2 TOOL NOMENCLATURE

British Standard 1296 : 1972[23] defines the angles on single point cutting tools in terms of the normal rake system. This defines two rake

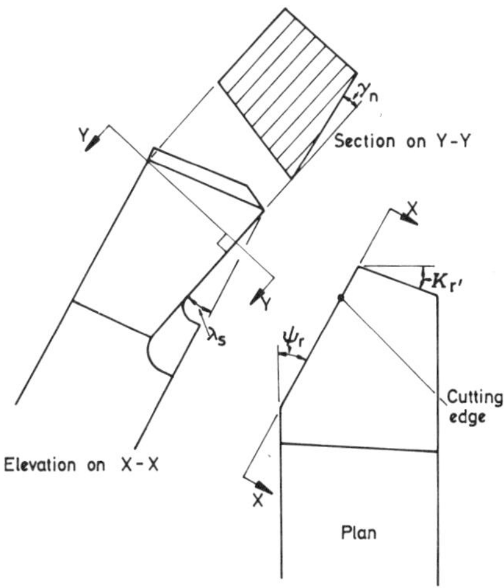

Fig. 7.1 Normal rake system

angles, as shown in Fig. 7.1. The angle λ_s is the back rake or cutting edge inclination, which is measured parallel to the cutting edge in the vertical plane; γ_n is the normal rake, measured in a plane perpendicular to the cutting edge. ψ_r is the tool approach angle and $K_{r'}$ is the tool minor cutting edge angle.

The normal rake system relates the important cutting angles directly to the cutting edge and not to some geometrical feature of the shank of the tool; it is therefore a more realistic method of nomenclature for studying chip formation. For multi-point tools the normal rake system again gives a realistic idea of tool geometry and is generally preferable to other systems.

7.3 DIRECTION OF CHIP FLOW

For most practical machining operations the cutting edge is presented obliquely to the cutting direction (Fig. 7.2). When designing cutting tools,

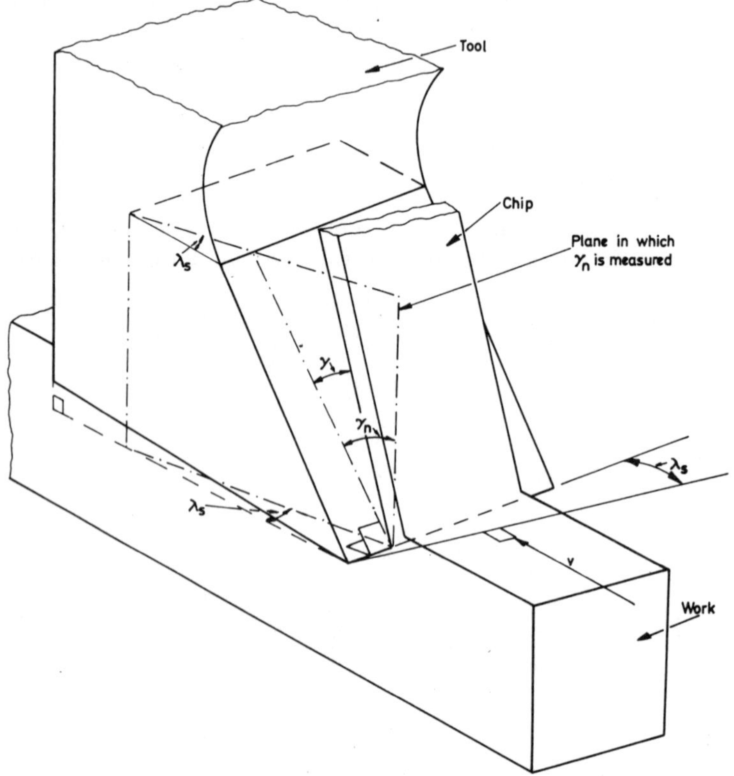

Fig. 7.2 Oblique machining

it is desirable to know the direction in which the chip will flow up the rake face. Experimental work by Stabler[24] has shown the direction to conform fairly closely to a simple law, which states that

$$\gamma = \lambda_s \tag{7.1}$$

The angle λ_s was referred to by Stabler as the angle of obliquity and corresponds to the cutting edge inclination, and γ is the angle made by the chip with the normal to the cutting edge along the tool rake face. The normal rake is denoted by the angle γ_n. The example shown in Fig. 7.2 has been simplified by assuming a tool approach angle of zero, but the result is equally applicable when the tool approach angle is not zero.

In oblique cutting, neither of the principal rake angles, λ_s and γ_n, is of much value in describing the chip geometry, since they do not lie in the plane of the chip flow. Stabler specified an effective rake angle measured in a plane containing both the cutting speed and chip flow vectors, β_e in Fig. 7.3

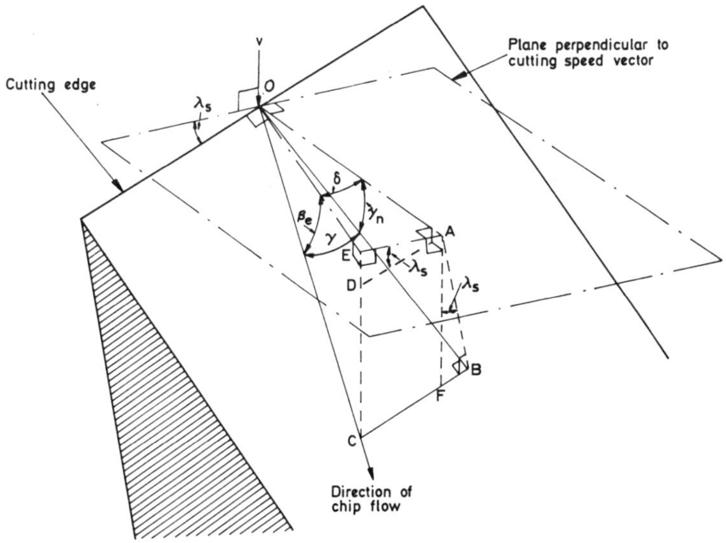

Fig. 7.3 Angular relationships for oblique cutting

$$\sin \beta_e = \frac{CE}{OC} = \frac{CD + ED}{OC} = \frac{AF + ED}{OC}$$

$$= \frac{AB}{\cos \lambda_s \, . \, OC} + \frac{AD \sin \lambda_s}{OC}$$

$$= \frac{AB \cos \gamma}{OB \cos \lambda_s} + \frac{(CB - BF) \sin \lambda_s}{OC}$$

$$= \frac{\sin \gamma_n \cos \gamma}{\cos \lambda_s} + \sin \gamma \sin \lambda_s - \frac{AB \tan \lambda_s \sin \lambda_s \cos \gamma}{OB}$$

$$= \frac{\sin \gamma_n \cos \gamma}{\cos \lambda_s} + \sin \gamma \sin \lambda_s - \sin \gamma_n \tan \lambda_s \sin \lambda_s \cos \gamma$$

$$= \sin \gamma_n \cos \gamma \left(\frac{1 - \tan \lambda_s \sin \lambda_s \cos \lambda_s}{\cos \lambda_s} \right) + \sin \gamma \sin \lambda_s$$

$$= \sin \gamma_n \cos \gamma \cos \lambda_s + \sin \gamma \sin \lambda_s$$

Assuming the truth of the flow law, $\lambda_s = \gamma$,

$$\sin \beta_e = \cos^2 \lambda_s \sin \gamma_n + \sin^2 \lambda_s \qquad (7.2)$$

This relationship enables the effective rake angle to be calculated from the principal rake angles. The nomogram, Fig. 7.4, enables this calculation to be performed.

7.4 CHIP FORMATION

Three types of chip are usually classified; they are: (*a*) discontinuous, (*b*) continuous, (*c*) continuous with a built-up edge. Unfortunately it is not possible to differentiate clearly between these groups because of the existence of semi-discontinuous chips and of discontinuous chips which produce a noticeable built-up edge. However, experience shows that certain combinations of cutting materials and cutting conditions are more likely to produce one sort of chip than another.

Discontinuous chips (Fig. 7.5) occur when the amount of deformation which the chips undergo is limited by repeated fracturing. As would be expected, brittle materials such as cast iron are most likely to give a discontinuous chip, although they may also be produced by ductile materials, particularly if the hydrostatic pressure near the cutting edge becomes tensile or the shear strain energy reaches a critical value, as demonstrated by Enahoro and Oxley,[25] and by Luk and Brewer.[26] The conditions under which discontinuous chips are likely to occur are low

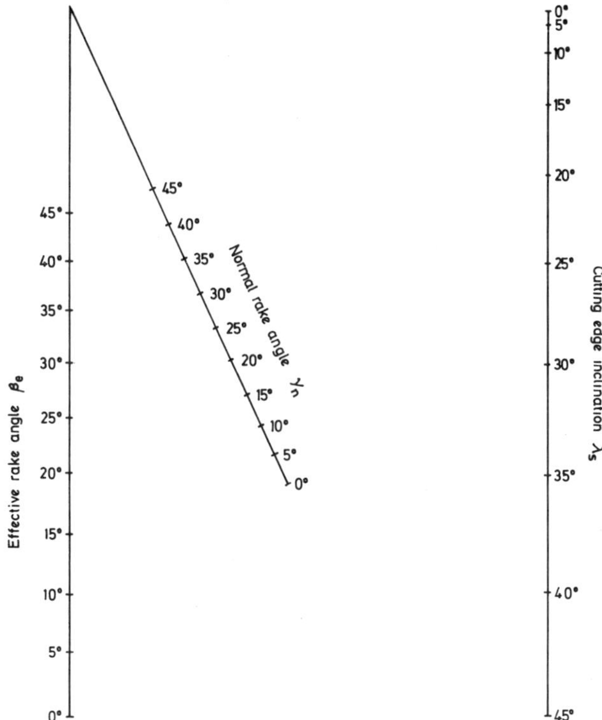

Fig. 7.4 Effective rake angle nomogram

cutting speeds, low rake angles, heavy feeds and high friction forces at the chip tool interface.

Continuous chips without built-up edge are most likely to be obtained when using ductile work materials, and under conditions the reverse of those causing discontinuous chips. Conditions leading to built-up edge formation are described in the next section.

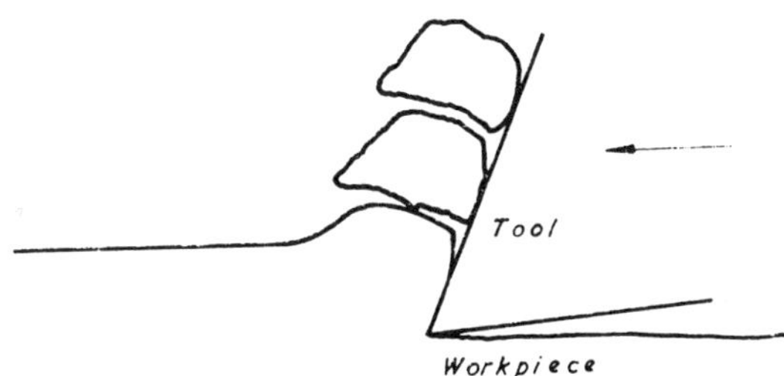

Fig. 7.5 Discontinuous chip formation

7.5 BUILT-UP EDGE

Over a large range of cutting speeds a built-up edge or nose exists on the cutting edge of the tool; this has the effect of changing the effective geometry of cutting. The form and size of such an edge depends largely on the cutting speed, and disappears almost entirely at very low and very high speeds. Since cutting speed directly affects the temperatures attained in the shear zone, it is reasonable to suppose there may be some relationship between temperature and build-up. Bisacre and Bisacre[27] suggested the idea of a dimensionless 'thermal number' which they considered should bear a relationship to all the cutting factors associated with the chip tool interface for a given tool. The thermal number Vf/h^2 is the product of cutting speed V and feed f divided by the thermal diffusivity h^2. It is, in fact, the coefficient of the first order term in the heat conduction equation (see Appendix 1).

Chao and Bisacre[28] assessed the size of the built-up edge by measuring its projected area A, while Heginbotham and Gogia[29] measured the length of the built-up edge L along the rake face of the tool. Graphs of A/f^2 and L/f plotted against thermal number for readings taken at different speeds and feeds showed that a unique relationship appeared to exist, although Heginbotham and Gogia found a better correlation by plotting against $Vf^{1 \cdot 5}/h^2$.

Two different theories exist to explain the initial formation of a built-up edge. The one most usually accepted is that the high friction forces existing on the rake face cause the material to reach its shear flow stress along a line inclined to the rake face; a velocity discontinuity occurs and wedge-shaped particles are left on the rake surface. The other theory, that of Palmer and Yeo,[30] assumes that the built-up edge is initiated by bluntness of the tool edge rather than by the high friction on the rake face.

By adopting a crystal-plotting technique, they constructed a slip-line field around an artificially blunted tool. This had a dead metal zone ABC (see Fig. 7.6). They predicted that the hydrostatic stress at A was more

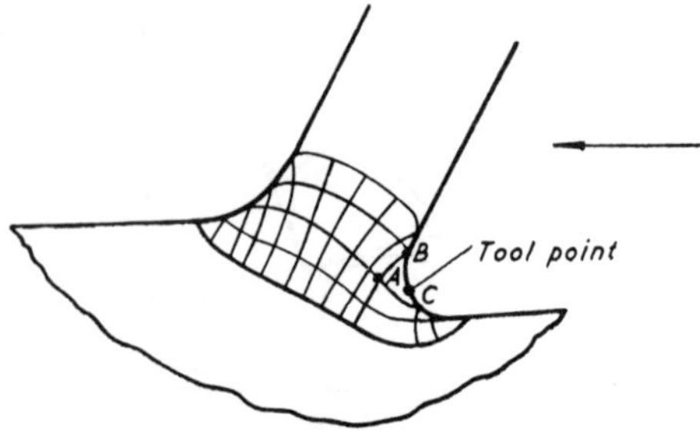

Fig. 7.6 Slip-line field around blunt tool

compressive than at B and C. The compression at A inhibits separation at this point, causing the dead metal zone to grow by advancing away from the tool cutting edge to form a built-up edge.

Qualitative experiments have demonstrated that the shape of the built-up edge depends mainly on cutting speed; typical built-up edges for three different cutting speeds are shown in Fig. 7.7. This variation in shape is explained by thermal softening, assuming the temperature to rise near the tip of the build-up as cutting speed increases. At speeds $>1{\cdot}5$ m s^{-1} (300 ft/min) Heginbotham and Gogia claim to have identified a very thin parallel built-up edge which persisted at much higher speeds. However,

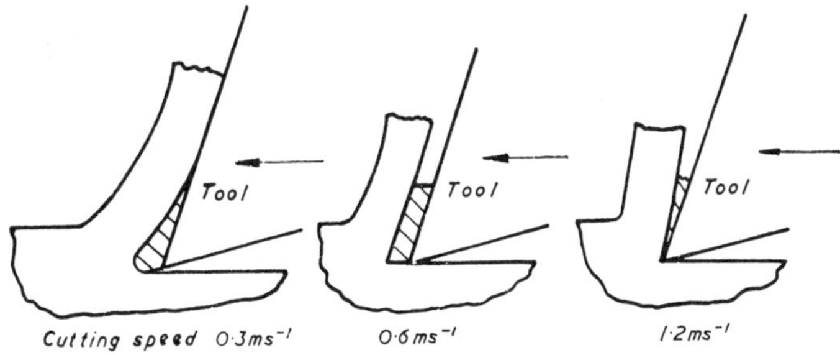

Fig. 7.7 Typical built-up edge formation for steel

other investigators are of the opinion that the built-up edge does not exist at high speeds.

The present state of knowledge of the built-up edge may be summarized as follows.

(*a*) It can be initiated in one of two ways. Wear at the point of the tool may result in the formation of a dead metal zone, and the subsequent growth in size of the build-up at any given speed depends on the work hardening properties of the material. Alternatively, a high friction force at the chip tool interface may cause adhesion to occur. Adhesion may be inhibited by using a polished tool, or under certain conditions by the application of a cutting lubricant.

(*b*) The shape of the built-up edge is a function of temperature, which is in turn determined by cutting speed.

(*c*) The built-up edge is apparently continuous with the chip and workpiece. Although a certain amount of diffusion welding occurs at the chip tool interface, the built-up edge on a tool will normally remain with the chip when the cut is 'frozen' by quick stopping.

(*d*) Positive rake angles on the tool lead to a decrease in the amount of build-up at low cutting speeds, but negative rakes result in a decrease in build-up at high cutting speeds.

(*e*) Increase in feed causes an increase in the dimensions of the built-up edge at low cutting speeds, but at higher speeds a value of feed is reached where maximum build-up occurs, and at higher feeds the build-up then decreases.
(*f*) A unique relationship appears to exist between the size of the built-up edge and the thermal number.
(*g*) The built-up edge appears always to exhibit a blunt form.
(*h*) Surface finish on the workpiece is dependent on the type of built-up edge and generally improves at higher speeds.

7.6 CUTTING TOOL MATERIALS

A large number of cutting tool materials has been developed to meet the demands of high metal-removal rates. The situation at the moment is that the latest developments, such as the use of refractory oxides, cannot be fully exploited because of the inadequacy of existing machine tools. The most important of these materials and their influence on cutter design, are described below.

7.6.1 High carbon steel. Historically, high carbon steel was the earliest cutting material used industrially, but it has now been almost entirely superseded since it starts to temper at about 220°C and this irreversible softening process continues as temperature increases (Fig. 7.8). Cutting speeds with carbon steel tools are therefore limited to about 0·15 m s^{-1} (30 ft/min) when cutting mild steel, and even at these speeds a copious supply of coolant is required.

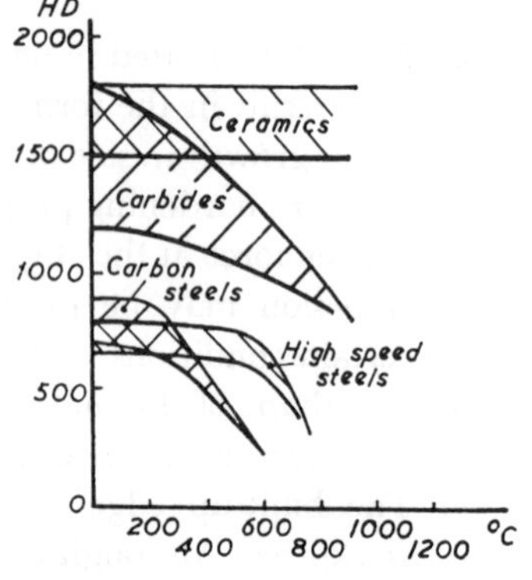

Fig. 7.8 Variation of tool hardness with temperature

7.6.2 High-speed steel. To overcome the low cutting speed restriction imposed by plain carbon steels, a range of alloy steels, known as high-speed steels, began to be introduced during the early years of this century. The chemical composition of these steels varies greatly, but they basically contain about 0·7% carbon and 4% chromium, with additions of tungsten, vanadium, molybdenum and cobalt in varying percentages. They maintain their hardness at temperatures up to about 600°C, but soften rapidly at higher temperatures. Experience shows that high-speed steels fail in a short time if used on mild steel at cutting speeds in excess of 1·8 m s^{-1} (350 ft/min), and many cannot successfully cut mild steel faster than 0·75 m s^{-1} (150 ft/min).

7.6.3 Sintered carbides. Carbide cutting tools, which were developed in Germany in the late 1920s, usually consist of tungsten carbide or mixtures of tungsten carbide and titanium or tantalum carbide in powder form, sintered in a matrix of cobalt or nickel. Because of the comparatively high cost of this tool material and its low rupture strength, it is normally produced in the form of tips which are either brazed to a steel shank or mechanically clamped in a specially designed holder. Mechanically clamped tool tips are frequently made as throw-away inserts. When all the cutting edges have been used the inserts are discarded, as regrinding would cost more than a new tip.

The high hardness of carbide tools at elevated temperatures enables them to be used at much faster cutting speeds than high-speed steel (of 3–4 m s^{-1} (600–800 ft/min) when cutting mild steel). They are manufactured in several grades, enabling them to be used for most machining applications. Their earlier brittleness has been largely overcome by the introduction of tougher grades, which are frequently used for interrupted cuts including many arduous face-milling operations.

Recently, improvements have been claimed by using tungsten carbide tools coated with titanium carbide or titanium nitride (about 0·0005 mm coating thickness). These tools are more resistant to wear than conventional tungsten carbide tools, and the reduction in interface friction using titanium nitride results in a reduction in cutting forces and in tool temperatures. Hence, higher metal removal rates are possible without detriment to tool life or alternatively longer tool lives could be achieved at unchanged metal removal rates.

The uses of other forms of coating with aluminium oxide and polycrystalline cubic boron nitride are still in an experimental stage, but it is likely that they will have important applications when machining cast iron, hardened steels and high melting point alloys.

7.6.4 Ceramics. The so-called ceramic group of cutting tools represents the most recent development in cutting tool materials. They consist mainly of sintered oxides, usually aluminium oxide, and are almost invariably in the form of clamped tips. Because of the comparative cheapness of ceramic tips and the difficulty of grinding them without causing thermal cracking, they are made as throw-away inserts.

Ceramic tools are a post-war introduction and are not yet in general factory use. Their most likely application is in cutting metal at very high speeds, beyond the limits possible with carbide tools. Ceramics resist the formation of a built-up edge and in consequence produce good surface finishes. Since the present generation of machine tools is designed with only sufficient power to exploit carbide tooling, it is likely that, for the time being, ceramics will be restricted to high-speed finish machining where there is sufficient power available for the light cuts taken. The extreme brittleness of ceramic tools has largely limited their use to continuous cuts, although their use in milling is now possible.

As they are poorer conductors of heat than carbides, temperatures at the rake face are higher than in carbide tools, although the friction force is usually lower. To strengthen the cutting edge, and consequently improve the life of the ceramic tool, a small chamfer or radius is often stoned on the cutting edge, although this increases the power consumption. Tool life is greatest for tools with negative rakes of about 15–20°. This is about twice the size of the largest negative rakes commonly used on carbide tools and except at very high speeds, where thermal softening of the work material predominates, results in noticeably higher power consumption.

7.6.5 Non-ferrous alloys. Several non-ferrous alloys containing varying percentages of cobalt, chromium, tungsten and carbon are sometimes used for machining hard metals at speeds slightly in excess of those used with high-speed steel tools. Their cutting properties are roughly intermediate between those of high-speed steel and tungsten carbide. The most important application of these alloys is in the drilling operation.

7.6.6 Diamonds. For producing very fine finishes of 0·05–0·08 μm (2–3 μin) on non-ferrous metals such as copper and aluminium, diamond tools are often used. The diamond is brazed to a steel shank. Diamond turning and boring are essentially finishing operations, as the forces imposed by any but the smallest cuts cause the diamond to fracture or be torn from its mounting. Under suitable conditions diamonds have exceptionally long cutting lives.

Synthetic polycrystalline diamonds are now available as mechanically clamped cutting tips. Due to their high cost they have very limited applications, but are sometimes used for machining abrasive aluminium–silicon alloys, fused silica and reinforced plastics. The random orientation of their crystals gives them improved impact resistance, making them suitable for interrupted cutting.

7.7 EFFECT OF CUTTING MATERIAL ON CUTTING ANGLES

The major consideration when determining the size of the angles ground on cutting tools is the ability of the tool material to withstand the cutting forces. The advent of hard sintered compounds with low tensile and impact strengths as cutting materials has demanded a more robust cutting edge than was necessary with the tougher carbon steel or high-speed steel tools. Hence the tendency has been to design sintered tools with smaller positive rake angles and lower clearance angles.

Chapter 8 shows the effect of rake angle on cutting force, and it is demonstrated that at low speeds the cutting forces associated with high rake angles are less than those associated with low positive rakes or negative rakes. However, at high speeds, thermal softening of the chip results in an increase in chip thickness ratio for negative rakes and a corresponding decrease in cutting force. Apart from the additional strength conferred upon the cutting edge, negative rakes cause the cutting force to be rotated so that the tool bits are held more effectively against their mountings.

8 Metal Cutting

8.1 ORTHOGONAL CUTTING

It is not yet possible to predict with any great accuracy the forces involved in metal cutting, in spite of a large number of theories which have been developed. This is largely due to the extreme complexity and the lack of geometrical constraint which is characteristic of metal cutting. Before explaining the effects of the many variables which are encountered, a description of the simplified theories based on orthogonal machining will be attempted.

In orthogonal machining, the tool approaches the workpiece with its cutting edge parallel to the uncut surface and at right angles to the direction of cutting. To prevent end effects, the tool is wider than the workpiece. Fig. 8.1 shows how a chip is removed under orthogonal conditions. All the theories of orthogonal machining imply or assume plane strain conditions, in that the width of the chip remains equal to the width of the workpiece.

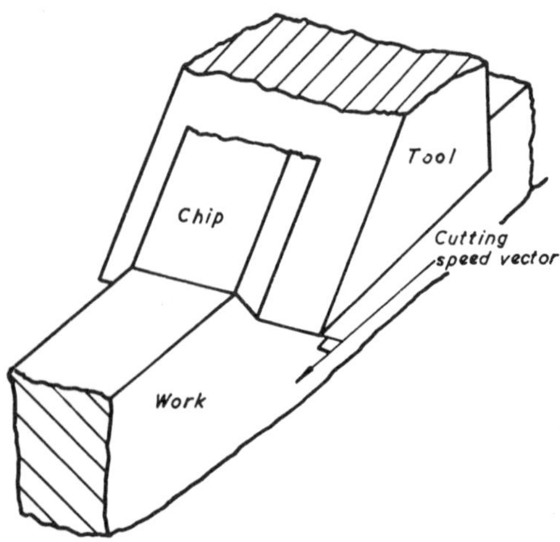

Fig. 8.1 Orthogonal cutting

If it is assumed that shearing in the primary deformation zone occurs across a narrow band (Fig. 8.2), the shear strain γ is dependent on the rake angle β_e and the shear angle ϕ. A particle entering the shear zone at A is deflected by the shearing action so that it leaves the shear zone at C.

$$\begin{aligned} \gamma &= BC/AD \\ &= (BD + CD)/AD \\ &= \cot\phi + \tan(\phi - \beta_e) \end{aligned}$$

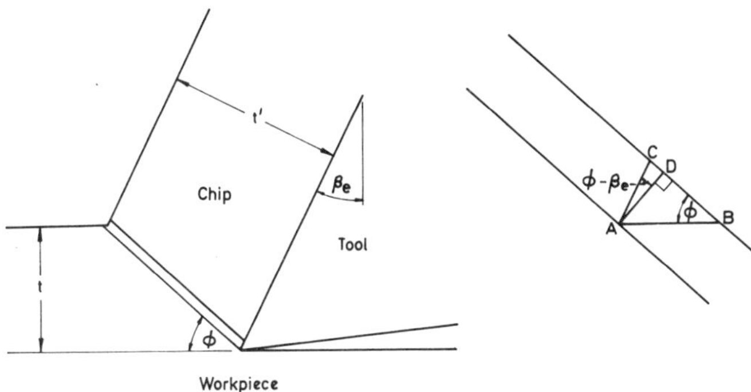

Fig. 8.2 Idealised chip formation

It follows that shear strain, and hence the energy of primary deformation, is reduced if β_e and ϕ are large. These angles determine the chip thickness ratio r_t which is the ratio of the uncut chip thickness t divided by the actual chip thickness t'. A large value of r_t therefore indicates a high efficiency in chip removal.

8.2 MERCHANT'S THEORY

This theory is usually attributed to Merchant,[31] although other investi-

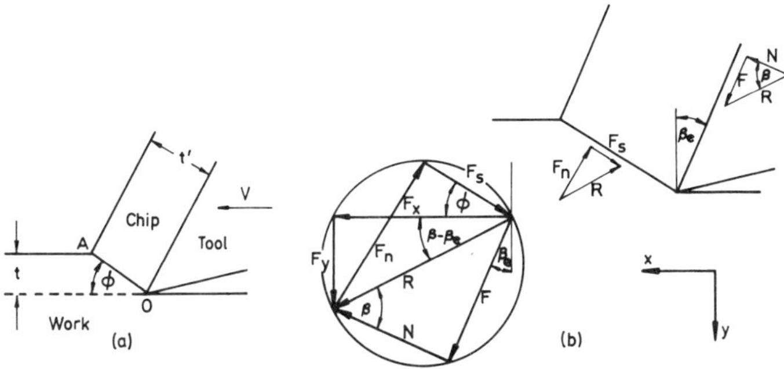

Fig. 8.3 Forces in orthogonal machining

gators have independently arrived at a similar conclusion.

The chip is assumed to shear continuously across a plane AO (Fig. 8.3 (*a*)), on which the shear stress reaches the value of the shear flow stress k. Alternatively the assumption that minimum work is done in shearing can be used.

From Fig. 8.3 (*b*),

$$R = F_x \,.\, \sec(\beta - \beta_e)$$

$$F_s = R \,.\, \cos(\phi + \beta - \beta_e) = F_x \,.\, \sec(\beta - \beta_e) \,.\, \cos(\phi + \beta - \beta_e)$$

$$F_n = R \,.\, \sin(\phi + \beta - \beta_e) = F_x \,.\, \sec(\beta - \beta_e) \,.\, \sin(\phi + \beta - \beta_e)$$

Given a chip width w, and uncut chip thickness t, shear stress on shear plane

$$\tau = \frac{F_x \,.\, \sec(\beta - \beta_e) \,.\, \cos(\phi + \beta - \beta_e) \,.\, \sin\phi}{t.w.}$$

$$F_x = \frac{\tau.t.w.}{\sec(\beta - \beta_e) \,.\, \cos(\phi + \beta - \beta_e) \,.\, \sin\phi}$$

Using Merchant's hypothesis that the shear angle adjusts itself to give minimum work, we can either seek to maximize τ or minimize F_x to find an equation for ϕ. The uncut chip thickness t, the tool rake angle β_e, and the width of the chip w are all constants for a given set of cutting conditions, and Merchant assumed that the angle of friction β was independent of ϕ. In the expressions for τ and F_x above, the term $\cos(\phi + \beta - \beta_e) \,.\, \sin\phi$ contains one variable, ϕ, the shear angle, and to maximize τ or minimize F_x then $\cos(\phi + \beta - \beta_e) \,.\, \sin\phi$ must itself be maximized.

If $y = \cos(\phi + \beta - \beta_e) \,.\, \sin\phi$

$$\frac{dy}{d\phi} = -\sin(\phi + \beta - \beta_e) \,.\, \sin\phi + \cos(\phi + \beta - \beta_e) \,.\, \cos\phi.$$

For maximum value of y,

$$\sin(\phi + \beta - \beta_e) \,.\, \sin\phi = \cos(\phi + \beta - \beta_e) \,.\, \cos\phi.$$

$$\tan(\phi + \beta - \beta_e) = \cot\phi = \tan\left(\frac{\pi}{2} - \phi\right)$$

$$\phi^\circ = 45^\circ - \tfrac{1}{2}(\beta - \beta_e)^\circ \tag{8.1}$$

If ϕ is plotted against $\beta - \beta_e$, it gives a linear relationship (as shown in

Fig. 8.4) which has an intercept value of 45° and a slope of $-\frac{1}{2}$. Experiments on a number of materials have shown that in most cases a linear

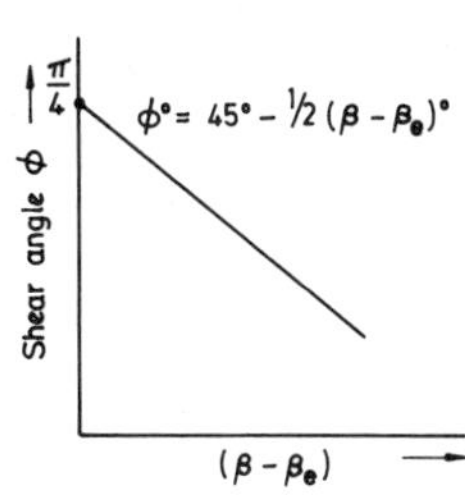

Fig. 8.4 Graphical representation of Merchant's relationship

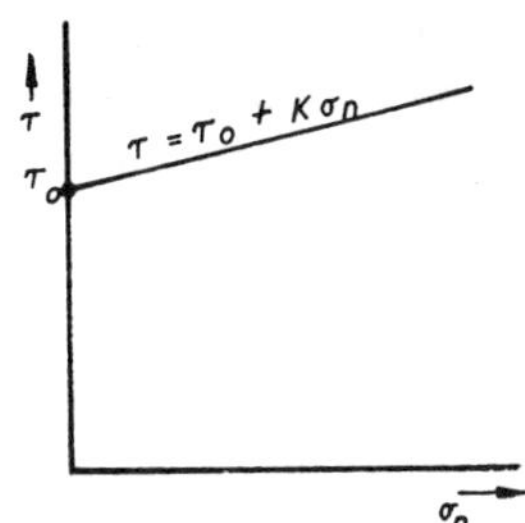

Fig. 8.5 Merchant's assumed relationship between shear flow stress and normal stress

relationship is obtained, but the slopes and intercept values have varied considerably according to the type of metal used. To explain these differences, Merchant called on the work of Bridgman[32] concerning the effect of hydrostatic pressure on the ultimate strength of a metal. He reasoned that if hydrostatic pressure increased the ultimate stress, it would also increase the yield stress. Merchant's assumed relationship between shear flow stress and normal stress is shown in Fig. 8.5.

$$\tau = \tau_0 + K \, . \, \sigma_n \text{ where } K \text{ is the slope.}$$

Normal stress on shear plane

$$\sigma_n = \frac{F_n}{A} = \frac{F_s}{A} \, . \tan(\phi + \beta - \beta_e)$$

$$= \tau \, . \tan(\phi + \beta - \beta_e)$$

where A = area of shear plane.
Substituting for τ in the equation for F_x,

$$F_x = \frac{\tau_0 \, . \, t \, . \, w}{\sec(\beta - \beta_e) \, . \cos(\phi + \beta - \beta_e) \, . \sin\phi \, . \, [1 - K \, . \tan(\phi + \beta - \beta_e)]}$$

Minimizing this equation,

$$\phi = \frac{\cot^{-1} K}{2} - \tfrac{1}{2}(\beta - \beta_e) \tag{8.2}$$

The modified expression for ϕ enables the line for the graph of ϕ against $\beta - \alpha$ to be moved vertically but does not alter its slope. For mild steel, the slope is fairly close to experimental results and the intercept value is about 32°. However, other materials such as aluminium have widely differing slopes from that predicted by Merchant's equation. Merchant's extension of Bridgman's theory has since been disproved, but this treatment has been included for historical completeness.

If the shear plane theory is considered from the point of view of time it will be realized that an infinite shear strain rate is implied. Much of the work published subsequent to Merchant's has assumed that instead of a plane, shearing takes place in a narrow plastic zone, where it is necessary to allow for work hardening.

8.3 LEE AND SHAFFER'S THEORY

Before leaving the shear plane theory the analysis of Lee and Shaffer[33] should be mentioned. They assumed, as did Merchant, that shearing occurred along a single plane, but took account of how the tool force is transmitted through the root of the chip to the shear plane.

The metal bounded by OAB in Fig. 8.6, is assumed to be rigid-plastic, and to have been stressed throughout to its yield point. The chip above OB is assumed to be stress-free and hence the slip lines meet OB at 45°. Shearing occurs along the entry slip line OA, which is a velocity discontinuity. Since the chip is assumed to be stress-free, the normal stress at *a* along the stress-free interface is zero and the Mohr stress circle is shown in Fig. 8.7.

The friction angle $\beta = \tan^{-1}(\tau_d/\sigma_d)$. Therefore the angle subtended at the centre of the circle between d and the horizontal axis is 2β.

$$2\beta + 2(\phi - \beta_e) = 90°$$

$$\phi = 45° - (\beta - \beta_e) \qquad (8.3)$$

From the above expression for ϕ it can be seen that the Lee and Shaffer solution can be represented by a line on the graph of ϕ against $\beta - \beta_e$, having an intercept of 45° and a slope of -1. For most metals this solution is even less accurate than the Merchant prediction, but to overcome its inadequacy an alternative field was proposed which allowed for the existence of a small built-up edge on the tool. This field is shown in Fig. 8.8, and progressive distortion of the metal occurs as the chip passes through the fan-shaped part of the field.

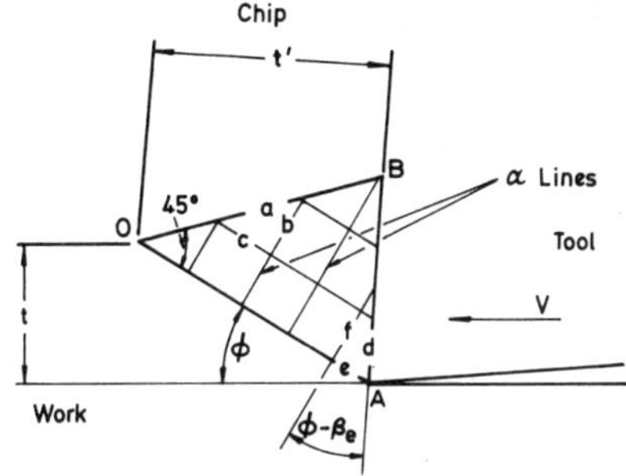

Fig. 8.6 Lee and Shaffer's slip-line field for machining

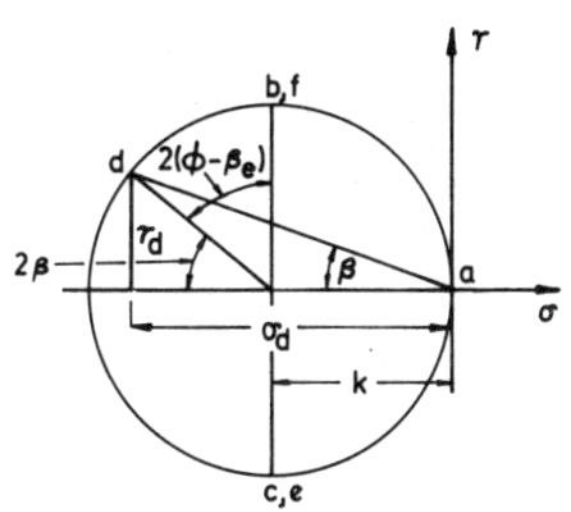

Fig. 8.7 Mohr stress circle for Lee and Shaffer's solution

From the Hencky equations it is possible to find the changes in compressive hydrostatic stress by tracing backwards from OC along the α lines to the entry slip line. No change occurs in hydrostatic stress until slip line OB is reached, as the lines are straight in this part of the field. In the zone OBA the compressive stress increases as the α lines here are curved.

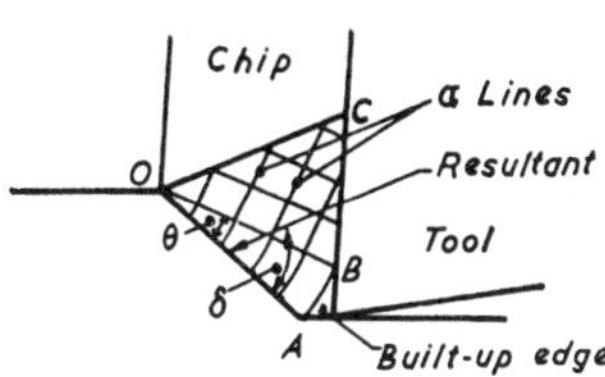

Fig. 8.8 Lee and Shaffer's built-up edge solution

Along an α line, $p - 2k\phi = \text{constant}$. If the angle between OB and OA is θ the increase in compressive stress is $2k\theta$, and the hydrostatic stress at the entry slip line is $k(2\theta + 1)$, the stress along OB being k. The angle δ between the line of action of the resultant force and the entry slip line is $\tan^{-1}(F_s/F_n) = \tan^{-1} k/[k(2\theta + 1)] = \tan^{-1}[1/(2\theta + 1)]$ and passes through the centre point of the entry slip line OA.

The built-up edge, in practice, is frequently of much greater size than is considered by this theory and in addition the work hardening of the deforming metal has been neglected.

8.4 CHRISTOPHERSON, OXLEY AND PALMER'S THEORY

Probably the most important recent development in the theory of metal cutting is the contribution by Christopherson, Oxley and Palmer[1], which allows for strain hardening by introducing an integral term to the modified Hencky equations (3.10/11). This means that, whereas with a rigid-plastic material the slip lines of one family must either be straight lines or all curve in the same direction, opposing curvature is now possible. Experimental work, generally conducted at low cutting speeds, has shown that the deformation zone has slip lines of this sort. A typical field for orthogonal machining with a sharp, smooth tool is shown in Fig. 8.9.

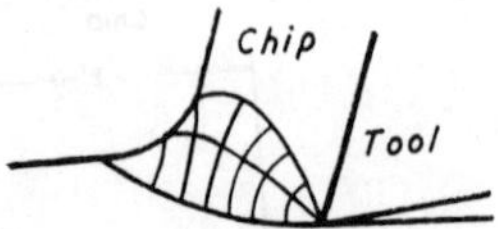

Fig. 8.9 Slip-line field for work hardening material

Unfortunately, a field of this type does away with a useful concept, the shear angle ϕ, and the curved boundary of the deformation zone causes considerable difficulty in predicting the forces involved. It has been suggested in a number of papers that at realistic cutting speeds the shear zone decreases in width and approximates to a narrow, rectangular band. Subsequent experimental work on the cutting of strain-hardening materials does not suggest that this is so, but more recent work by Oxley[34] uses this assumption as a starting point from which to estimate cutting forces.

Considering equilibrium of forces parallel to the deformation zone in the small element in Fig. 8.10.

$$(p + \delta p) \,.\, w + k\delta s = p.w + (k + \delta k) \,.\, \delta s$$

where p is a compressive stress.

$$\delta p = (\delta k/w) \,.\, \delta s = \text{constant} \,.\, \delta s$$

since $(\delta k/w)$ will be constant at any point along a slip line for a parallel sided shear zone. Therefore the hydrostatic stress varies linearly along the entry slip line, AB.

Total shear force on entry slip line, $F_s = k \,.\, (t/\sin\phi)$ assuming unit chip width.

Total normal force, $$F_N = \frac{p_A + p_B}{2} \cdot \frac{t}{\sin\phi}$$

If the resultant of F_s and F_N makes an angle γ with the shear lines,

$$\tan\gamma = \frac{F_N}{F_s} = \frac{p_A + p_B}{2k}$$

It should be noted that the hydrostatic stress becomes progressively less compressive as point B is approached, and in fact frequently becomes tensile near B.

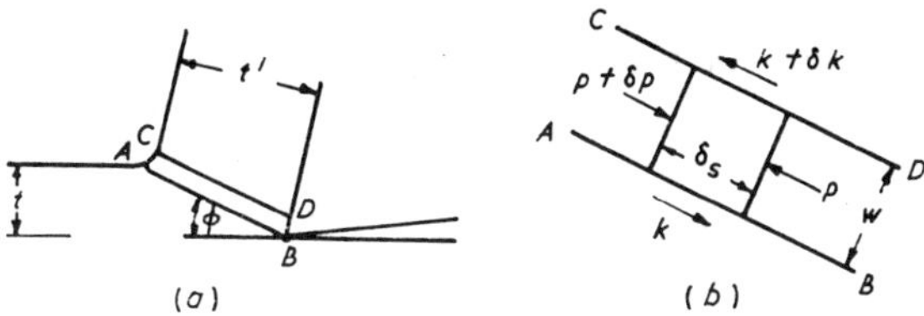

Fig. 8.10 Work hardening material with narrow rectangular shear zone

The idea of a rectangular zone must be modified slightly to allow for the fact that slip lines reach the free surface at 45°. In moving from the free surface to the point A in Fig. 8.11, the slip line turns through $(\pi/4) - \phi$ radians.

Using the modified Hencky equations,

$$p + 2k\phi + \int \frac{\partial k}{\partial \alpha} \cdot \mathrm{d}\beta = \text{constant along a } \beta \text{ line}$$

(where p is positive if tensile).

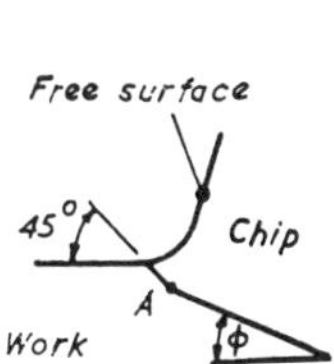

Fig. 8.11 Rotation of slip line at free surface

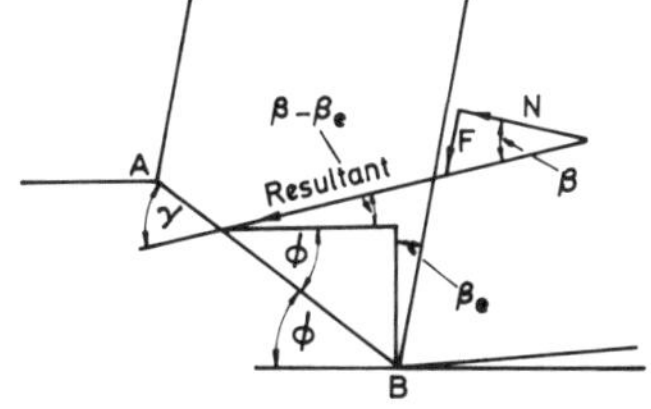

Fig. 8.12 Angular relationship of resultant force

Since the distance from the free surface to A is very small, the integral term can be ignored.

Compressive hydrostatic stress at A,

$$p_A = k\left[1 + 2\left(\frac{\pi}{4} - \phi\right)\right].$$

Since $\frac{\delta k}{w} \cdot \delta s = \delta p$, and $p_B = p_A +$ increase in hydrostatic compressive stress along a slip line,

$$p_B = p_A - \frac{\delta k}{w} \cdot \frac{t}{\sin \phi} \tag{8.4}$$

From Fig. 8.12, $$\gamma = \phi + \beta - \beta_e \tag{8.5}$$

Experiments show that for a large range of cutting conditions using a variety of materials the width of the shear zone increases in direct proportion to its length, and therefore the main factor affecting the variation of hydrostatic stress and the angle γ is the increase in shear flow stress due to work hardening. From this it follows that the greater the work hardening rate, the smaller becomes the angle γ, and the resultant force cuts the entry slip line nearer to the point A. Further, small values of γ lead to small values of ϕ for rapidly work hardening materials, a fact which is observed in practice.

This analysis does not provide a quantitative method of deducing the variables, but at least produces a qualitative basis for predicting trends in cutting force. Several attempts have been made to discover a dimensionless number for each material which would be of use in making predictions in metal cutting. One such number, suggested by Oxley and Welsh,[35] was derived from a static stress/strain curve. They began by assuming that, although a stress/strain curve at high strain rates was obviously different from one obtained at low strain rates, a material which exhibited a more rapid rate of work hardening at low strain rates than another would continue to do so at high strain rates. Therefore, a parameter derived from static stress/strain curves would be equally valid at high strain rates as a measure of comparison.

The parameter they used was m/k, where m is the average slope $d\sigma/d\varepsilon$ of the stress/strain curve (Fig. 8.13) above a strain of 0·2, and k is the shear flow stress corresponding to the yield stress at $\varepsilon = 0{\cdot}5$; i.e. $k = Y_{0\cdot5}/\sqrt{3}$. The higher the value of m/k, the lower the value of ϕ for a given value of $\beta - \beta_e$.

It is known from experiment that the value of ϕ increases with increase in cutting speed. This is in accordance with Oxley's theory, from the knowledge that as strain rate is increased m tends to decrease and k increases.

8.5 FRICTION IN METAL CUTTING

When cutting without a built-up edge, several investigators have found, on examining the underside of a 'frozen' chip obtained by quick stopping methods, that a short length of chip measured from the cutting edge bears unmistakable impressions left by the grinding marks on the tool. This indicates that the normal stress on the interface due to the cutting force reaches such a magnitude that yielding occurs and the chip material flows

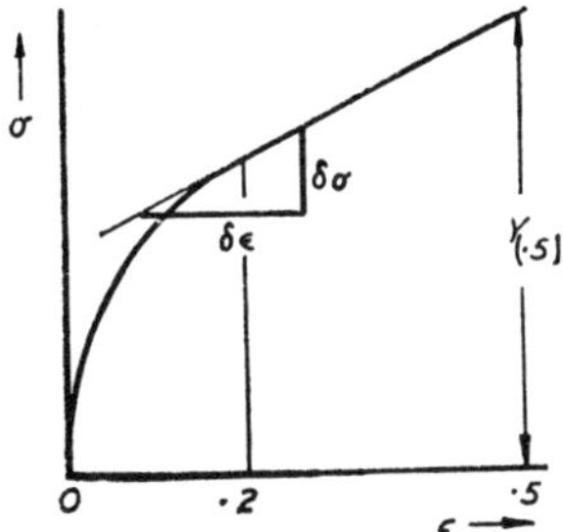

Fig. 8.13 True stress/strain curve for work-hardening material

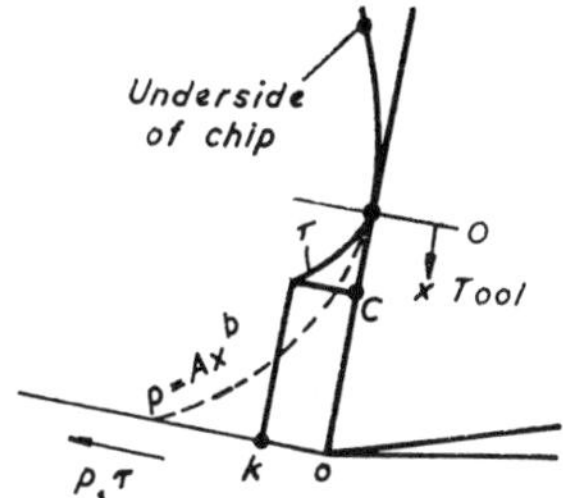

Fig. 8.14 Shear stress and normal pressure distribution along rake face (after Zorev)

into the asperities on the ground face of the tool. The frictional force on the rake face over the area where yielding occurs is then equal to the force necessary to cause shearing of either the body of the chip or the surface contaminants. Increase in normal stress over this area will therefore have no further effect, and Coulomb's laws no longer apply.

Zorev[36] assumed that the normal stress at a distance x from the end of contact obeyed the law $p = Ax^b$. The shear stress is then as shown in Fig. 8.14, so that for part of the length of contact it obeys the Coulomb laws and then at C, having reached the shear flow stress, it remains constant for the remainder of the distance to the cutting edge.

The forces measured with a tool force dynamometer give poor agreement with Zorev's model, and in some cases the normal stress may actually decrease in the secondary deformation zone near the tool point. The model implies that shearing occurs parallel to the rake face, but if the slip lines meet the rake face at an angle greater than zero, the shear stress parallel to the rake face can fall considerably below the value of the shear flow stress.

If the contact length between the tool and the chip is artificially restricted by relieving the rake face, the shear flow stress of the material may be reached across the entire contact area, and the whole interface is subject to secondary deformation. The shear force on the interface is less than that experienced with conventional flat-top tools and the compressive force is also reduced.

To balance the rake face forces, the shear angle increases with the amount of contact length restriction, resulting in a larger chip thickness ratio and reduced primary shear strain. It follows that the specific cutting energy is reduced, resulting in a cooler chip and improved tool life when operated under suitable conditions. The reduced cutting forces enable the metal removal rate to be increased if spindle power is a limiting factor.

8.6 CUTTING FLUIDS

When cutting metals at speeds below about 0·7 m s^{-1} (140 ft/min) cutting forces can be considerably reduced and the formation of a built-up edge can often be inhibited by the use of suitable cutting lubricants. For lubrication a mineral oil containing sulphur or chlorine additives is generally believed to be most effective. This is popularly believed to be because the high interface temperatures which occur in metal cutting cause a breakdown of the cutting oil which chemically reacts with the underside of the chip material to produce low shear strength chlorides or sulphides, thereby reducing the shear stress. Shaw,[37] however, suggested that for good lubricating action a high shear strength film is desirable.

Observations of lubricated and unlubricated cutting within this speed range indicate that a lubricant improves the cutting action by increasing the shear angle (and hence the chip thickness ratio), and also induces chip curl by reducing the rake contact length. Childs[38] demonstrated that the shear stress along the rake face near the cutting edge was unaltered by lubrication, but that the stress gradient away from the cutting edge was increased. Thus it appears that there is no effective penetration of lubricant to the cutting edge, and any reduction in cutting force is effected by reducing the stresses in the elastic zone.

At higher cutting speeds, fluids are used for cooling purposes, usually when using carbon steel or high-speed steel cutting tools. Hardened and tempered carbon steels lose their hardness above about 250°C and high-speed steels start to soften at about 650°C; it is therefore essential that tool temperatures should be kept below these figures. Heat in the cutting process is generated by plastic deformation and by chip/tool friction on the rake and flank faces. Some of this heat flows into the body of the workpiece and causes a temperature rise, so that when the tool cuts the surface on a second or subsequent pass the metal is already heated above room temperature. It is unrealistic to suppose that a coolant could remove heat generated by plastic deformation or friction before it flowed into the tool. However, the workpiece can be effectively cooled by a cutting fluid, allowing a greater cutting speed to be achieved. If cutting speed is progressively increased, heating approaches adiabatic conditions: heat flow by conduction has a smaller effect, and the coolant ceases to be effective in cooling the tool. With carbon steel or high-speed steel tools, softening will occur, followed by rapid failure.

For efficient cooling, a liquid possessing a high specific heat and good metal wetting properties is necessary. These requirements are best met by using water soluble mineral oils. Recently sulphurized and chlorinated soluble oils have been introduced to give the combined benefits of lubrication and cooling.

It is well known that when cutting with a lubricant at low speeds the surface finish obtained is superior to that obtained when cutting dry. This is due to the inhibition of the built-up edge which is the main factor in determining variation in accuracy and surface finish. When using tungsten carbide or ceramic tools, it is customary to machine at high speeds at which the built-up edge is seldom a problem. Under these conditions lubricants and coolants are little used, as they have a detrimental effect on tool life by causing thermal cracking, and anyway the softening temperatures of these tool materials are much higher than those for steels.

Apart from the lubricating and cooling effects of cutting fluids, they are also sometimes used to assist the clearance of swarf. A typical example is deep-hole drilling, where fluid is pumped to the cutting edges at pressures of about 5·5 N mm^{-2} (800 lbf/in^2).

8.7 FORCES IN METAL CUTTING (LARGE CHIP PROCESSES)

The foregoing sections show that the variability inherent in the metal cutting operation makes the accurate prediction of cutting forces impossible. Although such accurate predictions are usually unnecessary, some estimate of cutting force is desirable to enable cutting speeds to be kept within the available power of the machine tool, or to prevent unacceptable deflexions of the machine or workpiece. The following paragraphs consider the main factors affecting cutting forces.

8.7.1 Cutting speed and cutting fluid. At low cutting speeds, due to the unpredictability of built-up edge formation, the cutting forces vary in a manner which is seldom repeatable in consecutive tests, giving one or more maxima and minima before achieving a relatively constant value at high speeds where built-up edge is insignificant.

The lubricating effect of cutting fluids at low cutting speeds helps to inhibit built-up edge formation and at speeds below about 0·7 m s^{-1} (140 ft/min) the resulting forces may be significantly reduced. At higher speeds with or without lubricant cutting forces are not appreciably different. Fortunately, most metal cutting on steel can now be performed at speeds where the built-up edge is of little consequence, and cutting force coefficients can be specified which enable the forces to be estimated with reasonable accuracy.

8.7.2 Effective rake. As will be shown in subsequent sections of this chapter one of the factors affecting cutting force is determined by the effective rake in oblique cutting. This factor C in the cutting force equation (8.6), can be found from orthogonal cutting tests which enable the rake to be varied independently of the other parameters.

At cutting speeds used for machining steel with high speed steel tools, 0·4–1·4 m s^{-1} (80–280 ft/min), forces reduce considerably as rake angles increase in the positive sense (see Fig. 8.15). The reason for this decrease in

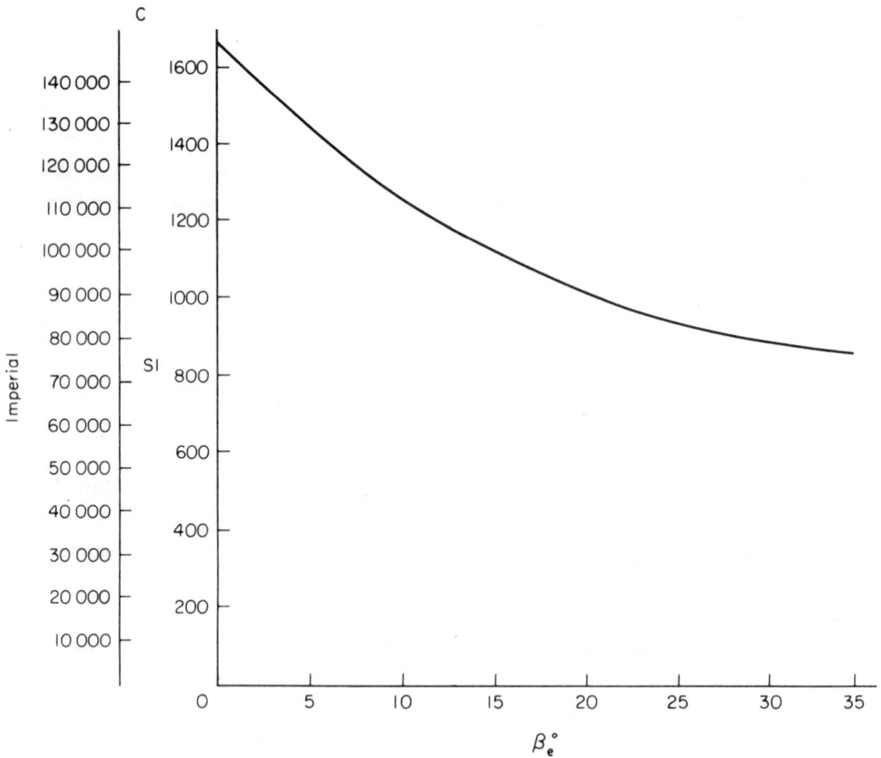

Fig. 8.15 Variation of cutting force constant with effective rake (ENIA)

force at high rakes can be ascribed largely to the greater shear angle at higher rakes and the consequent reduction in the shear plane area. At these cutting speeds very high cutting forces result from the use of negative rakes, and hence high-speed steel tools almost invariably have positive rakes.

Tungsten carbide and ceramic tools are frequently used with negative rake angles to overcome the inherent brittleness of these materials. Due to their resistance to thermal softening, they are used at much higher speeds,

2–6 m s^{-1} (400–1200 ft/min). At these speeds the chip geometry undergoes a considerable change, as shown in Fig. 8.16, resulting in a decrease in the area of the shear plane. Although the change in geometry has never been fully explained, it is probably mainly caused by the higher strain rates and by the greater thermal softening effect of the work material due to the closer approach to adiabatic conditions as heat conduction in the chip and workpiece is reduced at high speeds. The sensitivity of cutting force to rake angle is thus less at high speeds than at low speeds, and the cutting forces using negative rakes are seldom more than 10% greater than with positive rakes at speeds in excess of about 4 m s^{-1} (800 ft/min).

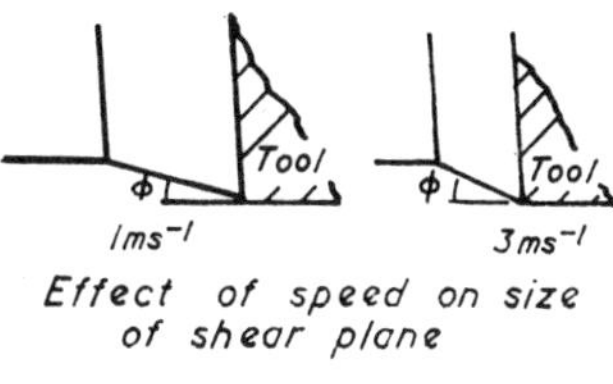

Fig. 8.16

Rake angle affects the rate of flank wear, and the high rakes which give low cutting forces at low cutting speeds also produce rapid tool wear when used on hard materials. Generally speaking, the harder the work material the lower the rake selected. Typical rake angles for high speed steel tools are shown in Table 8.1.

TABLE 8.1

Material	*Rake angle* (degrees)
Aluminium and soft alloys	30–40
Free cutting steels	20–25
Medium carbon steels	10–15
Hard alloy steels, hard cast iron	0–5

Therefore, although it is known that rake angle has a considerable effect on force at low speeds, rake is predetermined by considerations of tool life and cannot generally be regarded as a variable parameter when calculating cutting forces. Where negative rakes are used they are usually between $-8°$ and $-10°$.

8.7.3 Forces in orthogonal cutting. When machining under orthogonal conditions, i.e. with zero tool approach angle and zero cutting edge inclination, the cutting force can be reasonably accurately expressed by the formula:

$$F = Ct^x w \tag{8.6}$$

where F = cutting force, N (lbf)
C = a material constant at a given cutting speed and rake angle (derived from Fig. 8.14)
t = uncut chip thickness (or f in the turning operation) mm rev^{-1} (in/rev)
w = width of cut (or d in the turning operation, mm (in))
x = constant for the material machined.

The exponent x is less than unity and varies slightly with rake angle. When machining mild steel at 1·5 m s^{-1}, x increases from 0·8 at zero rake to about 0·9 at a positive rake angle of 25°, and decreases again at higher rakes. For practical purposes it can be taken as 0·85 for all steels over the usual ranges of rake angle. Observed values when machining cast iron are rather lower, in the order of 0·7. As feed increases, the shear angle also increases, giving rise to a non-linear increase in cutting force. This accounts largely for the non-unity exponent of feed.

The values of C given in Table 8.2 have been proposed by Brewer from experimental results, based on the normally used positive rake angles. Where negative rakes are used at high cutting speeds the values of C should be increased by about 10%.

TABLE 8.2

Material		*C (Imperial)*	*C(SI)*
Free machining carbon steel	120 HB	70 000	980
	180 HB	85 000	1190
Carbon steels	125 HB	115 000	1620
	225 HB	160 000	2240
Nickel-chrome steels	125 HB	104 000	1460
	270 HB	150 000	2100
Nickel molybdenum and chrome molybdenum steels	150 HB	114 000	1600
	280 HB	140 000	1960
Chrome vanadium steels	170 HB	130 000	1820
	190 HB	170 000	2380
Flake graphite cast iron	100 HB	35 000	635
	263 HB	73 000	1330
Nodular cast irons	annealed	61 000	1110
	as cast	68 000	1240

8.7.4 Forces in oblique cutting. Orthogonal cutting is little used in practice, and it is necessary to extend the simple formula (8.6) to allow for the other variables of tool approach angle and cutting edge inclination.

Cutting force in oblique cutting when using a pointed tool is dependent on the length of the cutting edge in engagement and on the uncut chip thickness measured in the direction of chip flow (i.e. in the plane of the effective rake angle).

Equation (8.6) then modifies to:

$$F = Ct''^{x} d'' \tag{8.7}$$

where C = cutting force constant associated with the effective rake angle of the tool

t'' = uncut chip thickness measured in the direction of chip flow

d'' = length of cutting edge in engagement.

From Fig. 7.3 it can be seen that the plan angle between the normal to the cutting edge and the chip flow direction is found from

$$\cos \delta = \frac{\cos \lambda_s \cos \gamma_n}{\cos \beta_e}$$

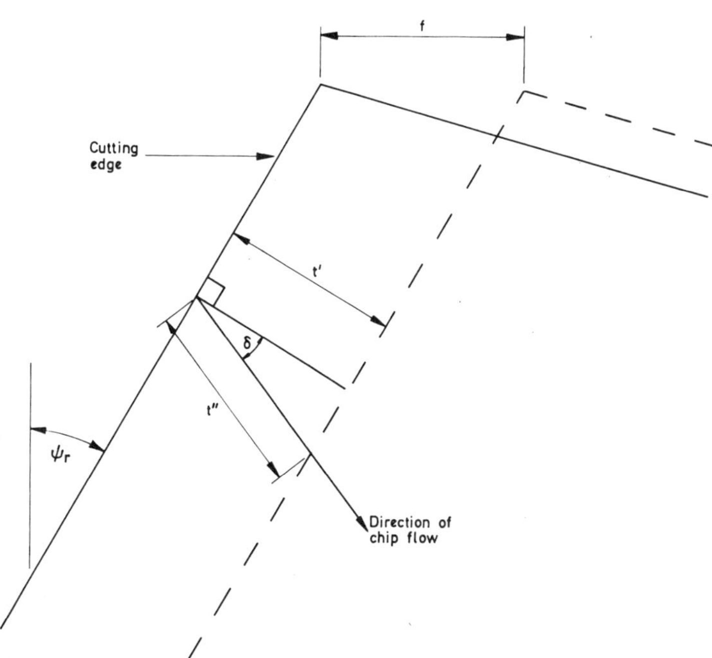

Fig. 8.17 Plan view of single point tool

Also, from Fig. 8.17

$$t'' = f\frac{\cos\psi_r}{\cos\delta} = f\frac{\cos\psi_r\cos\beta_e}{\cos\lambda_s\cos\gamma_n} \tag{8.8}$$

where f = feed/rev in turning
ψ_r = tool approach angle.

Allowing for the effects of tool approach angle and cutting edge inclination, the length of the cutting edge in engagement d'' is given by

$$d'' = \frac{d}{\cos\psi_r\cos\lambda_s} \tag{8.9}$$

Equation (8.7), expressed in terms of feed/rev and depth of cut becomes

$$F = Cf^x d(\sec^{1-x}\psi_r \cos^x\beta_e \sec^{1+x}\lambda_s \sec^x\gamma_n) \tag{8.10}$$

The range of values assumed by ψ_r λ_s and γ_n in the turning operation is fairly restricted, and taking extreme values of these variables the product of the trigonometrical terms in equation (8.10) seldom exceeds 1·06. This implies that for a pointed lathe tool equation (8.6) rarely underestimates cutting force by more than 6%.

The nomogram, Fig. 8.18, enables the principal cutting force to be evaluated provided the value of C is known. The value of cutting force so obtained may be increased by as much as 25% when cutting with worn tools. Most turning tools have a nose radius which increases cutting force slightly at small depths of cut, but has a negligible effect for deep cuts.

8.7.5 Power consumption of helical roller milling cutters. The angles describing the geometry of roller milling cutters are the helix angle and the radial rake, β'. It can be seen that the helix angle is the cutting edge inclination λ_s, and the normal rake is found from the radial rake using the formula

$$\tan\gamma_n = \tan\beta'\cos\lambda_s \tag{8.11}$$

The helix angle of roller milling cutters is frequently large, an angle of 45° not being unusual. For this reason equation (8.6) is not a reasonable approximation of the instantaneous cutting force on a milling cutter tooth, and the trigonometrical terms in equation (8.10) must be included, but are simplified by the fact that ψ_r is zero for roller milling.

Section 9.1.3 shows how the cutting power P_0 of a roller milling cutter with zero helix angle can be calculated. To calculate the power for a roller milling cutter having helical teeth, the value of P_0 for a cutter whose radial rake is equal to the effective rake of the cutter under consideration must first be found.

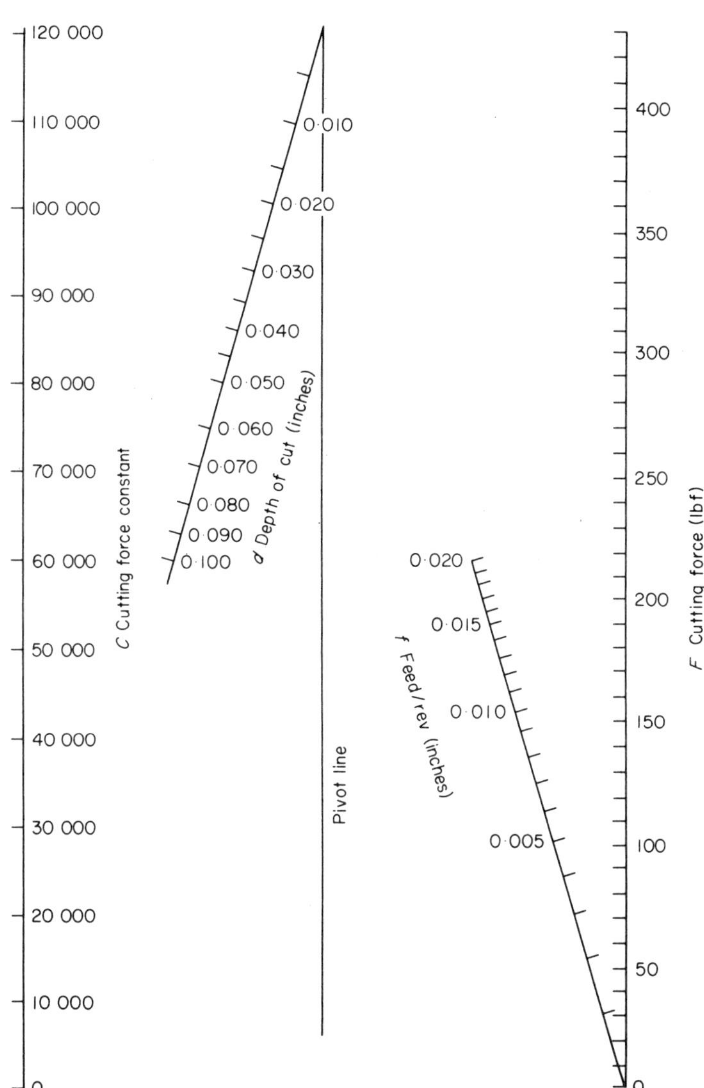

Fig. 8.18 Nomogram to evaluate cutting force

The required cutting power is then found from

$$P = P_0 \sec^{1+x}\lambda_s \cos^x\beta_e \sec^x\gamma_n \tag{8.12}$$

Taking $x = 0{\cdot}85$ for steel and putting

$$X = \sec^{1\cdot 85}\lambda_s \cos^{0\cdot 85}\beta_e \sec^{0\cdot 85}\gamma_n$$
$$P = P_0 X \tag{8.13}$$

where X is calculated from the nomogram Fig. 8.19.

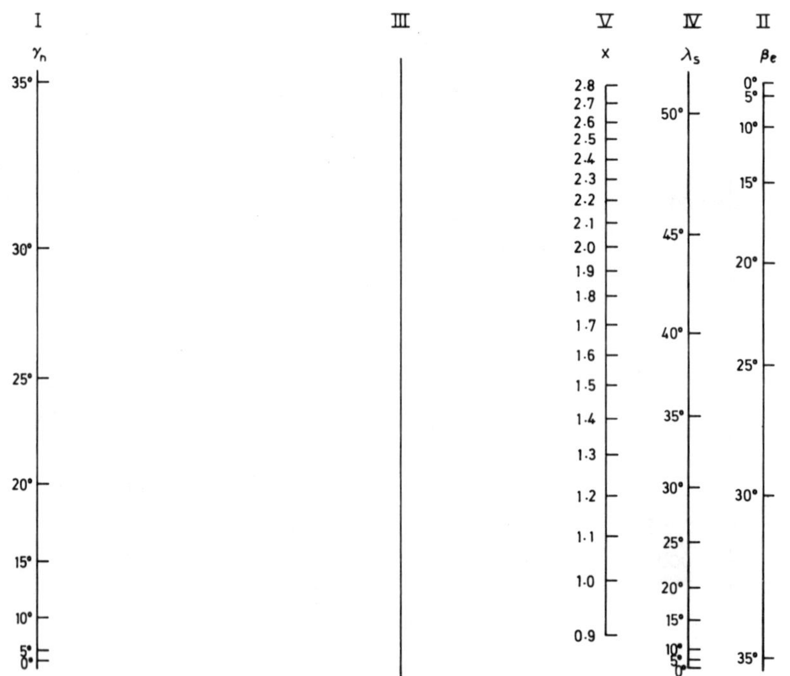

Fig. 8.19 Nomogram to evaluate X

8.8 TOOL WEAR AND TOOL LIFE

Tool life can be defined as the time a newly sharpened tool will cut satisfactorily before it becomes necessary to remove it and regrind or replace. Usually, this is based on the amount of wear which occurs before the cutting edge becomes so worn that catastrophic failure becomes imminent, although other considerations, such as dimensional accuracy or poor surface finish, may occasionally demand earlier replacement.

Tool failure can be due to a variety of reasons. These have been investigated by Trent,[39] who observed six types of deterioration with tungsten carbide and tungsten–titanium carbide tools; these are:

(*a*) cratering wear on rake face of tool;
(*b*) crumbling of cutting edge due to built-up edge and associated factors;
(*c*) deformation of tool due to high stresses and temperatures;
(*d*) cracking due to thermal shock;
(*e*) chipping of tool edge due to mechanical impact;
(*f*) wear on flank or clearance face.

8.8.1 Cratering at high speeds. This type of wear is a major cause of failure when using tungsten carbide tools to cut steel, but is less prevalent when using high-speed steels or tungsten–titanium carbide. A crater forms on the rake face of the tool behind the cutting edge and continues to enlarge until the cutting edge is eventually weakened. It appears that the high temperatures generated along the chip/tool interface at high cutting speeds cause the steel to alloy with the tool material and the resultant alloy is carried on the underside of the chip. This process may be due to diffusion or to actual fusion.

Cratering occurs with high-speed tools when machining near the maximum cutting speeds, although it seldom causes them to fail. The addition of titanium carbide to tungsten carbide greatly reduces cratering on carbide tools, and it is believed that this is because titanium carbide less readily alloys with steel at high temperatures.

8.8.2 Crumbling of the cutting edge at low speeds. The speed dependency of built-up edge formation has been explained earlier. At low speeds the build-up can reach large proportions and is frequently unstable, breaking away with the underside of the chip. At the interface temperature, diffusion bonding can occur between the tool and the built-up edge, and when the particles of built-up edge break away they can cause particles of tool to break away with them.

This crumbling type of wear occurs especially with brittle tool materials such as carbides. Flaking can also occur when the tool cools, because the greater coefficient of expansion of the built-up edge induces tensile forces at the surface of the tool. A solution to this sort of wear is to increase the cutting speed to a point where the built-up edge diminishes in size and ceases to present a problem.

8.8.3 Deformation of the cutting edge. Generally, this form of failure manifests itself at high interface temperatures which occur when cutting under arduous conditions at high speeds. It is really a form of creep, causing the geometry of the cutting edge to change. Once deformation starts, the stresses increase further and failure eventually occurs due to the cracking of the cutting edge. Because of the critical nature of the temperatures inducing creep, this kind of failure is likely to occur rapidly once a certain speed has been exceeded. Large nose radii result in smaller stresses at the tool point and correspondingly less creep.

8.8.4 Cracking due to thermal shock and mechanical impact. Thermal cracking occurs when there is a steep temperature gradient. It is usually associated with intermittent cutting such as milling, where rapid heating and cooling recur continuously. Thermal shock in the cutting process is inseparable from the mechanical shock suffered when the cutter makes violent contact with the workpiece. For this reason the cracks which occur are probably due to both thermal and mechanical shock. Appropriate selection of the cutting parameters can do much to reduce this form of failure, and the use of tougher tool materials can also help.

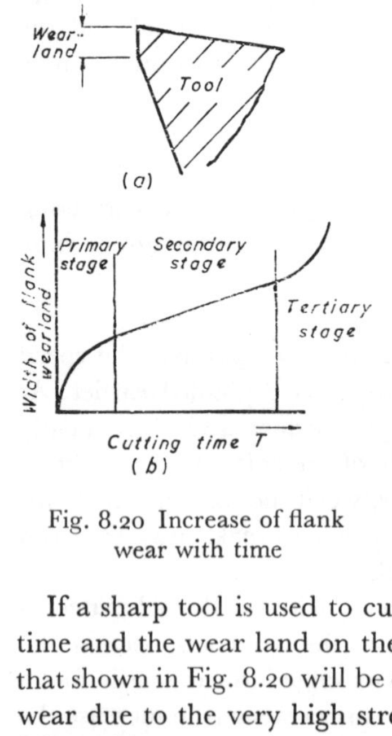

Fig. 8.20 Increase of flank wear with time

8.8.5 Flank wear. The five modes of failure previously discussed can usually be effectively reduced by changing speed, feed or depth of cut. The sixth mode, wear on the flank face, is a progressive form of deterioration which will ultimately result in failure whatever precautions are taken. This sort of failure was recognized by Taylor[40] in the early years of this century. He succeeded in specifying an exponential law relating cutting speed to tool life, which is still applicable.

If a sharp tool is used to cut at a given speed for a specified period of time and the wear land on the flank face is measured, a curve similar to that shown in Fig. 8.20 will be obtained. The primary stage is one of rapid wear due to the very high stresses at the point of the sharp tool. This is followed by a secondary stage of relatively linear wear rate, until a third stage occurs when the wear rate increases rapidly and frequently results in catastrophic failure. The third stage presumably occurs when the wear land has reached such proportions that friction on this face of the tool causes thermal softening.

Fig. 8.21 (*a*) shows the effect of increasing the cutting speed from V_1 to V_4 with four tools of identical geometry operating under identical conditions of feed and depth of cut. In each case the tertiary stage of wear occurs at approximately the same land size, about 1 mm (0·040 in) for high-speed steel tools and 0·75 mm (0·030 in) for carbide or ceramic tools.

If tool life corresponding to these wear lands is plotted against cutting speed, a graph similar to that shown in Fig. 8.21 (*b*) will be produced, leading to a tool life equation of the form VT^n = constant. When machining steel, n takes the value of 0·125 if high-speed steel tools are used, 0·25 for positive rake carbide tools, 0·20 for negative rake carbide tools, and 0·38 for ceramic tools.

The equation VT^n = constant is specific to a given tool with a particular feed and depth of cut. It is not easy to qualify the effects of changes in tool geometry and cutting conditions. Intuitively, it might be expected that increase in either feed or depth of cut would decrease tool life, due to the greater rate of metal removal. Increase in plan approach angle or nose radius might be expected to improve tool life, as there is an increase in the effective length of the cutting edge, allowing the cutting force to be supported along a greater length. Therefore, a generalized equation of

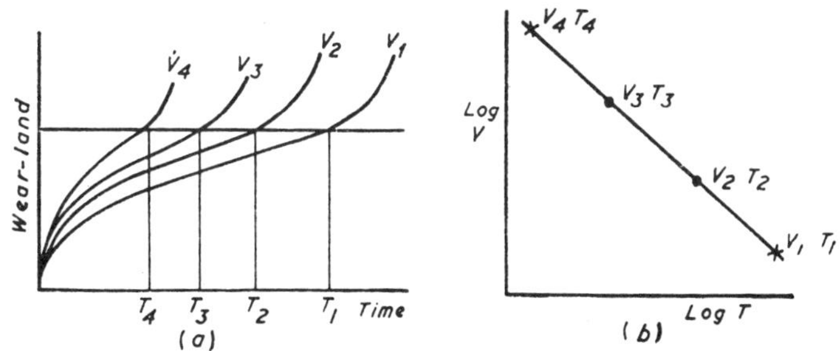

Fig. 8.21 Logarithmic relationship of tool life and cutting speed

the form $V \cdot T^n \cdot g(f, d, r, \psi_r)$ = constant could be expected, where g is a function of feed (f), depth of cut (d), nose radius (r) and tool approach angle (ψ_r).

Such an equation would be unwieldy, and Brewer and Rueda[41] sought to simplify it by finding a unique relationship connecting these four

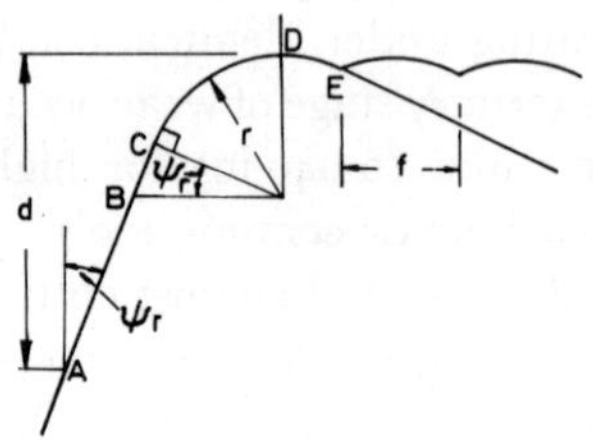

Fig. 8.22 Plan view of tool

variables. They reasoned that tool life is a function of the temperature at the cutting edge, θ, and the steady state temperature depends on a balance being struck between the heat generated and heat removed, that is, $\theta \propto$ (heat generated/heat removed.)

It was assumed that heat generated was proportional to the area of the cut, $f \,.\, d$, and that heat removed from the cutting edge was proportional to the length of the cutting edge. Provided the depth of cut exceeds r (Fig. 8.22) and ignoring the short length DE, the length of cutting edge

$$l_e = \frac{d - r(1 - \sin \psi_r)}{\cos \psi_r} + \frac{90 - \psi_r}{180} \,.\, \pi \,.\, r$$

Hence

$$\theta \propto \frac{f \,.\, d}{[d - r(1 - \sin \psi_r)]/\cos \psi_r + [(90 - \psi_r)/180] \,.\, \pi \,.\, r} = b_e = f \,.\, G \tag{8.14}$$

From dimensional considerations b_e is a length which is known as the equivalent chip thickness, and, as has been shown, includes the four variables f, d, r and ψ_r. This enables a generalized tool life equation to be written as follows:

$$V \,.\, T^n \,.\, b_e^m \,. = V \,.\, T^n \,.\, (f \,.\, G)^m = \lambda \tag{8.15}$$

where λ is a constant for the material.

The index m appears to be dependent only on the material being cut. Brewer suggested a value of 0·45 for cast iron, and more recent work by PERA,[42] varying feed only, indicates a value of 0·37 for steel.

The dependence placed on equivalent chip thickness as a valid parameter of tool life is based on incomplete experimental results, and the authors suspect that at extreme values of the variables the results may be considerably in error.

8.9 SURFACE FINISH

Assuming initially that cutting conditions do not affect the finish of the workpiece, it will then be only the plan geometry of the cutting tool that determines the smoothest surface which can be achieved. If a smooth finish is required, a pointed tool would not be used, so the geometry of the cut surface will be as shown in Fig. 8.23.

Provided the surface consists of circular arcs it can be shown that the centre line average height

$$R_a \simeq \frac{f^2}{18\sqrt{(3)}r} \tag{8.16}$$

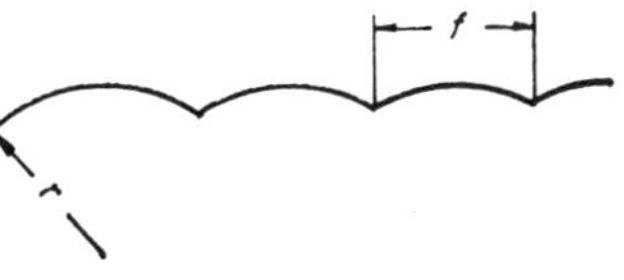

Fig. 8.23 Section through turned surface

where R_a, f and r are in compatible units (for a description of centre line average see Chapter 17). Thus the ideal surface roughness is directly dependent on the square of the feed and inversely dependent on the size of nose radius.

In practice, other factors adversely affect the surface finish produced. These are mainly associated with the formation of a built-up edge and it is therefore not surprising that the most important is cutting speed. Fig. 8.23 shows that at high cutting speeds, where the effect of built-up edge is small, the surface roughness approaches the ideal value. Similarly, rake angle has a noticeable effect at low speeds, but its effect is small at speeds used for finish machining. Cutting fluid, in so far as it inhibits build-up, also has some effect at low cutting speeds. Depth of cut has a very slight effect, provided it is not sufficiently large to cause chatter.

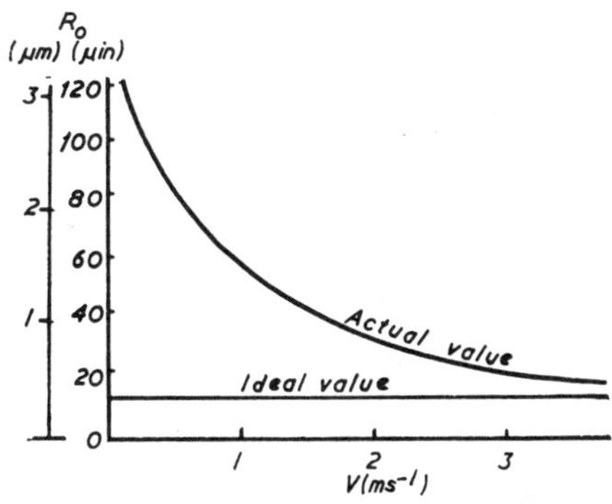

Fig. 8.24 Variation of surface roughness with cutting speed

8.10 TORQUE IN DRILLING

There does not appear to be an accurate method of calculating the magnitude of forces involved in drilling, but a semi-empirical approach provides a reasonably satisfactory formula for calculating torque.

It can be assumed that the feed/rev affects the torque in the same manner as it affects force in the turning operation, i.e. torque $\propto f^x$ where x is usually about 0·85. Assuming that a given volume of metal requires an approximately constant amount of energy to remove it, the torque can be expected to vary directly as the area of the hole being drilled, i.e. torque $\propto D^2$, where D is the drill diameter. In practice a better approximation appears to be $T = Cf^{\cdot 85}\, D^{1\cdot 9}$, where T is the torque, N mm (lbf in), f = feed/rev, mm (in), D = drill diameter, mm (in) and C is a constant which depends on work materials. Average values of C are as follows:

	Mild steel	High carbon steel	Grey cast iron	Aluminium	Brass
C (SI)	240	370	130	75	100
C (non-metric)	15 500	24 000	8 300	4 800	6 500

One factor not accounted for in the formula is the thickness of web left between the flutes. Thick webs considerably increase the cutting force. The web thickness increases as the drill is progressively shortened by grinding, and hence the forces acting on an old drill are considerably greater than those acting on a new one unless the web is thinned.

8.11 ACCURACY OF MACHINED SURFACES

The accuracy of machined parts is affected by static, steady state machining and dynamic machining considerations. Static alignment tests for most standard machine tools are well documented,[43] and can be extended to allow for the testing of non-standard and special purpose machines. These tests enable machines to be levelled within generally acceptable limits of accuracy and also enable the alignments of slideways and spindles to be checked.

Steady-state effects are due to the elastic forces set up in the machine structure and workpiece when cutting. To reduce the effect of these forces it is necessary to increase the rigidity of the tool–workpiece combination. This can be achieved by selection and setting of cutting tools to minimize deflexion due to overhang, and by adequate clamping and support of the workpiece, e.g. by fitting workpiece support steadies on a centre lathe.

Dynamic effects in most cases are either due to forced vibrations or to induced vibrations brought about by the interaction of the structural response of the machine tool and the cutting process. Forced vibrations are caused by unbalanced rotating masses or by periodic force vibrations, as when the teeth of a milling cutter engage the workpiece. These vibrations are troublesome when the cyclical force variation corresponds with resonant frequencies of the machine structure, and can usually be reduced by either an increase or decrease in spindle speed.

Self-induced vibrations are more difficult to cure as they may be due to a number of imperfectly understood mechanisms of chip removal. These vibrations, generally referred to as chatter, have been extensively investigated and have yielded much useful information for the machine tool designer. However, the purpose of this book is not to consider the design of machine tools, but their uses, and the ensuing remarks are directed towards the practical methods of chatter elimination.

8.11.1 Cutting tool and workpiece rigidity. The first approach should aim at achieving the maximum rigidity of both tool and workpiece, thereby increasing the natural frequencies to values which are unlikely to cause chatter. Effective clamping and support of the cutting tool and workpiece can do much to increase rigidity and natural frequencies. On horizontal milling machines a similar effect can be achieved by fitting ties between the overarm support and the table of the machine.

8.11.2 Cutting tool geometry. A large nose radius on a cutting tool will improve surface finish and tool life, but is likely to promote chatter. For this reason cutting tools designed for removing large quantities of metal seldom have nose radii larger than 1 mm. Large side cutting edge angles also improve tool life, but they too are likely to encourage chatter, and for practical purposes seldom exceed 30°. It is frequently observed that blunt tools are less likely to chatter than sharp ones. This is presumably because bluntness increases the resistance of the tool to penetration of the workpiece, and thereby damps vibration normal to the workpiece surface. For this reason the sharp cutting edge is sometimes deliberately dulled before use. Very little appears to have been reported concerning the effect of rake angles on machining stability, but in general high positive rakes increase the shear angle, giving a more efficient cutting action, and it is likely that stability is marginally improved by increasing rake.

8.11.3 Width of cut. Steady-state and dynamic cutting forces are proportional to the width of cut, and hence stable cutting can be achieved at the cost of metal removal by reducing the width, which is analogous to the depth of cut in the turning process.

8.11.4 Cutting speed. When the other parameters are kept constant, chatter can be frequently overcome by either an increase or decrease in cutting speed. Since a reduction of speed decreases metal removal rates it is obviously preferable to attempt to stabilize the process by a speed increase rather than a decrease.

In general, it is to be expected that the cutting speed is initially selected either to give a reasonable tool life or to restrict the cutting power to that available at the spindle. A significant increase in cutting speed without altering the other cutting parameters would therefore result either in unacceptably short tool life or in stalling the spindle.

8.11.5 Feed. This is another factor whose influence on chatter is not well documented. Pearce and Richardson[44] showed that, when cutting at a constant cutting speed and width of cut, it is sometimes possible to pass from an unstable to a stable condition by increasing the feed.

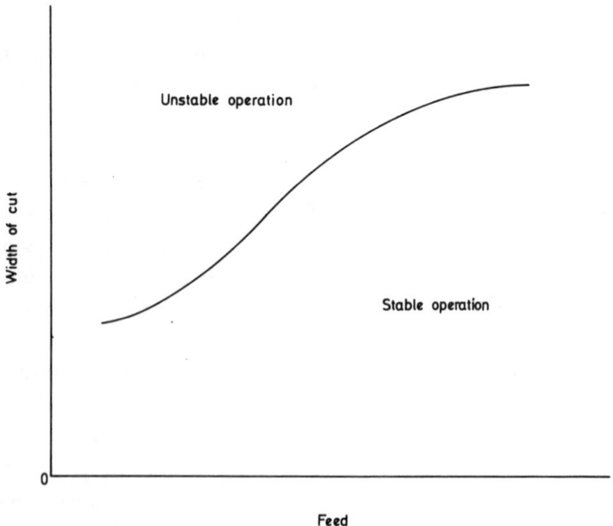

Fig. 8.25 Effect of feed on stable operation

The shape of the stability boundary in Fig. 8.25 shows that the greatest benefit from increasing feed occurs at relatively low feeds. At higher feeds the curve becomes flatter and stabilization by this method becomes impractical. However, it should be noted that, whereas a large feed may not give stable machining, a small feed (i.e. reduced metal removal rate) is always likely to promote chatter.

8.11.6 Tool contact length. If the chip-tool contact length is artificially restricted by relieving the rake face, the shear angle is increased and the cutting forces decrease.

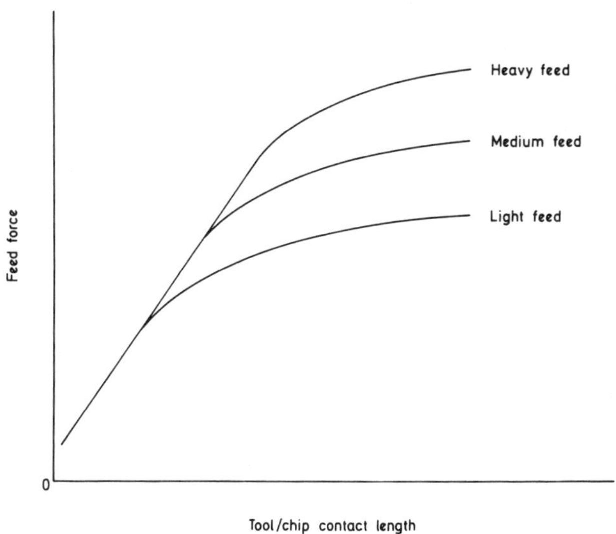

Fig. 8.26 Effect of contact length on feed force

Fig. 8.26 shows a typical steady state relationship of feed force and contact length for a zero rake tool. For short contact lengths the force is invariant over a range of feeds, so that the effect of a surface wave on the workpiece or of a tool vibration normal to the surface would not produce a variation in feed force. There is therefore no dynamic cutting force attributable to steady state effects to help sustain a vibration when using a restricted contact tool. Richardson and Pearce[45] showed that when applying a controlled vibration to such a tool a small dynamic force does, in fact, exist but it leads tool displacement and is therefore likely to contribute a stabilizing effect.

Restricted contact tools have been shown to have a marked stabilizing effect when machining a range of steels with contact lengths between 1·5 and 3·0 times the feed. They also reduce the specific energy of metal removal, thereby increasing the metal removal capacity of a machine tool, and give increased tool life as a consequence of their lower operating temperatures.[46]

8.11.7 Chatter in boring bars. Chatter is frequently encountered when boring holes where the length/diameter ratio is large. Tlusty[47] showed that such vibration is often due to mode coupling and can sometimes be eliminated by machining flats on opposite sides of the boring bar so that it has unequal flexibilities about the two principal axes. An alternative device for reducing chatter in boring bars incorporates a heavy metal slug surrounded by a viscous fluid in a cylindrical coaxial cavity machined in the end of boring bar near the tool. The damping of such devices has been shown to have a dramatic effect in improving bored surfaces.

8.11.8 Variable helix roller milling cutters. The cyclical pattern of force variation in roller milling can be varied by machining the teeth with varying helix angles. These cutters are very effective in preventing chatter vibrations.

9 Milling and Broaching

9.1 MILLING

At first sight milling and broaching may appear to be vastly dissimilar processes, but this is not so, since surface broaching is somewhat similar to peripheral milling with a cutter of infinite diameter. In fact, surface broaching is being used increasingly in place of milling as a cutting process.

Milling processes are diverse and the cutters used cover a wide range of shapes and sizes. However, only two basic methods of metal removal, peripheral and face milling, will be considered. As peripheral and face milling are generally used to remove large volumes of metal, considerable economies can be achieved in these by optimum selection of machining parameters. Other applications, such as the milling of slots, forms and helices, are essentially variants of the basic methods and each is specialized to such an extent that a detailed treatment would be inappropriate here.

9.1.1 Peripheral milling. The cutter teeth are machined to give cutting edges on the periphery. They may be gashed either axially or spirally. Fig. 9.1 shows a spirally gashed cutter where the spiral angle λ_s corresponds to the cutting edge inclination (see Fig. 7.1). The radial rake β' is not the normal rake angle γ_n, but from Fig. 7.3 it will be seen that $\tan \gamma_n = \tan \beta' \,.\, \cos \lambda_s$ where $\beta' = \mathrm{A\hat{O}F}$. Also, using the formula derived in Chapter 7 for the effective rake angle, β_e

$$\sin \beta_e = \cos^2 \lambda_s \,.\, \sin \gamma_n + \sin^2 \lambda_s$$

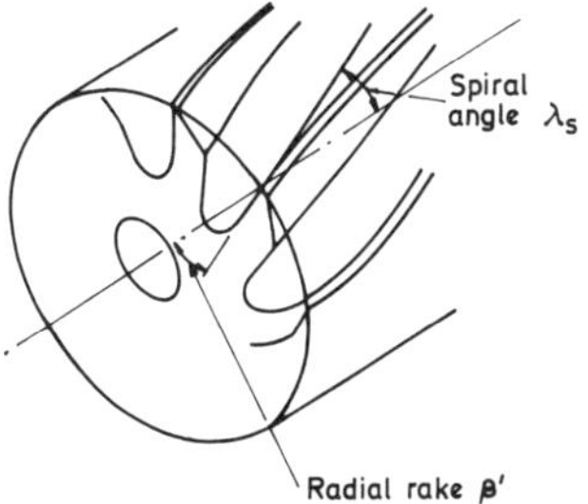

Fig. 9.1 Rake angles on spirally gashed cutter

It follows from the above equation that a large effective rake can be produced on a milling cutter by introducing a high spiral angle, without unduly weakening the tooth by having a large radial rake. This is of considerable importance when designing milling cutters since, as in turning, the cutting force per unit length of cutting edge reduces rapidly with increase in rake.

Another advantage of a helical cutter is the more even distribution of cutting force since the cutting edges engage with the workpiece progressively as the cutter rotates. A disadvantage is the end thrust component of the cutting force, which can sometimes be balanced or reduced by mounting two cutters of opposing helix angles on the same arbor.

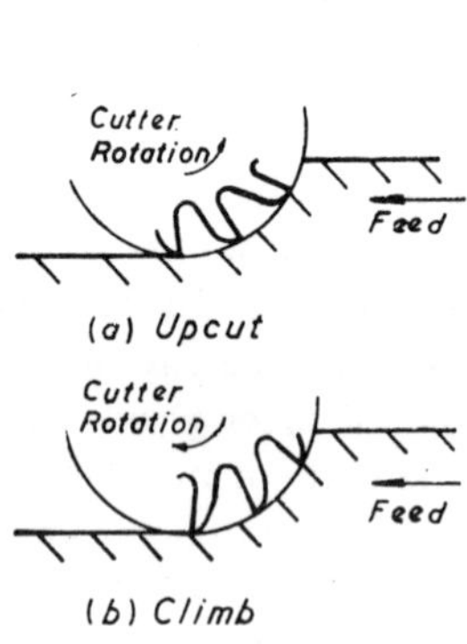

Fig. 9.2 Upcut and climb milling

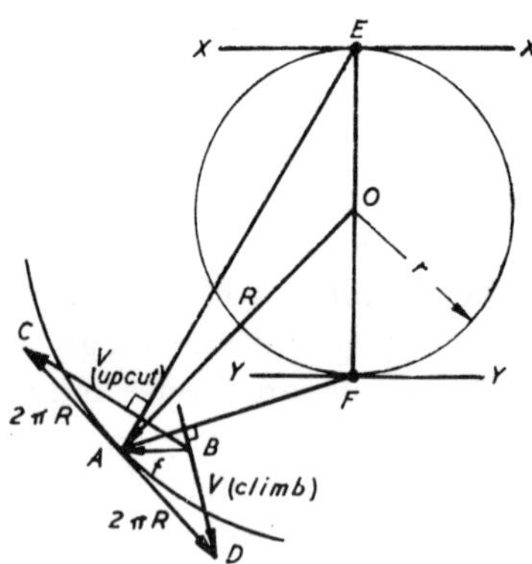

Fig. 9.3 Velocity diagrams for upcut and climb milling

There are two methods of chip removal, one known as upcut milling and the other as downcut or climb milling. The essential features of each are shown in Fig. 9.2. Although upcut milling is generally used, downcut milling is preferable, as will be seen later, provided that the machine is fitted with an adequate backlash eliminator.

9.1.2 Chip formation in peripheral milling. The relative motion of the cutting edge and workpiece is the vector sum of the cutter rotation and the feed. Fig. 9.3 shows the velocity diagrams for upcut and climb milling with a cutter of radius R and a feed f mm rev^{-1} (in/rev). Triangles ABC and AOE are similar;

hence
$$\frac{\text{OE}}{\text{AB}} = \frac{\text{OA}}{\text{AC}}$$

$$\frac{r}{f} = \frac{R}{2\pi R}$$

$$r = \frac{f}{2\pi}$$

ABD and AFO are also similar. The relative velocity in upcut milling, in direction BC, is perpendicular to AE, and the instantaneous centre is at point E. The cutting path is a curve whose instantaneous centre is the point of tangency of the circle of radius r as it rolls without slipping on XX. Similarly, the instantaneous centre for climb milling is the point F, the point of tangency when the circle of radius r rolls without slipping on the line YY.

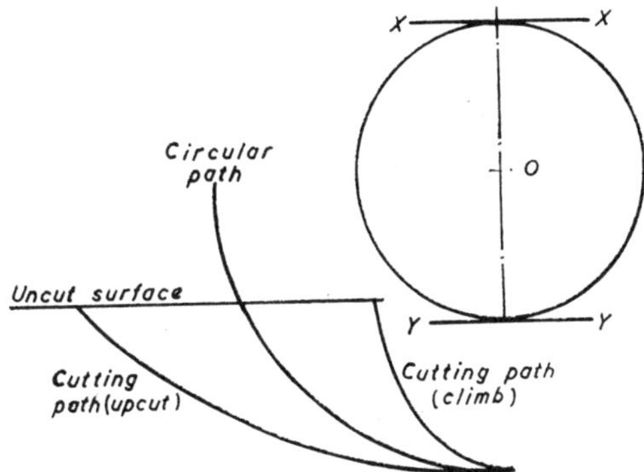

Fig. 9.4 Cutter paths in upcut and climb milling

Fig. 9.4 shows how the cutting path for the two methods differs from the circle swept by the cutter, climb milling giving a shorter cutting path than upcut milling for the same feed rate. Since the same metal-removal rate applies in both cases, it follows that climb milling gives a higher average uncut chip thickness and lower cutting speed. In practice the feed rate is very small compared with the cutter speed and a circular path may be assumed in either case with little loss of accuracy.

The depth of cut influences the length of cutting path, assuming a circular path, in the following manner. Fig. 9.5 shows that

$$\cos \theta = (R - d)/R = 1 - (d/R)$$

Fig. 9.6 shows how the length of cutter path increases at a decreasing rate as the depth of cut d increases. Since it is reasonable to assume that tool life is partly dependent on the total length of chip cut per tooth between regrinds, it is clear that the amount of metal removed between regrinds

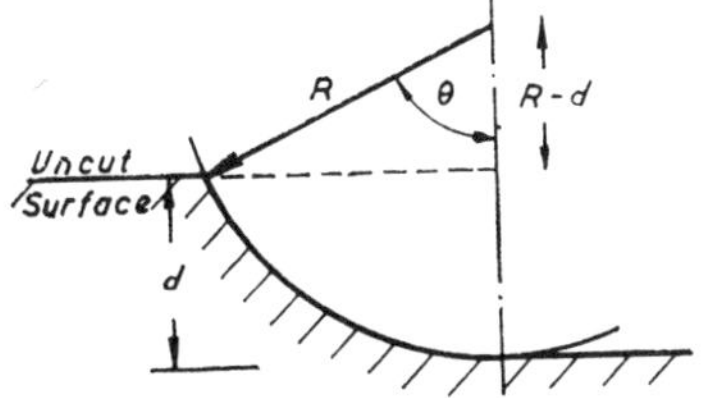

Fig. 9.5 Influence of depth of cut on length of cutter path

will be increased if the surplus metal is removed in one cut. The economic sense of this argument is further reinforced when the additional time required for making more than one pass is considered.

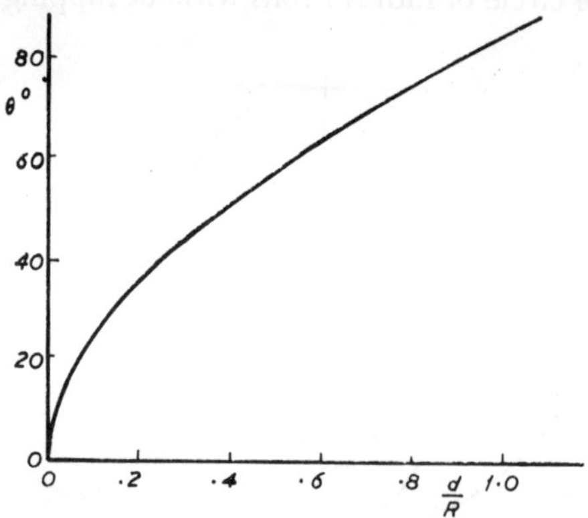

Fig. 9.6 Influence of depth of cut on length of cutter path

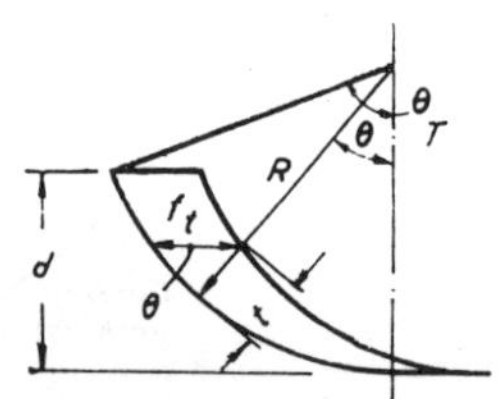

Fig. 9.7 Variation of uncut chip thickness with angle of rotation

9.1.3 Power required for peripheral milling. It is assumed that the principal cutting force relationship $F = C \, . \, f^x \, . \, d$ developed in Chapter 8 for turning also applies to milling. Since f has a different significance in this chapter it will be preferable to change the symbols as follows:

$$F = C \, . \, t^x \, . \, w \tag{9.1}$$

where t is the uncut chip thickness, mm (in)
w is the width of the chip, mm (in).

Considering a cutter having zero helix angle, Fig. 9.7 shows that t approximately equals $f_t \sin \theta$, where f_t is the feed/tooth.

Work done in removing one chip $W_{\text{tooth}} = R \, . \, w \, . \, C \, . \int_0^{\theta_T} t^x \, \mathrm{d}\theta$

$$W_{\text{tooth}} = R \, . \, w \, . \, C \, . \, f_t^x \, . \int_0^{\theta_T} (\sin \theta)^x \, \mathrm{d}\theta$$

$$= R \, . \, w \, . \, C \, . \, f_t^x \int_0^{\theta_T} \left(\theta - \frac{\theta^3}{3!} + \frac{\theta^5}{5!} \cdots \right)^x \mathrm{d}\theta$$

$$\simeq R \, . \, w \, . \, C \, . \, f_t^x \int_0^{\theta_T} \left(\theta - \frac{\theta^3}{3!} \right)^x \mathrm{d}\theta \text{ since } \theta \text{ is usually} < 1 \text{ rad.}$$

$$\simeq R \, . \, w \, . \, C \, . \, f_t^x \left[\frac{\theta_T^{\;1+x}}{1 + x} - \frac{x \theta_T^{\;3+x}}{6(3 + x)} - \frac{x(1 - x) \theta_T^{\;5+x}}{72(5 + x)} + \cdots \right]$$

The second and subsequent terms can generally be ignored.

$$W_{\text{tooth}} \simeq R \,.\, w \,.\, C \,.\, f_t^x \,.\, \frac{\theta_T^{1+x}}{1 + x} \tag{9.2}$$

$$\cos \theta_T = \frac{R - d}{R} = 1 - \frac{d}{R}$$

$$= 1 - \frac{\theta_T^2}{2} + \frac{\theta_T^4}{24} - \frac{\theta_T^6}{720} + \ldots \simeq 1 - \frac{\theta_T^2}{2}$$

$$\theta_T \simeq \sqrt{\frac{2d}{R}} \tag{9.3}$$

$$\text{Power } P_0 = \frac{2^{(1+x)/2}}{2\pi} \times \frac{1}{1 + x} \times V . S . w . C . \frac{f_t^x \, d^{(1+x)/2}}{R^{(1+x)/2}}$$

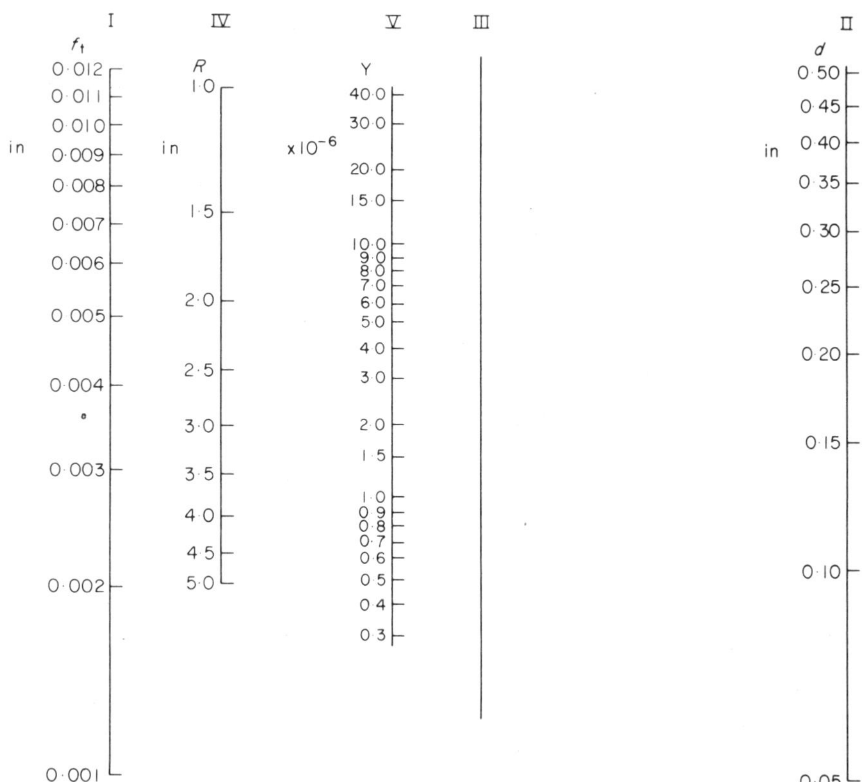

Fig. 9.8 (*a*) Nomogram to evaluate *Y* (Imperial Units)

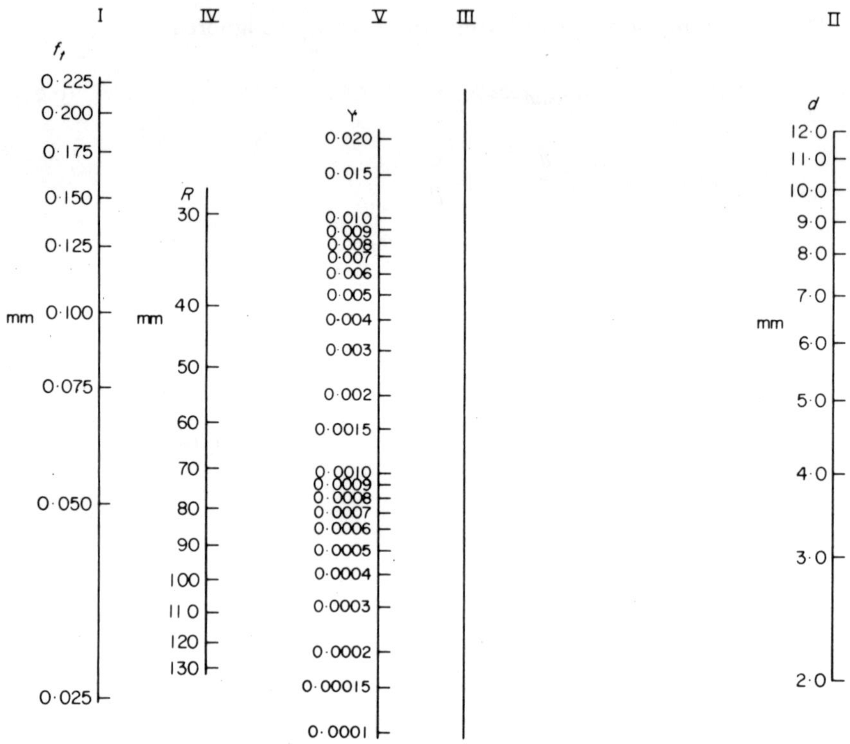

Fig. 9.8 (*b*) Nomogram to evaluate *Y* (Metric Units)

Where V = cutting speed, m s^{-1} (ft/min)
S = number of cutting teeth
R = cutter radius, mm (in).

For steel $x \simeq 0{\cdot}85$ and for cast iron $x \simeq 0{\cdot}7$.

When cutting steel,

$$P_0 = 0{\cdot}164\ V \,.\, S \,.\, w \,.\, C \,.\, \frac{f_t^{0\cdot85}\,\mathrm{d}^{0\cdot925}}{R^{0\cdot925}} \text{ watt (ft lbf/min)} \qquad (9{\cdot}4)$$

To facilitate the calculation of P_0 when cutting steel, two nomograms (Fig. 9.8 (*a*) and (*b*)) have been compiled. These nomograms incorporate the feed, depth of cut and cutter radius terms to give a value for Y such that $P_0 = VSwCY$ watts, using either Imperial units (Fig. 9.8 (*a*)) or metric units (Fig. 9.8 (*b*)).
When cutting cast iron,

$$P_0 = 0{\cdot}169\ V \,.\, S \,.\, w \,.\, C \,.\, \frac{f_t^{0\cdot7}\,\mathrm{d}^{0\cdot85}}{R^{0\cdot85}} \text{ watt (ft lbf/min)} \qquad (9{\cdot}5)$$

9.1.4 Forces in peripheral milling. Unfortunately the discontinuous nature of chip formation in milling does not enable a simple assessment of radial cutting force to be made. This can be most easily understood if the formation of a chip by the upcut method is considered. Initially the cutting edge rubs the surface cut by the previous tooth, and only when the elastic limit of the workpiece has been reached does the cutting edge penetrate the surface. This causes a peak value of radial force to occur before it settles down to a fairly steady value.

The variations in tangential and radial forces in upcut milling are shown in Fig. 9.9 (*a*). When machining rapidly work hardening materials, such as the austenitic stainless steels, the radial force can reach very high values, which lead to the rapid failure of the cutting edge. The surface finish produced on these materials is usually poor because of rubbing of the milling cutter.

Fig. 9.9 (*b*) shows the tangential and radial forces in climb milling. The tangential force is similar to that obtained in upcut milling except that, due to the reverse rotation of the cutter, the force rapidly reaches a maximum and then decreases; with upcut milling the reverse obtains. It will be seen that the cut achieves its maximum depth almost immediately and consequently there is no appreciable rubbing. The radial force

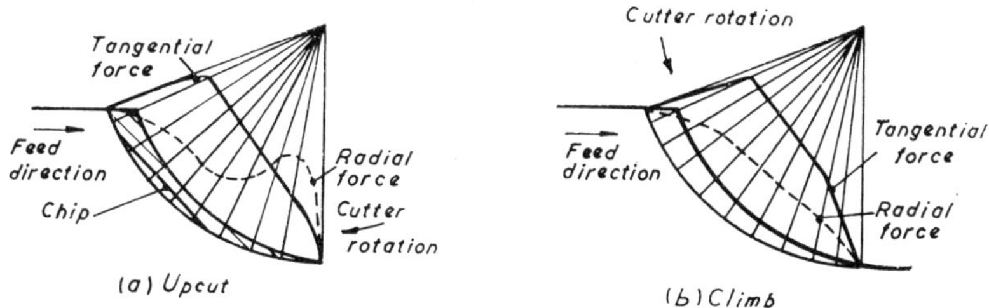

Fig. 9.9 Cutting forces in peripheral milling

in climb milling does not reach the same high peak value as in upcut milling, and this enables higher radial rakes to be used without detriment to tool life. High rakes result in lower tangential forces, as shown in Chapter 8, and a reduction in the power consumed.

The resultant force in upcut milling opposes the feed force, whereas in climb milling it assists the feed force. Thus, any backlash in the leadscrew causes the table to vibrate, leading to rapid wear of the machine and the cutter. For this reason an efficient backlash eliminator is essential when climb milling.

9.1.5 Face milling. When producing a surface by peripheral milling the smoothness of the surface obtained is dictated by the radius of the cutter and the feed per tooth. In addition, the resultant surface will be affected by any eccentricity. of the arbor holding the cutter.

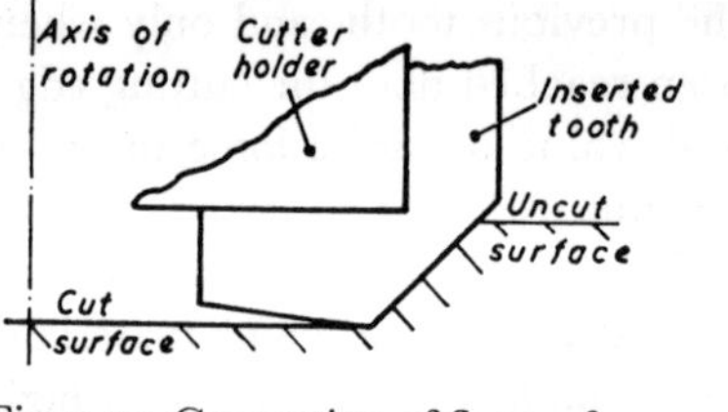

Fig. 9.10 Generation of flat surface with face milling cutter tooth

Face milling, as its name implies, uses the face of a multi-point cutter to produce a smooth finish (Fig. 9.10). If the trail angle on the cutting tooth was zero, a perfectly flat surface would result, provided the axis of rotation was at right angles to the plane of the slideways.

The geometry of face milling cutters is somewhat complicated. For simplicity, consider a single cutting tooth (Fig. 9.11). The corner angle corresponds to the plan approach angle in turning. To give strength to the cutter and to improve the surface finish it is usual also to have a small chamfer at the point of the tool, and a finishing land along which the trail angle is zero. The rake angles may be both positive (as shown) or,

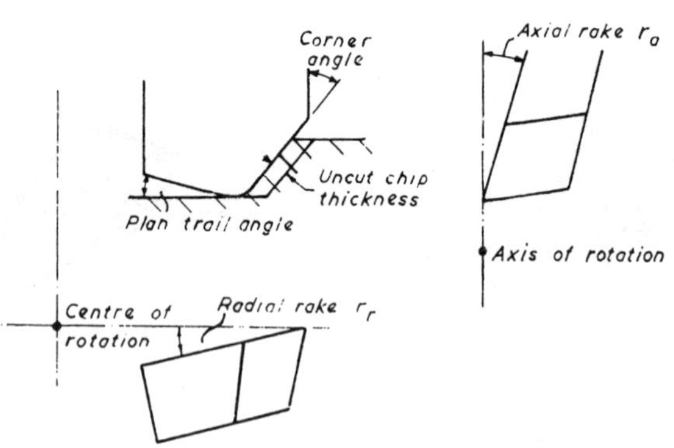

Fig. 9.11 Geometry of face milling cutter tooth

when using carbide tipped tools, one or both of the rake angles may be negative.

A large corner angle gives improved cutter life, but increases the likelihood of chatter; it also induces a heavy end thrust on the spindle. An angle of 45° is commonly used.

Fig. 9.12 shows the cutting edge of a face mill tooth, where r_a is the

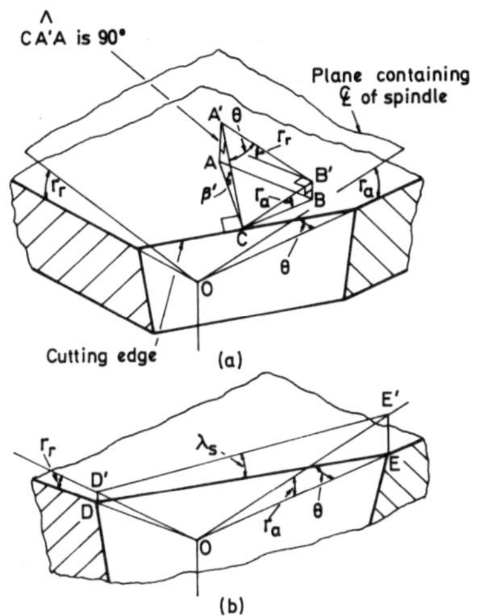

Fig. 9.12 Rake angles on face milling cutter

axial rake, r_r is the radial rake (both being positive), θ is the corner angle and λ_s the cutting edge inclination.

$$\tan \lambda_s = \frac{\text{E'E} - \text{D'D}}{\text{D'E'}} = \frac{\text{EO} \tan r_a - \text{DO} \tan r_r}{\text{D'E'}}$$

$$\tan \lambda_s = \cos \theta \,.\, \tan r_a - \sin \theta \,.\, \tan r_r \qquad (9.6)$$

The angle β' is measured in the vertical plane at right angles to the cutting edge, and corresponds to the side rake in the German (DIN) Standard.

$$\tan \beta' = \frac{\text{AA'}}{\text{A'C}} = \frac{\text{B'C} \,.\, \tan r_a + \text{A'B'} \,.\, \tan r_r}{\text{A'C}}$$

$$\tan \beta' = \sin \theta \,.\, \tan r_a + \cos \theta \,.\, \tan r_r \qquad (9.7)$$

To find the normal rake γ_n, from Fig. 7.3, where $\beta' = \text{A}\hat{\text{O}}\text{F}$

$$\tan \gamma_n = \tan \beta' \,.\, \cos \lambda_s \qquad (9.8)$$

and to evaluate the effective rake β_e,

$$\sin \beta_e = \cos^2 \lambda_s \sin \gamma_n + \sin^2 \lambda_s \qquad (7.2)$$

Table 9.1 shows how the effective rake is affected by variation in the axial and radial rakes.

TABLE 9.1

Cutter	γ_a°	γ_r°	θ°	λ_s°	γ_n°	β_e°
A	+10	+10	45	0	+14	+14
B	−10	−10	45	0	−14	−14
C	+ 5	−12	45	12	− 5	− 2½
D	+ 5	−18	45	16	− 9¼	− 4¼

Using tools with geometry similar to C and D it is possible to derive the benefits of negative rake angles whilst retaining the lower axial thrust associated with positive rakes.

It is difficult to estimate accurately the forces in face milling because of the considerable variations in chip geometry which may occur. Fig. 9.13 shows three possible cutter/workpiece arrangements in face milling, each giving different chip geometry. Under these conditions a fair approximation of power required can be obtained by making use of the concept of specific cutting pressure. It is assumed that the energy required to remove a unit volume of a specific material is a constant. (This is in fact a very rough approximation since, as we have seen, the variation of cutting conditions affects cutting forces appreciably.)

Let volume of metal removed in unit time $= X$

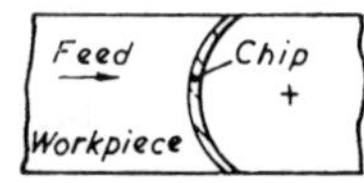

If V is the cutting speed,

$$\text{Average area of cut} = \frac{X}{V}$$

$$\text{Specific cutting pressure } u = \frac{\text{cutting force}}{\text{area of cut}}$$

$$= \frac{FV}{X} \qquad (F \text{ is cutting force})$$

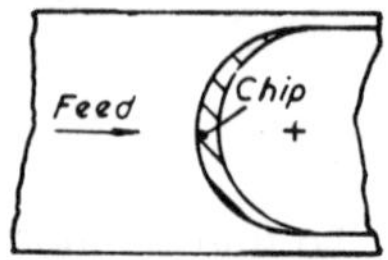

But cutting power $= FV$

∴ Power required to remove unit volume of

$$\text{metal in unit time} = \frac{FV}{X} = u$$

Fig. 9.13 Differing chip geometry due to cutter/workpiece arrangement

Values of u have been tabulated for various metals and some of the more important values are quoted in Table 9.2.

TABLE 9.2

Work material	hp min in^{-3}	Jmm^{-3}
Steel 100 HB	1·20	3·30
400 HB	2·00	5·50
Cast iron	0·70–1·20	1·90–3·30
Aluminium	0·40	1·10
Magnesium	0·28	0·75
Brass	0·55	1·50
Bronze	0·72	1·98
Monel	1·70	4·70

It should be noted that the power needed in cutting is often only a small percentage of the total power supplied to the machine. The other power is required to overcome frictional losses. In addition the backlash eliminator is a source of power loss which detracts slightly from the cutting advantages of using climb milling.

A recent development in face milling is the PERA face milling cutter[48], which combines the advantages of a high rate of metal removal, good surface finish, and rapid indexing of throw-away tool tips. There are several variants of the cutter, but they all contain a number of hexagonal roughing tips with radiused corners, together with a single finishing tip which can be either circular or hexagonal with radiused sides, as shown in Fig. 9.14. The finishing tip is unconventionally mounted, so that the

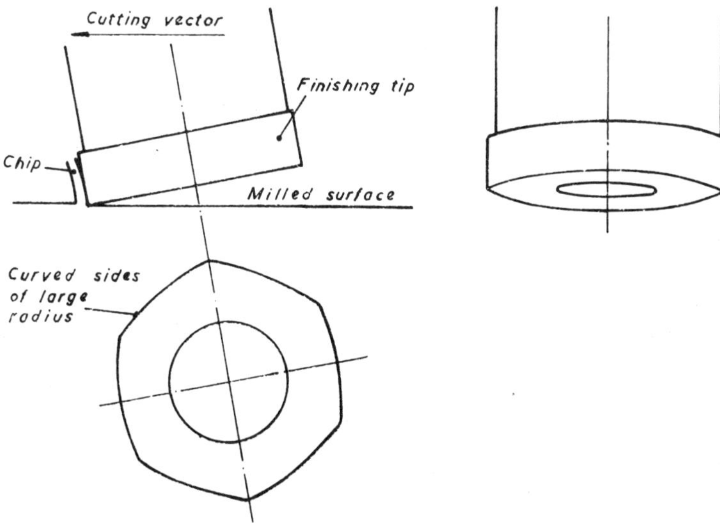

Fig. 9.14 Mounting for hexagonal finishing tip in PERA face milling cutter

side of the tip, and not the flat surface, acts as the rake face. This method of mounting gives the cutting edge of the finishing tip an almost flat geometry when presented to the workpiece, and results in a much reduced cusp height on the machined surface.

9.2 BROACHING

Broaching consists of passing a tool known as a broach across part of the surface of a component to give it a desired form. The broach consists of a series of cutting teeth, the cutting edge of each tooth standing slightly proud of the preceding tooth.

Fig. 9.15 Section of broach cutting work

A small part of a broach is shown in Fig. 9.15; the uncut chip thickness t is a few thousandths of an inch. The teeth are designed to accommodate the chip, which curls like a watch spring.

Most of the early applications of broaching were to produce internal forms such as splined holes, which would have been difficult or impossible to produce by other means. The broaches used for these purposes are generally pulled through a hole drilled in the workpiece, although when small quantities are required a push-type broach is sometimes used under the ram of a hydraulic press. Push broaches are limited in length to prevent buckling. Pull broaches are not limited in this sense and are sometimes up to 1·5 m (5 ft) long. This allows a large number of teeth to be used and consequently a large change of internal form is possible.

Surface broaching is becoming increasingly popular as an alternative to milling. The form is produced on the outside surface of the component by a translatory motion between the broach profile and the workpiece. In surface broaching the broach is rigidly located, whereas in internal broaching the drilled hole acts as its own location. Surface broaching is a more rapid and accurate method of metal removal than milling, but due to the high cost of broaches it is economical only when large quantities of parts are required. Broaches are usually made from high-speed steel, but for very large runs carbide inserted teeth may be used. The teeth of surface broaches are usually gashed on an angle to minimize peak loads.

It is essential for smooth operation that at least two or three teeth are cutting simultaneously. A suitable rule of thumb is that the tooth spacing shall be about $1{\cdot}76\sqrt{l}$ mm or $0{\cdot}35\sqrt{l}$ in, where l is the broached length (mm or in). The cut per tooth is usually between 0·05–0·09 mm (0·0020–0·0035 in) and cutting speeds between 0·1–0·4 m s^{-1} (20–80 ft/min) are generally used.

Machines are almost invariably hydraulically operated and may be either horizontal or vertical. Vertical machines are favoured because of the smaller space required, but for internal broaching, where the broach must be disconnected to pass through the workpiece, a horizontal machine is sometimes easier to operate.

The broaching load per tooth can be evaluated from the formula $F = C \,.\, t^x \,.\, w$, where t is the feed/tooth and w is the mean periphery of a tooth on an internal broach, or the width of the broached surface for surface broaching.

The speed range in which broaching is performed is likely to produce a severe built-up edge which spoils the surface finish of the work and shortens broach life. To minimize the incidence of this a plentiful supply of cutting lubricant is used. Sulphurized or chlorinated E.P. oils are normally used for steels, and soluble oils for non-ferrous materials.

10 Economics of Metal Removal

10.1 OPTIMIZATION OF PARAMETERS IN TURNING

The cost of producing components is made up of the raw material, and machining costs and factory overheads. In many machining operations the percentage of raw material utilized in the finished product is extremely low, and in these cases considerable savings can often be made by adopting metal forming or casting rather than metal cutting techniques. By using value analysis, which is an uninhibited approach to reducing product costs, drastic reductions can sometimes be obtained, by changing either the design or the manufacturing process, or both.

Despite this, machining may still be the most economical method of production and it is often necessary to machine away large volumes of material. It is surprising, therefore, that this is usually done with little concern for the economic selection of machining parameters. The turning process has been investigated fairly exhaustively by Brewer and Rueda[40] and PERA[41], but the other machining processes do not appear to have received much attention.

Once the process has been selected raw material cost will presumably be fixed, so in compiling the cost equation the only elements considered are those relating to machining cost and overhead. These elements are as follows:

(*a*) set-up and idle time cost/piece, K_1;
(*b*) machining cost/piece, K_2;
(*c*) tool changing cost/piece, K_3;
(*d*) tool re-grinding cost/piece, K_4;
(*e*) tool depreciation cost/piece, K_5.

If throw-away tips are used, tool regrinding can be ignored.

For turning, assume a cylindrical component of length L mm (in) and diameter D mm (in) is turned, using a feed f mm rev^{-1} (in/rev) and a cutting speed of V m s^{-1} (ft/min). The cost/min of labour and overhead is k_1 and cost of regrinding a tool is made up of two components, k_2 and k_3, where k_2 is the setting cost and k_3 is the cost per mm (in) of tool ground.

Set-up and idle time cost

$K_1 = k_1 . t_s$, where t_s is the set-up and idle time/piece.

Machine cost

$K_2 = k_1 . t_m$, where $t_m = (L . \pi . D)/f . V$, the machining time/piece, assuming the metal is removed in a single cut.

Tool changing cost

$K_3 = k_1 . t_c \times$ no. of tool changes/component, where t_c is the time required to change a tool.

The number of tool changes/piece $= \dfrac{t_m}{T}$, where $T =$ tool life.

Using equation (8.15) for a specified amount of tool wear,

$$VT^n(f . G)^m = \lambda$$

$$K_3 = k_1 . t_c . \frac{L . \pi . D}{f . V}\left(\frac{V}{\lambda}\right)^{1/n}(f . G)^{m/n}$$

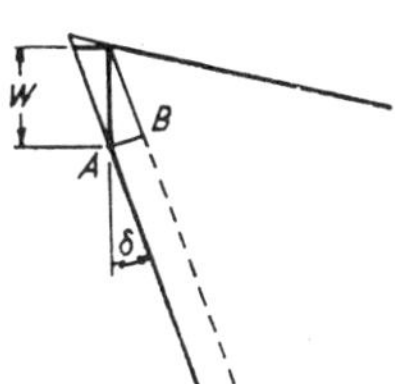

Fig. 10.1 Flank wear on cutting tool

Tool regrinding cost

Assuming that the permitted size of the wear land on the flank face before regrinding is W, it is seen from Fig. 10.1 that the minimum amount reground perpendicular to the clearance face of the tool is AB $= W \sin \delta$, where δ is the flank clearance angle.

$$K_4 = (k_2 + k_3 . W . \sin \delta) \times \text{tool changes/piece}$$

$$= (k_2 + k_3 \quad W \quad \sin \delta) \frac{L . \pi . D}{f . V}\left(\frac{V}{\lambda}\right)^{1/n} (f . G)^{m/n}$$

Tool depreciation cost

If the total amount which can be ground off the flank face of a new tool before it is no longer of use is A, the permitted number of regrinds is $A/W \sin \delta$, and hence the total number of tool lives/tool is $1 + (A/W \sin \delta)$.

The depreciation cost per tool life is $N/[1 + (A/W \sin \delta)]$, N being the cost of a new tool.

$$K_5 = \text{tool changes/piece} \times \frac{N}{1 + (A/W \sin \delta)}$$

$$= \frac{N}{1 + (A/W \sin \delta)} \cdot \frac{L \,.\, \pi \,.\, D}{f \,.\, V} \left(\frac{V}{\lambda}\right)^{1/n} (f \,.\, G)^{m/n}$$

Total cost

$$\text{Total cost } K = k_1 \,.\, t_s + k_1 \frac{L \,.\, \pi \,.\, D}{f \,.\, V} + \frac{L \,.\, \pi \,.\, D}{f \,.\, V} \left(\frac{V}{\lambda}\right)^{1/n} \,.\, (G \,.\, f)^{m/n}$$

$$\left[k_1 \,.\, t_c + k_2 + k_3 \,.\, W \,.\, \sin \delta + \frac{N}{1 + (A/W \sin \delta)}\right] \quad (10.1)$$

The terms in the square bracket are a constant for a given tool, denoted by H, assuming an optimum value for flank wear.

$$K = k_1 \,.\, t_s + k_1 \,.\, \frac{L \,.\, \pi \,.\, D}{f \,.\, V} + \frac{L \,.\, \pi \,.\, D}{f \,.\, V} \,.\, \left(\frac{V}{\lambda}\right)^{1/n} \,.\, (G \,.\, f)^{m/n} \,.\, H. \quad (10.2)$$

10.1.1 Optimization of cutting variables. In equation (10.2) the variables are G, f and V. From equation (8.14),

$$G = \frac{d}{[d - r(1 - \sin \psi_r)]/\cos \psi_r + [(90 - \psi_r)/180] \,.\, \pi \,.\, r}$$

The value of G is minimized by using large nose radius and tool approach angle, but these factors are both limited by chatter considerations, and sometimes by the geometry of the workpiece. Depth of cut is determined by the required reduction in diameter and by the number of passes of the tool to give the desired size. Although multi-pass machining may be more economical in some cases, in the vast majority of situations the single-pass method is likely to give minimum cost and the present analysis is confined to a single-pass method.

We are then left with the possible variation of feed and cutting speed. It is found that when partial derivatives of K with respect to f and V are equated to zero to find minimum total cost, the results obtained are mutally inconsistent. This means that there is no single value of f and V which gives minimum cost.

$$\frac{\partial K}{\partial V_{(f\ \text{fixed})}} = -\frac{k_1 . L . \pi . D}{f . V^2} + \left(\frac{1}{n} - 1\right) . f^{(m/n)-1} . \frac{L . \pi . D}{\lambda^{1/n}} . V^{(1/n)-2} . G^{m/n} . H$$

Equating to zero to obtain minimum cost for a specified value of f,

$$f = \frac{1}{G}\left(\frac{k_1^n \lambda}{[(1/n) - 1]^n . V . H^n}\right)^{1/m} \tag{10.3}$$

$$\frac{\partial K}{\partial f_{(V\ \text{fixed})}} = \frac{-k_1 . L . \pi . D}{f^2 . V} + \left(\frac{m}{n} - 1\right) . f^{(m/n)-2} . \frac{L . \pi . D}{\lambda^{(1/n)}} . V^{(1/n)-1} . G^{m/n} . H$$

Equating to zero to give minimum cost for a specified value of V,

$$f = \left(\frac{k_1^n . \lambda}{[(m/n) - 1]^n . V . H^n}\right)^{1/m} \times \frac{1}{G} \tag{10.4}$$

10.1.2 Factors limiting feed. Fig. 10.2 shows cost/piece plotted against f and V, with the minimum value of K becoming progressively smaller as f is increased. In practice f is limited by the maximum feed available on the lathe, the surface finish obtainable, or the force on the tool. Also, a very high feed is associated with a very low cutting

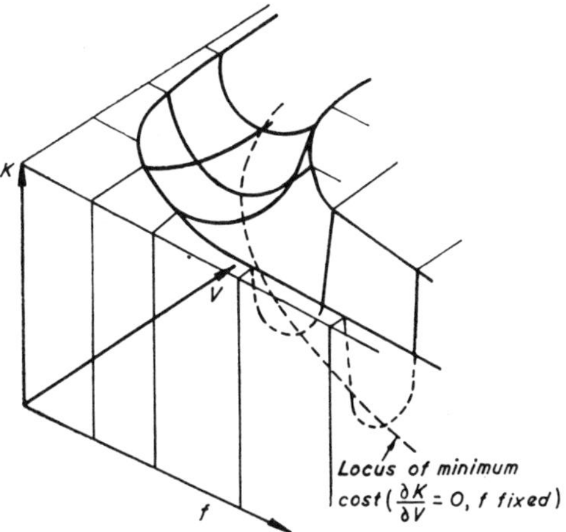

Fig. 10.2 Variation of cost with feed and speed (*after Brewer*)

speed, and at low speeds Taylor's tool life equation no longer applies since the tool fails from crumbling of the cutting edge, rather than flank wear.

Often the maximum practical feed and its associated cutting speed for minimum cost conditions may require power in excess of that obtainable from the lathe motor: a lower and less economic speed has therefore to be selected.

Using equation (8.6)

$$\text{Cutting force} = C \, . \, f^x \, . \, d$$

$$\text{Power } P = \text{cutting force} \times \text{cutting speed}$$

$$\text{Power} = C \, . \, f^x \, . \, d \, . \, V_p$$

Maximum cutting speed due to power available

$$V_p = \frac{P}{C \, . \, f^x \, . \, d} \qquad (10.5)$$

Substituting for V_p in the equation for total cost (10.1) it can be shown that the larger the value of f the smaller the cost per piece.

Fig. 10.3 shows a likely practical situation where over the higher speed range the minimum cost locus, represented by the line $\partial K/\partial V = 0$, determines the optimum selection of V, whereas at lower speeds the value of V is limited by the power constraint. When throw-away tips are used H can be simplified to $[k_1 \, . \, t_c + N/X]$ where X is the number of cutting edges on the tip.

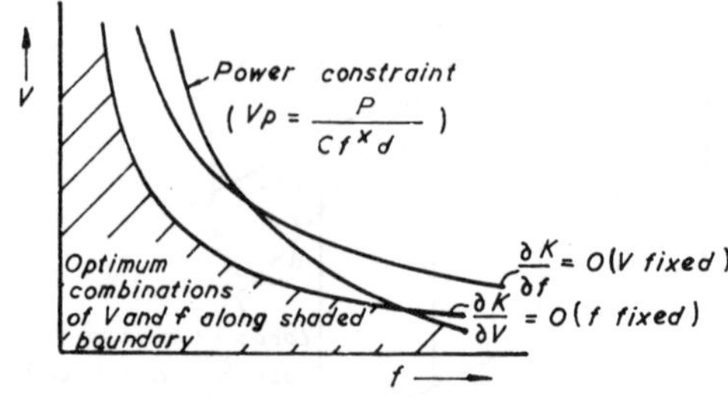

Fig. 10.3 Effect of power constraint on optimisation

10.1.3 Use of nomograms to find optimum cutting speed. Brewer and Rueda, realising that the approach outlined required a considerable amount of tedious arithmetic to find V, compiled nomograms

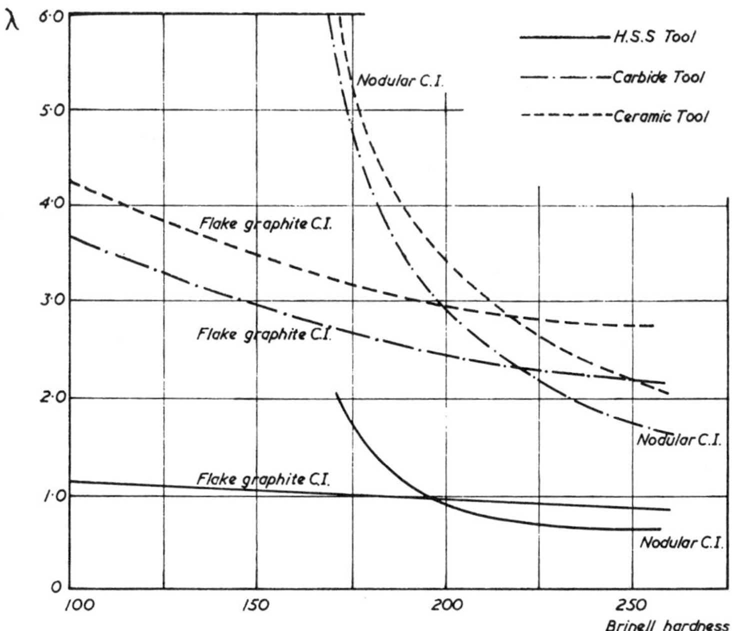

Fig. 10.4 Variation of hardness for cast iron (metric units)
See Appendix 3 for values of λ in Imperial units

from which the optimum values of V can be read. Further, lack of sufficient tool life and cutting force data for machining non-ferrous materials at present limits the application of this analysis to steels and cast irons.

Fig. 10.4 shows the variation of λ for cast irons of differing hardness when using high-speed steel, carbide and ceramic tools. Fig. 10.5 shows some typical values of λ for steel.

When the necessary constants are not available it has been suggested that, without great loss of accuracy, the following approximation can be adopted. From Fig. 10.2 it is seen that K is smallest when f is greatest; therefore the maximum practical feed should be used to give lowest total cost.

Steel	Hardness (HV30)	λ for carbide tools (SI units)	(non-metric)
EN3A	118	7·10	425
EN8	140	4·78	285
EN16	310	3·25	195
EN24	280	3·47	205
EN35B	204	3·95	235

Approximate values of λ for H.S.S. Tools 0·33 × value for carbide

Approximate values of λ for ceramic tools 1·3 × value for carbide

Fig. 10.5 Values of λ for steel when using carbide tools

Considering Taylor's tool life equation and equation (8.15),

$$V\,.\,T^n = C$$

and

$$V\,.\,T^n\,.\,(f\,.\,G)^m = \lambda$$

Then,

$$\lambda = (f\,.\,G)^m\,.\,C.$$

Also, considering the constant H, this can be simplified to give a component of cost due to the time required to change a tool, $k_1\,.\,t_c$, and a component of cost to allow for tool depreciation, k_g, where

$$k_g = k_2 + k_3\,.\,W\,.\,\sin\delta + \frac{N}{1 + (A/W\sin\delta)}$$

$$H = [k_1\,.\,t_c + k_g]$$

The new total cost equation now becomes

$$K = k_1 t_s + k_1\,.\,\frac{L\,.\,\pi\,.\,D}{f\,.\,V} + \frac{L\,.\,\pi\,.\,D}{f\,.\,C^{1/n}}\,.\,V^{(1/n)-1}[k_1\,.\,t_c + k_g]$$

$$\frac{\partial K}{\partial V} = -k_1\,.\,\frac{L\,.\,\pi\,.\,D}{f\,.\,V^2} + \left(\frac{1}{n} - 1\right)\frac{L\,.\,\pi\,.\,D}{f\,.\,C^{1/n}}\,.\,V^{(1/n)-2}[k_1\,.\,t_c + k_g]$$

Equating to zero for minimum cost conditions,

$$V = C\,.\left(\frac{k_1}{k_1\,.\,t_c + k_g}\right)^n\left(\frac{n}{1-n}\right)^n \tag{10.6}$$

The term $[k_1/(k_1t_c + k_g)]$ is of the order of 0·1 for solid or brazed tools, or 0·25 for throw-away tips, and since it is raised to a low power the error in assuming these values is small. Hence,

$$V = \left(\frac{0{\cdot}1n}{1 - n}\right)^n C = X\,.\,C \text{ for solid brazed tools.}$$

$$V = \left(\frac{0{\cdot}25\,n}{1 - n}\right)^n C = X\,.\,C \text{ for throw-away tips.}$$

Values of X are plotted against n in Fig. 10.6. The value of C can be obtained from tool life tests where data are unobtainable, or from the nomograms in Figs. 10.7 and 10.8, where steel or cast iron is machined.

Example. A brazed carbide tool with positive rake is used to turn a carbon steel component of EN24. The tool approach angle is 15°, nose radius is 1·00 mm and depth of cut is 4·00 mm. The largest practical feed is taken as 0·75 mm. Find the cutting speed which gives minimum total cost. The value of G is found from Fig. 10.7 to be 0·86, and λ is found from Fig. 10.5 to be 3·47. Using these values, Fig. 10.8 gives C as 4·00. The value of X is obtained from Fig. 10.6 as 0·425, using a value of 0·25 for n with positive rake carbide tools, as suggested in Chapter 8.

Optimum cutting speed $V = X\,.\,C = 1{\cdot}70 \text{ ms}^{-1}$ (330 ft/min)
If throw-away tips had been used the optimum cutting speed would have been raised to $2{\cdot}18 \text{ m s}^{-1}$ (430 ft/min).

Experiments show that, when cutting under nominally identical conditions, the standard deviation of tool life is about 10 per cent of the expected life. To allow for this variability when cutting, it is therefore advisable to reduce the computed cutting speed by a factor of about 10 per cent.

An interesting point arising from this analysis is that the optimum selection of cutting parameters produces shorter tool lives than are usually considered economic in industry. However, in recent years there has been a persistent shortage of skilled labour such as setters. This leads many employers to work on the basis of maximum output/setter rather than the output giving minimum cost/piece, and has resulted in an increase in the number of machines looked after by each setter and the consequent use of low cutting speeds to reduce the amount of resetting. The analysis shown above has therefore achieved little popularity in the United Kingdom.

10.1.4 Comparison of cutting speeds giving maximum output and minimum cost. It is possible to compile an equation to give the output from a machine for a particular metal cutting operation, and

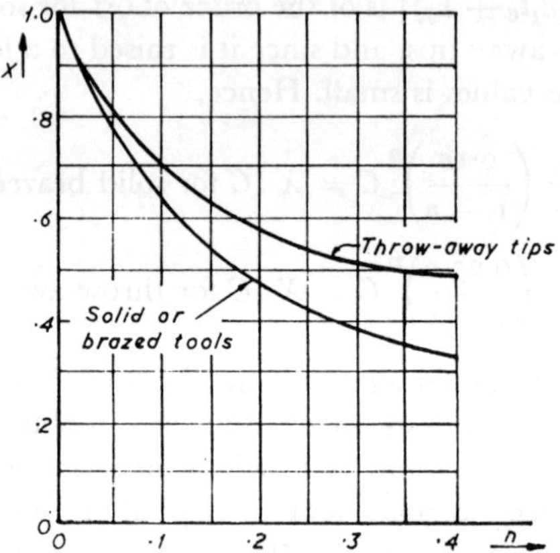

Fig. 10.6 Variation of X with n

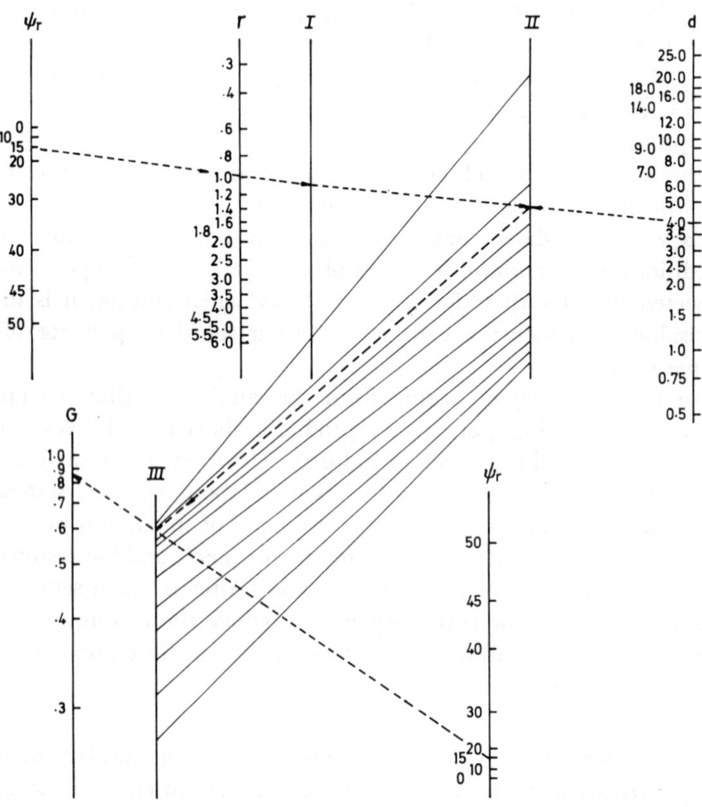

Fig. 10.7 Nomogram for determining value of G

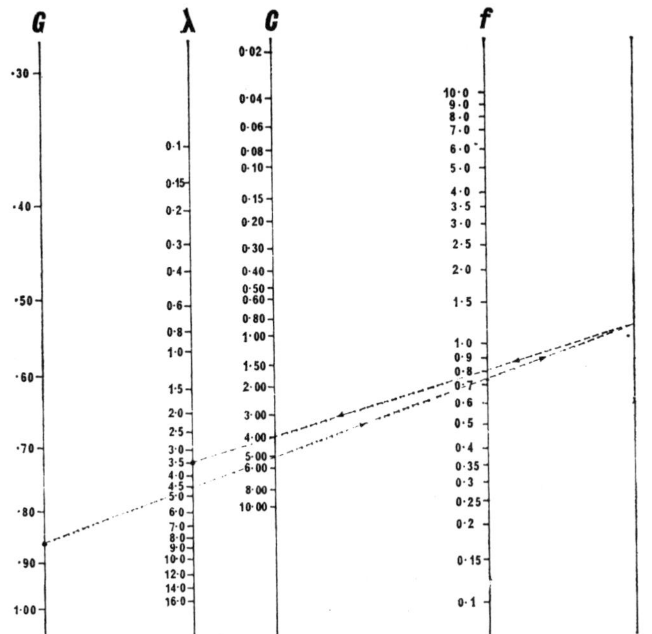

Fig. 10.8 (*a*) Nomogram for determining tool life constant *C* for steel

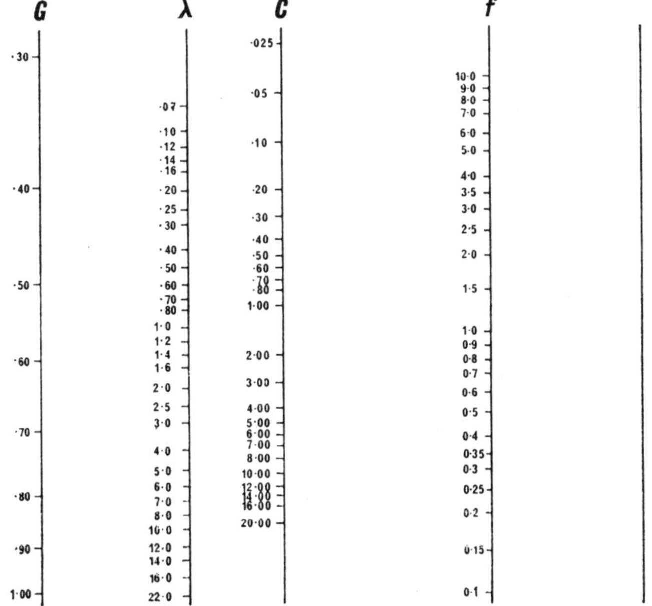

Fig. 10.8 (*b*) Nomogram for determining tool life constant *C* for cast iron

hence to find the combination of feed and cutting speed for maximum production from a machine.

Time to produce one component, $M = t_s + t_m + t_c \times$ tool changes/piece

$$= t_s + \frac{L \cdot \pi \cdot D}{f \cdot V} + t_c \times \frac{L \cdot \pi \cdot D}{f \cdot V}\left(\frac{V}{\lambda}\right)^{1/n}(f \cdot G)^{m/n}$$

Differentiating with respect to V and equating to zero to give minimum time,

$$V = \frac{\lambda}{[(1/n) - 1]^n (f \cdot G)^m} \times \left(\frac{1}{t_c}\right)^n \text{ for maximum output} \quad (10.8)$$

From equation (10.3),

$$V = \frac{\lambda}{[(1/n) - 1]^n (f \cdot G)^m} \times \left(\frac{k_1}{H}\right)^n \text{ for minimum cost per piece} \quad (10.9)$$

From previous work it is seen that $1/t_c > k_1/H$. The cutting speed to give maximum output from a particular machine will thus always be greater than that to give minimum cost per piece.

10.1.5 Pre-set tooling. To reduce machine down-time during changeover and to reduce the re-setting time when individual tools are changed, it is possible to fit tool holders to lathes which allow for pre-setting away from the machine. A number of advantages can be claimed for this approach to setting:

(1) By reducing changeover times it is possible to visualize production of smaller work batches without significantly reducing output.
(2) Operators who lack setting skills could change tools on the machine, thereby increasing job satisfaction and releasing existing setters for pre-setting.
(3) Optimum tool lives would be still further reduced, due to the reduction in tool changing time, and productivity further increased.

10.2 MATERIALS CAUSING DIFFICULTY IN MACHINING

Technical advances have demanded the use of materials which are increasingly difficult to machine. These include materials developed to retain their strength at elevated temperatures, titanium alloys used in the aerospace industries, and radioactive metals associated with nuclear power. Detailed analyses of optimum cutting conditions, such as those described for steel and cast iron, do not exist for these metals, but our present machining experience enables general recommendations to be made so that satisfactory cutting conditions may be selected.

The first group includes stainless steels, nimonic alloys and refractory metals such as molybdenum, tungsten and tantalum. Stainless steels rapidly work harden, and light cuts must be avoided as the rubbing action promotes skin hardness. For the same reason climb milling is necessary with most stainless steels. Although high-speed steel tools and a plentiful supply of chlorinated or sulphurized cutting fluid can be used to cut stainless steels, dry cutting with carbide tools is to be preferred. Since cutting speeds are necessarily low, (about 0·5–2 m s^{-1} (100–400 ft/min) with tungsten carbide) positive rake tools are essential.

The nimonic alloys consist basically of nickel and chromium, and are used in gas turbines because of their resistance to creep at high temperatures. These alloys vary considerably in their machining properties, but with the more difficult, cutting speeds using carbide tools are frequently restricted to less than 0·5 m s^{-1} (100 ft/min).

Tungsten is another material which presents problems in machining. Current practice favours the use of carbide tools operated at feeds of about 0·25 mm (0·010 in), although the feed may need to be reduced to prevent chipping of the workpiece. Zero or negative rake tools are recommended, with turning speeds of 1·0 m s^{-1} (200 ft/min) using soluble oil, and milling speeds of 0·4 m s^{-1} (75 ft/min) using chlorinated cutting oil.

Molybdenum and TZM molybdenum alloy can be cut satisfactorily with carbide tools at speeds of 1·8 m s^{-1} (350 ft/min) using soluble cutting oil. For turning, 20° positive rake tools can be operated at a feed of 0·25 mm (0.010 in) and for face milling zero rake tools are recommended at reduced feeds.

Tantalum and 90Ta–10W tantalum alloy can be cut with carbide or high speed steel tools having 20° positive rake. With either type of tool soluble oil should be used with cutting speeds of 0·4 m s^{-1} (75 ft/min) and 0·25 mm (0·010 in) feeds.

Titanium alloys cause rapid tool wear, but are best cut with carbide tools. The reasons for the high rate of wear are the high interface temperatures, resulting from the poor thermal conductivity of titanium and its chemical affinity for most metals at high temperatures. Cutting speeds of 0·25–0·5 m s^{-1} (50–100 ft/min) are recommended for titanium alloys, with a preference for large feeds and low speeds.

The use of metals which are difficult to machine has led to several developments in casting and metal forming in order to reduce the amount of stock removal necessary. Notably, investment casting is used in the production of nimonic alloy turbine blades.

When machining some of the hard materials such as nimonic alloys and Stellite, considerable improvements can be achieved by thermally softening the material with a plasma torch positioned ahead of the cutter. The pre-heat temperature is of the order of 600°C and the resulting chip temperature is correspondingly increased. Carbide tools are generally unsuitable for turning at these temperatures and ceramic tips are used. It is possible to machine some of the more intractable nimonic alloys by this method at cutting speeds of 8·0 m s^{-1} (1500 ft/min) although current practice is to use lower speeds and correspondingly higher feeds.

10.3 ECONOMICS OF MILLING

Although the economic aspects of milling do not appear to have been investigated to the same depth as those of turning, the analysis for turning, coupled with the remarks in Chapter 9 on the milling process, indicate that high feeds per tooth should be used, and cutting speeds should be adjusted accordingly. When a tooth first starts to cut, the resulting impact causes a variation of arbor torque. To keep this variation within reasonable limits the maximum feed per tooth is restricted to a fairly low value. The recommended feeds for cutters are usually quoted in the manufacturers' literature.

In the absence of a more accurate economic analysis it is usual to adjust the cutting speed, using the recommended feeds, to give a power consumption near to the maximum available from the drive motor. Should this procedure result in excessively high cutting speeds which would cause rapid tooth failure, it will be necessary to limit the speed to a lower value based on experience of the machine tool and the type of cutter used.

11 Abrasive Machining

11.1 Abrasive machining uses hard non-metallic particles to cut the workpiece. Processes within this group include grinding, honing, superfinishing, or abrasive belt machining and lapping. The first three use abrasive particles (often called grits), rigidly held in a wheel, stone or belt, whereas in lapping the particles are contained in a fluid. Unlike most other major machining operations this group of processes can shape workpieces harder than 400 Vickers Hardness Number (HV). Abrasive machining produces smooth surface finishes and enables close control to be maintained over the amount of workpiece material removed; in consequence it is mainly used for finishing operations.

11.2 GRINDING PROCESSES

Grinding is one of the major methods of metal machining, ranking with turning and milling. Most shapes can be produced by using an appropriate grinding wheel with either a surface grinder, a cylindrical grinder, or a special-purpose thread or gear grinder. Examples of some of the more important grinding operations are shown in Fig. 11.1.

The grinding operation is difficult to analyse because of the random shape of the abrasive particles, the minute chips produced, and the rapidity with which they are formed. One of the first papers dealing with the mechanics of grinding was by Guest[49] in 1915, but despite a considerable amount of subsequent work there is still no generally accepted theory of grinding.

11.2.1 Grinding wheels. Grinding wheels are made up of a large number of abrasive grains held together by a bonding agent. The abrasive material can be aluminium oxide, silicon carbide or diamond. Diamonds, which may be natural or synthetic, are the hardest of the materials used (6500 Knoop); they are also the most resistant to wear. However, because of their high cost, diamond wheels are normally confined to the grinding of very hard materials, such as tungsten carbide. Silicon carbide (2500

Knoop) is the next hardest, but it is liable to fairly rapid chemical wear when cutting steels and its use is generally restricted to machining non-ferrous materials and non-metals. The general purpose abrasive is aluminium oxide (2000 Knoop). Within these three main abrasive groups are sub-groups caused by the presence in the abrasive of impurities, imperfections in the crystal structure, and differences in crystal size. A

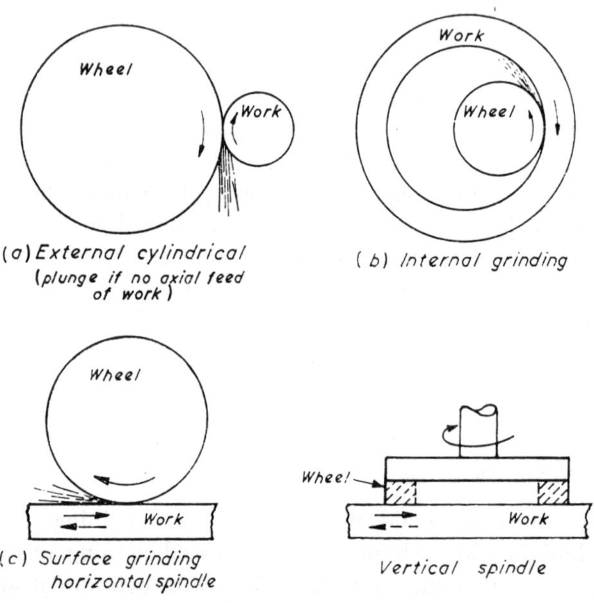

Fig. 11.1 Grinding operations

recently introduced abrasive material is cubic boron nitride (4700 Knoop). It has excellent wear resistance and despite its present extremely high cost has proved economical in certain precision grinding applications.

The size of abrasive grains used in a particular wheel is normally constant, although mixed sizes are sometimes used. Standard sizes of abrasive grain vary in steps from 8 to 600 (coarsest to finest). The larger grains are used for rapid metal removal and produce a comparatively rough finish, while small grains are used for fine surfaces; a typical general-purpose size is 46.

A variety of bonding materials is available. By far the most common is the vitrified bond, which is manufactured from felspar, with refractories and fluxes added. Wheels with vitrified bonds are produced by mixing abrasive particles, bonding material and an organic filler, the filler being used to adjust the structure of the wheel. Moulds are employed to obtain the wheel shape, and after drying the wheels are baked in a furnace for several days at about 1200°C. The bonding material melts and by surface tension forms bond posts between adjacent grains. The organic filler is burnt out, with the gases escaping through the network of interconnecting voids.

Thermosetting resin (resinoid) bonds enable thin flexible wheels to be produced, these are suitable for cutting-off operations and the grinding

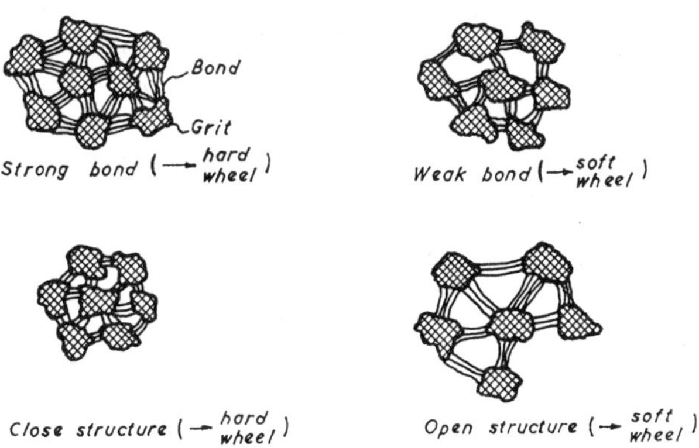

Fig. 11.2 Variation of bond and structure

of rough surfaces with hand-held grinders. The high strength of resinoid bonds enable them to operate at speeds up to 80 m s^{-1} (16000 ft/min), almost treble those for vitrified bonds. Rubber bonds are widely used for feed wheels on centreless grinding machines and for very thin cutting-off wheels. Shellac bonds are rarely used although they are suitable for fine finishing operations. Diamond wheels employ resinoid, vitrified or metallic bonds.

The relationship between both bond and structure as it affects wheel strength is illustrated in Fig. 11.2. Bond strength is graded by a letter code, the strength increasing alphabetically from A to Z. Weakly bonded wheels are referred to as soft and those with strong bonds as hard wheels. Wheel structure is indicated numerically (0–14), close structures having low numbers and open structures high numbers.

Wheel speeds conventionally used for cylindrical and surface grinding machines are in the order of 25 m s^{-1} (5000 ft/min); however, machines capable of surface speeds in the range of 60 m s^{-1} (12000 ft/min) to 90 m s^{-1} (18000 ft/min) are now being produced. As surface speeds increase the forces between the wheel and work are correspondingly reduced;

hence if forces are kept unchanged considerably higher metal removal rates can be achieved. Specially designed vitrified wheels, capable of withstanding the greatly increased bursting forces are available. Much improved wheel guarding must be provided to protect the operator in the event of a wheel burst.

11.2.2 Geometry of the grinding process. Shaw and his fellow workers considered grinding as a micro-milling process.[50–52] Although this was an oversimplification it enabled approximate expressions for depth of cut, length of chip and other parameters to be obtained.

Plunge grinding, an operation in which there is no longitudinal feed of the work, is shown in Fig. 11.3. It is assumed that the cutting grains are equally spaced round the periphery of the wheel, analogous to a milling cutter with a very large number of teeth. The depth of cut t can then be obtained from the expression

$$t = \frac{f}{KN} \text{ mm(in)} \qquad (11.1)$$

where f is the rate of infeed of work, mm min^{-1} (in/min)

N is the rev/min of the grinding wheel,
K is the number of grains in line/rev.

The term K can be found from the relationship

$$K = \pi DCb' \qquad (11.2)$$

where D is the wheel diameter, mm (in),
C is the number of cutting grains mm^{-2} (/in^2)
b' is the average grain width of cut, mm (in)

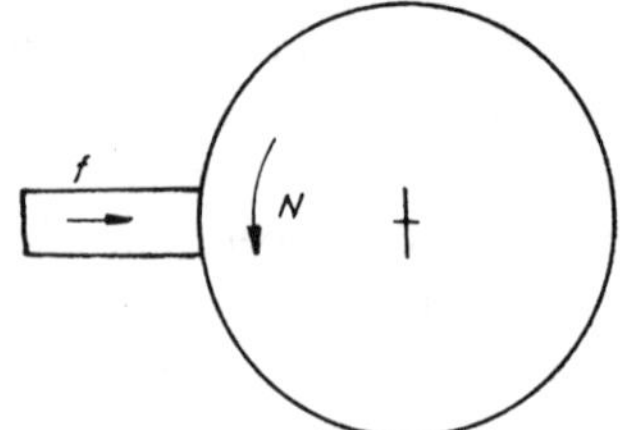

Fig. 11.3 Plunge grinding

The ratio of width to depth of cut is represented by r_p where $r_p = b'/t$

The terms C and r_p are somewhat difficult to evaluate. The number of cutting grains mm^{-2}(/in^2), C, is found by rolling the wheel over a soot-blackened glass and counting the points where the soot has been removed. A taper section of the ground surface (Fig. 11.4) can be used to find the width:depth ratio. The tapered surface is used to provide a magnification of the groove depth t.

If plunge grinding is considered, the chip thickness will equal t, the depth of groove. By substitution for b' in equation (11.2) and using the result to substitute for K in (11.1)

$$t = \sqrt{\left(\frac{f}{\pi\, DNCr_p}\right)} \tag{11.3}$$

Surface grinding on a horizontal spindle machine can be represented by Fig. 11.5(a) where the depth of cut has been greatly exaggerated,

Fig. 11.4 Taper sections

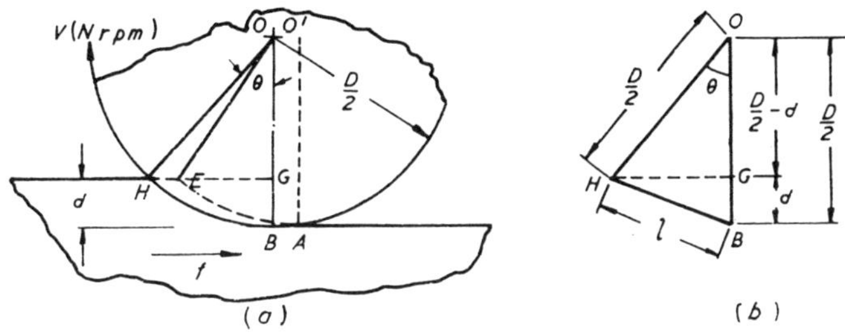

Fig. 11.5 Surface grinding (depth of cut exaggerated)

a typical value being only 0·0001 × wheel diameter. The path ABH traced by the tip of a cutting grain is a trochoid generated by a combination of the circular movement of the wheel and the horizontal movement of the work.

The undeformed chip length l and the maximum chip thickness t can be found by reference to Fig. 11.5.

$$l = \overline{\mathrm{ABH}} \simeq \overline{\mathrm{BH}} \text{ (Actually } \overline{\mathrm{BH}} \gg \overline{\mathrm{AB}}\text{)}.$$

As θ is very small when grinding

$$\text{curve } \overline{\mathrm{BH}} \simeq \text{chord } \overline{\mathrm{BH}} \text{ (Fig. 11.5 } (b)\text{)}$$

$$\therefore\ l = \overline{\mathrm{BH}} = \sqrt{(\overline{\mathrm{HG}}^2 + d^2)} = \sqrt{\left\{\left[\left(\frac{D}{2}\right)^2 - \left(\frac{D}{2} - d\right)^2\right] + d^2\right\}}$$

$$l = \sqrt{Dd} \tag{11.4}$$

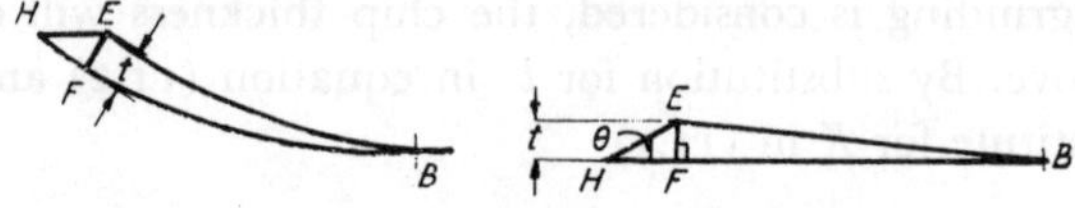

Fig. 11.6 Chip thickness

To find maximum chip thickness t, the chip is flattened as shown in Fig. 11.6.

$$t = \overline{\mathrm{EF}} = \overline{\mathrm{HE}} \sin \theta$$

$$\sin \theta = \frac{\overline{\mathrm{HG}}}{D/2} \qquad \text{(Fig. 11.5 } (b))$$

but $$\overline{\mathrm{HG}} = \sqrt{\left[\left(\frac{D}{2}\right)^2 - \left(\frac{D}{2} - d\right)^2\right]} = \sqrt{(Dd - d^2)}$$

hence $$t = \overline{\mathrm{HE}}\,\frac{2}{D}\sqrt{(Dd - d^2)} = 2\overline{\mathrm{HE}}\sqrt{\left[\frac{d}{D} - \left(\frac{d}{D}\right)^2\right]}$$

and as d/D is small $(d/D)^2$ can be neglected.

$$\therefore \qquad t = 2\overline{\mathrm{HE}}\sqrt{\left(\frac{d}{D}\right)}$$

$\overline{\mathrm{HE}}$ is work feed per grain. If the grains are equally spaced and there are K in line/rev,

$$\overline{\mathrm{HE}} = \frac{f}{K\,.\,N}$$

where f is rate of work feed, mm min^{-1} (in/min)

N is rev/min of wheel.

Substituting for $\overline{\mathrm{HE}}$

$$t = \frac{2f}{K\,.\,N}\sqrt{\frac{d}{D}} \tag{11.5}$$

In surface grinding, unlike the plunge grinding example previously considered, the chip thickness increases from zero to a maximum t. It is now more convenient to consider the b'/t ratio at $t/2$ instead of at t. This ratio is indicated by r, not r_p as in plunge grinding.

Hence $$b' = \frac{tr}{2} \tag{11.6}$$

From equation (11.2) $K = \pi DCb' = \pi DC\frac{t}{2}r$

Substituting for K in equation (11.5)

$$t = \frac{4f}{\pi DCtrN}\sqrt{\frac{d}{D}}$$

and

$$t = \left[\frac{4f}{\pi DNCr}\sqrt{\frac{d}{D}}\right]^{\frac{1}{2}} \qquad (11.7)$$

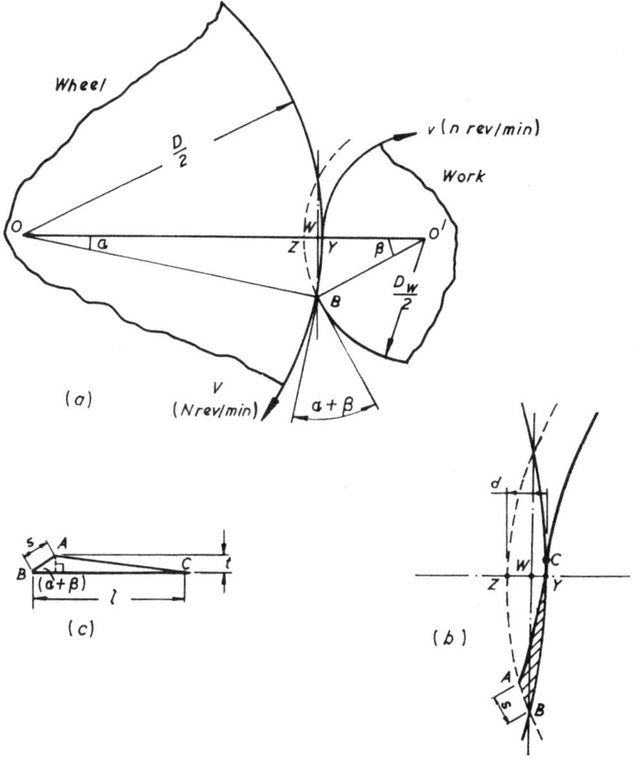

Fig. 11.7 Cylindrical grinding

The geometry of external cylindrical grinding is shown in Fig. 11.7 (*a*) and again for clarity the depth of cut has been greatly exaggerated. In grinding cylindrical shapes longer than the wheel width the work is reciprocated; if short cylindrical shapes or profiles are required, the work is plunge ground by feeding the wheel radially into the work. If

Figs. 11.7 (*a*), (*b*) and (*c*) are considered, expressions can be obtained for the chip length l and the undeformed chip thickness t.

$$\overline{YZ} = \text{radial work feed} = \frac{f}{n} \text{ mm rev}^{-1} \text{ (in/rev)}$$

where f is rate of radial work feed, mm min^{-1} (in/min)
n is rotational speed of work, rev min^{-1}

$$\overline{YZ} = \overline{WY} + \overline{WZ}$$

$$\frac{f}{n} = \overline{WY}\left(1 + \frac{\overline{WZ}}{\overline{WY}}\right)$$

$$\overline{WY} = \frac{f}{n[1 + (\overline{WZ}/\overline{WY})]}$$

For small angles of contact normal in grinding

$$\frac{\overline{WZ}}{\overline{WY}} \simeq \frac{\beta}{\alpha} \simeq \frac{D}{D_w}$$

$$\therefore \qquad \overline{WY} = \frac{f}{n(1 + D/D_w)}$$

In triangle OWB $\overline{WB}^2 = \overline{OB}^2 - \overline{OW}^2$

$$\overline{WB}^2 = \overline{OB}^2 - (\overline{OY} - \overline{WY})^2$$

Substituting for $\overline{OB}$, $\overline{OY}$ and $\overline{WY}$

$$\overline{WB}^2 = \left(\frac{D}{2}\right)^2 - \left\{\frac{D}{2} - \frac{f}{n[1 + (D/D_w)]}\right\}^2$$

$$\overline{WB}^2 = \frac{Df}{n[1 + (D/D_w)]} - \left\{\frac{f}{n[1 + (D/D_w)]}\right\}^2$$

As $f/n \ll 1$ $$\overline{WB}^2 \simeq \frac{Df}{n[1 + (D/D_w)]}$$

But $\overline{WB} \simeq l$ and $f/n = d$ (radial feed/rev)

$$\therefore \qquad l = \sqrt{\frac{Dd}{1 + (D/D_w)}} \qquad (11.8)$$

For small angles

$$\sin \alpha \simeq \alpha = \frac{\overline{\mathrm{WB}}}{D/2} = \frac{2}{D}\sqrt{\left(\frac{Dd}{1 + (D/D_{\mathrm{w}})}\right)} = \sqrt{\left(\frac{4d}{D[1 + (D/D_{\mathrm{w}})]}\right)}$$

$$\beta \simeq \frac{D\,\alpha}{D_{\mathrm{w}}} = \frac{D}{D_{\mathrm{w}}}\sqrt{\left(\frac{4d}{D[1 + (D/D_{\mathrm{w}})]}\right)}$$

Considering the flattened chip (Fig. 11.7 (*c*))

$$t = s \sin (\alpha + \beta) = s\left\{\left(1 + \frac{D}{D_{\mathrm{w}}}\right)\sqrt{\left(\frac{4d}{D[1 + (D/D_{\mathrm{w}})]}\right)}\right\}$$

since $\sin (\alpha + \beta) \simeq \alpha + \beta$ and $s = v/KN$

$$t = \frac{v}{KN}\sqrt{\left(\frac{4d[1 + (D/D_{\mathrm{w}})]}{D}\right)}$$

But from equations (11.2) and (11.6) $K = \pi DCb'$ and $b' = tr/2$

$$\therefore \qquad K = \frac{\pi DCtr}{2}$$

and

$$t = \frac{2v}{\pi DCtrN}\sqrt{\left(\frac{4d[1 + (D/D_{\mathrm{w}})]}{D}\right)}$$

$$N = \frac{V}{\pi D}$$

$$\therefore \qquad t = \left[\frac{4v}{VCr}\sqrt{\left(\frac{d[1 + (D/D_{\mathrm{w}})]}{D}\right)}\right]^{\frac{1}{2}} \qquad (11.9)$$

11.2.3 Grinding forces. The rate of metal removal in grinding is low, and in consequence the cutting forces are of the order of a few pounds, compared with several hundred pounds usual in milling and turning. In contrast to most other machining operations the radial force F_{n} in grinding is larger than the tangential force F_{t}; Fig. 11.8 shows the direction of these forces in surface grinding where the ratio $F_{\mathrm{n}}/F_{\mathrm{t}} \simeq 2$. As might be expected from the high radial force, grinding is an inefficient metal cutting process when judged on the basis of specific energy (energy required to remove unit volume of the workpiece). Specific energy will vary with wheel and grinding conditions, a typical value for mild steel being 48 J mm^{-3} (7×10^6 in lbf/in^3) this is about 20 times greater than that

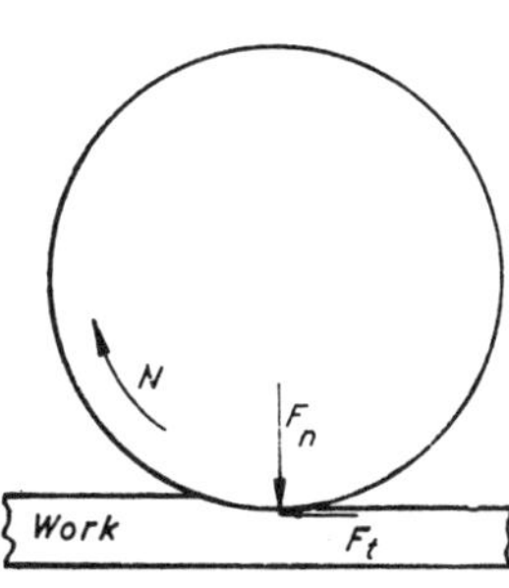

Fig. 11.8 Forces on work in surface grinding

required to cut similar material with a single-point tool. Compared with other metal cutting tools the abrasive grain has an inefficient shape; most grains have large negative rake and rapidly develop flats where they contact the workpiece. Often the area of these flats covers such a large proportion of the wheel surface that it has a glazed appearance, and in this state cutting frequently becomes impossible.

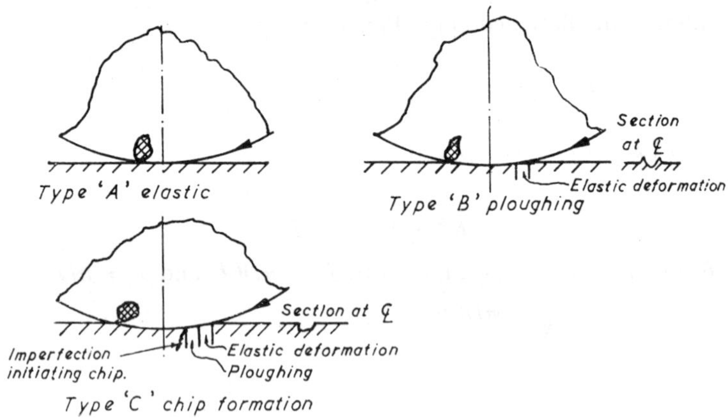

Fig. 11.9 Types of surface deformation (*after Hahn*)

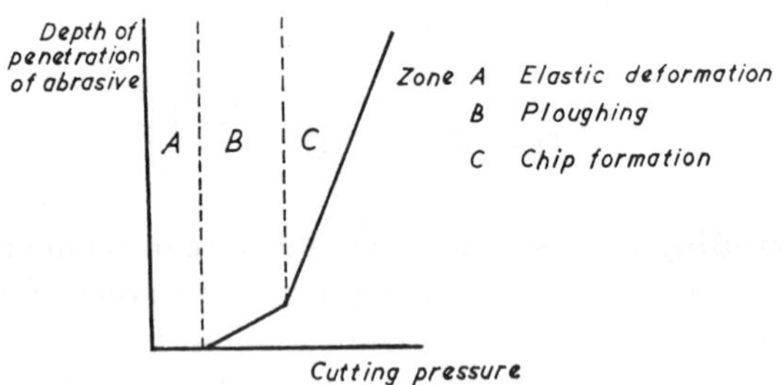

Fig. 11.10 Relationship between depth of abrasive penetration and cutting pressure (*after Hahn*)

Hahn[53] has examined the effect of different depths of cut on workpiece deformation (Fig. 11.9). At very small depths there is only elastic deformation of the metal surface, but as the setting is increased the surface is plastically deformed and metal is pushed up on either side of the furrows ploughed by the abrasive grains. Eventually as the depth of cut is further increased a chip is formed, after some initial elastic and plastic deformation of the work. The transition from ploughing to cutting is indicated by a diminution in the force increment required to secure an increase in depth of metal removed (Fig. 11.10). Hahn proposed that chips were initiated by minute imperfections in the metallic structure of the work surface and suggested that this explained the freer grinding properties of certain steels. He also indicated that the presence of a flat on the grit inhibited chip formation, as the interference between grit flat and work reduced the intensity of pressure at the critical chip-forming stage (Fig. 11.11).

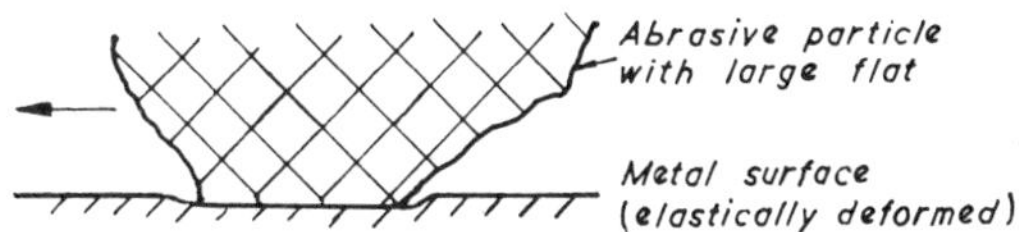

Fig. 11.11 Worn abrasive unable to cut

The magnitude of grinding forces can be related to the specific energy.

Power $= F_t V = bdvu$

$$F_t = \frac{bdvu}{V}$$

where F_t is tangential force on work,
V is wheel speed,
b is wheel width,
d is depth of cut.
v is workspeed (or rate of work feed f, where appropriate),
u is specific energy of the material.

The specific energy for grinding has been found by Backer, Marshall and Shaw[48] to vary as follows:

when maximum chip thickness $t > 1\ \mu m$ (40×10^{-6} in), $u = k_1/t^{0.8}$ (11.10)

when maximum chip thickness $t < 1\ \mu m$ (40×10^{-6} in), $u = k_2$ (11.11)

The value of u for grinding can be related to the value of specific energy for turning. For instance, $k_1 = 0.0092\ u_t$ and $k_2 = 30\ u_t$, where u_t is the specific energy for turning similar material using a feed of 0.25 mm rev^{-1} (0.010 in/rev) and a 15° tool rake angle; for mild steel the value of u_t is 2.27 J mm^{-3} (0.33×10^6 in lbf/in^3).

11.2.4 Effective wheel hardness. Perhaps the most generally recognized characteristic of wheel hardness is the ability of the wheel to retain dulled abrasive grains: the duller the retained grains, the harder the wheel.

Colwell, Lane and Soderlund[54] have pointed out that the hardness of a wheel in service can be different from its graded hardness. When a wheel is changed to a less rigid machine it appears to cut in a harder manner as it is now a relatively stiffer component in a system of overall lower stiffness. Resinoid wheels are harder in use than similar vitrified wheels, despite their greater elasticity (E for a resinoid wheel is approximately a quarter that for a vitrified one). This apparent paradox can be explained when it is realized that the cutting tool is the individual abrasive grain, not the wheel. The grain becomes relatively stiffer when set in a resinoid bond, as it is now operating in a system of lower all-over rigidity than when part of a stiffer vitrified wheel.

The ability of a wheel to retain its grains is influenced by the magnitude of the forces acting on it. The mean force F_c acting on a chip can be found from

$$F_c = \frac{\text{work done/chip}}{\text{chip length}}$$

$$\text{Work done/chip} = \frac{\text{Work done/min}}{\text{Number of chips/min}} = \frac{uvbd}{VCb}$$

and

$$F_c = \frac{uvd}{VCl} \tag{11.12}$$

From equations (11.8) and (11.9)

$$l = \sqrt{\left(\frac{Dd}{1 + (D/D_w)}\right)} \text{ and } t = \left[\frac{4v}{VCr}\sqrt{\left(\frac{d[1 + (D/D_w)]}{D}\right)}\right]^{\frac{1}{2}}$$

Hence $l = \dfrac{4vd}{VCrt^2}$

Substituting for l in equation (11.12), $F_c = \dfrac{urt^2}{4}$

Equations (11.10) and (11.11) show that for chip thickness $> 1\ \mu m$ (40×10^{-6} in) $u \propto (1/t^{0.8})$ but for smaller thicknesses, u is unaffected by t.

$$\therefore \text{ when } t < 1\ \mu m\ (40 \times 10^{-6} \text{ in})\ F_c \propto t^2 \tag{11.13}$$

$$t > 1\ \mu m\ (40 \times 10^{-6} \text{ in})\ F_c \propto t^{1.2} \tag{11.14}$$

It will be seen from equations (11.13) and (11.14) that the forces on the grains are considerably influenced by chip thickness. An increase in t will make the wheel act softer since the dislodging and fracturing forces on grains will be greater. Reference to equation (11.9), i.e.

$$t = \left[\frac{4v}{VCr}\sqrt{\left(\frac{d[1 + (D/D_w)]}{D}\right)}\right]^{\frac{1}{2}}$$

shows that a wheel can be made to cut softer by:

(*a*) increasing work speed,
(*b*) increasing depth of cut,
(*c*) reducing wheel speed,
(*d*) decreasing the number of active cutting grains, i.e. by using a larger grain size or a wheel of more open structure,
(*e*) decreasing the diameter of the workpiece. A given grade of wheel will appear to be harder in surface grinding than in external cylindrical grinding and hardest in internal grinding (negative curvature). For this reason comparatively soft wheels are used for internal grinding.

11.2.5 Temperature in grinding. The temperature reached by the tip of the abrasive particle when cutting is extremely high, and is thought to be in excess of the melting point of steel (1500°C). However, no melting occurs because of the brief time of contact, often less than 100×10^{-6} s. There is no serious heating of the wheel, as the temperature gradient in the cutting grains is very steep and there is ample cooling time between cuts (ratio of contact to cutting time 1:400 for surface grinding with 200 mm (8 in) wheel and 12·5 μm (0·0005 in) depth of cut). The ground surface can be affected to a depth of 100 μm or more by the thermal and mechanical shock it receives from successive grains. Residual tensile stresses are developed at the metal surface. If severe, these can cause grinding cracks in hardened steels, cemented carbide and high-temperature alloys. The heat from grinding can also change the microstructure of the material. If the temperature is high enough, a layer of austenite will be formed. This is rapidly quenched to form martensite by the mass of cold around it. The thermal and mechanical stresses produced by grinding can also cause warping of thin workpieces.

The approximate theoretical mean chip/tool interface temperature θ_t is given by

$$\theta_t = Ku\sqrt{\left(\frac{Vt}{k\rho C}\right)} \qquad (11.15)$$

where K is a constant
u is the specific energy of grinding
V is wheel speed
t is grain depth of cut
k is thermal conductivity of the work
ρC is volume specific heat of workpiece.

By using values of u from equations (11.10) and (11.11) it can be deduced from equation (11.15) that

$$\theta_t \propto V^{0.5} t^{0.5} \text{ when } t < 1\ \mu\text{m } (40 \times 10^{-6} \text{ in})$$

$$\theta_t \propto \frac{V^{0.5}}{t^{0.3}} \text{ when } t > 1\ \mu\text{m } (40 \times 10^{-6} \text{ in})$$

Hence for fine grinding, $t < 1\ \mu$m (40×10^{-6} in), the chip/tool temperature can be reduced by decreasing both wheel speed and chip thickness. For normal grinding, $t > 1\ \mu$m (40×10^{-6} in), temperature can be reduced by lowering wheel speed but not by decreasing chip thickness, in fact thermal damage can result from light finishing cuts.

A fluid is used in most grinding operations, although off-hand grinding and some surface grinding is done dry. Pahlitzsch[55] found considerable advantage in supplying oil radially outwards through the pores of the grinding wheels and a water-based coolant to the outside of the wheel. Not only did wheel wear decrease but there was less loading of the wheel by workpiece material, which substantially reduced the frequency of wheel dressing.

Fluids will also reduce workpiece temperature but they cannot prevent surface damage to the workpiece owing to the rapidity with which the heat generated at the work surface travels inwards.

11.2.6 Surface finish. The surface finish obtained by grinding is usually considerably better than that resulting from turning or milling. Usually the surface finish is between 0·75–0·1 μm (30 and 4 μ in) although smoother finishes are possible. The smoother the finish the more expensive it is to produce; hence the coarsest acceptable surface finish should be specified by designers.

The topography of a ground surface will vary with the direction in which it is measured (Fig. 11.12). The distribution of cutting grains on the wheel surface will have a major effect on surface finish; with small abrasive size and dense structure, the spacing will be close and likely to produce a smoother surface. For a given wheel, the finish can be improved by reducing work speed and increasing wheel speed.

Apart from geometrical considerations, if a good surface finish is needed much depends on using a rigid machine, a well designed spindle bearing, and an accurately balanced wheel. When very low micro-inch surface finishes are required, effective coolant filtration is necessary to prevent the recirculation of abrasive fragments, which are liable to be picked up by the wheel and severely scratch the surface of the work. Wheel dressing can also affect the roughness of the ground surface. Slow traversing of the dressing tool, small dressing depths, and a blunt diamond should be used if a fine surface is required on the workpiece.

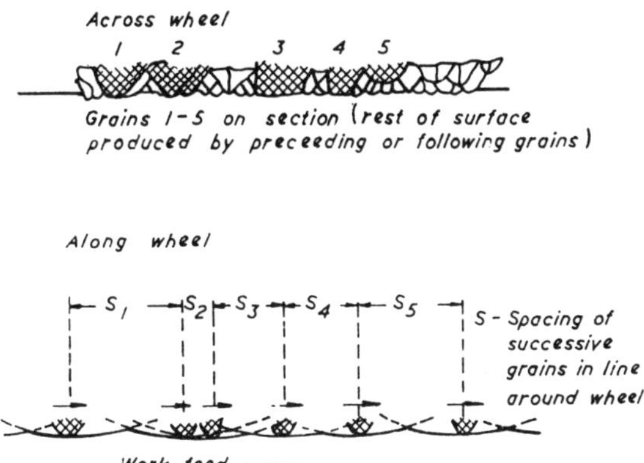

Fig. 11.12 Production of ground surfaces

11.2.7 Wheel wear. Grinding wheels wear more rapidly than most other metal cutting tools, although they normally suffer far greater size reduction by dressing than by wear in contact with the work. The generally accepted parameter of wheel wear is the grinding ratio, which is the ratio of volume of metal removed per unit volume of wheel worn away. The wheel wear curve (Fig. 11.13) shows three distinct phases. In the first the newly dressed wheel wears quickly as the sharp edges of the abrasive grains are rapidly worn away. The second and major phase is one of steady and slower wear, and in the final phase the rate of wear is accelerated as whole grains are lost from the wheel. The third stage will not be reached if the wheel glazes in phase two, as often happens with hard wheels.

A grinding wheel may wear from a variety of causes, these are bond fracture, grain fracture and gradual wear. Gradual wear occurs continuously in grinding and causes flats to develop and grow on the abrasive grains. These flats reduce the cutting ability of the wheel, and may eventually result in grains ceasing to cut. There are likely to be many causes of gradual wear. One of those isolated is the chemical reaction occurring between the tip of the grain and the metal being cut. Although abrasive and workpiece materials are not reactive at normal temperatures, Goefert and Williams[56] have shown that diffusion occurs when they are heated to the temperatures reached in grinding. Particularly severe chemical wear occurs when titanium is ground, but it has been found by Yang and Shaw[57] that this can be reduced considerably by the choice of a suitable grinding fluid. Bokuchava[58] considers that gradual grain wear is caused by plastic flow of the abrasive and that this type of wear depends on the temperature at the abrasive/chip interface and the heat penetration into the body of the grain. The depth of penetration increases considerably with contact time, the contact time being inversely proportional

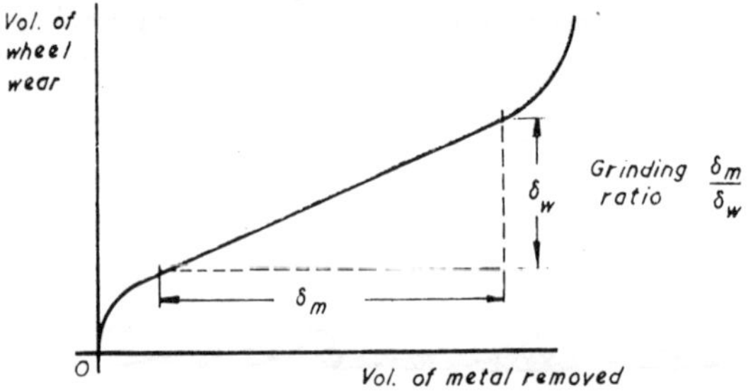

Fig. 11.13 Wheel wear plotted against metal removed

to wheel speed. Due to the higher conductivity of silicon carbide, the heat penetration is 30% greater than with aluminium oxide.

Wear also occurs by grain fracture, but unlike gradual wear this produces new cutting edges and improves the cutting ability of the abrasive grains. Grain splitting does not occur in any set pattern. Sometimes the grain splits into several fragments, perhaps leaving only a small proportion of the original grain in the wheel, or at the other extreme only a very small fragment is lost, leaving the grain shape substantially unaltered. Apart from the natural tendency of some abrasives to split more readily than others when subjected to mechanical stresses, splitting is also induced by thermal shock. It has been reported by Harada and Shinozaki[59] that grain fracture is more likely to occur in wet grinding than in dry, and the more effective the coolant the greater the likelihood of fracture.

The third type of wheel wear is bond post fracture in which whole grains are forced from the wheel. Yosikawa[60] assumed brittle fracture in both bond and grain of vitrified wheels and showed that the probability of both types of failure was approximately proportional to $t'e^{f}$, where t' is the grain workpiece contact time and f the force on the abrasive.

11.2.8 Dressing abrasive wheels. Dressing removes the outer layers of the cutting surface of the wheel, thereby creating a new cutting surface from a previously blunted or loaded surface. Truing, which is normally combined with dressing, produces an accurate grinding wheel profile. Too much dressing unnecessarily increases grinding wheel usage and lengthens production time; too little dressing is likely to reduce the rate of metal removal, cause thermal damage to the component and adversely affect its accuracy. The ideal dressing frequency is affected by the following factors:

(*a*) Wheel hardness—soft wheels rapidly lose their shape and need to be dressed to re-establish their geometry. Hard wheels are liable to glaze and to restore their cutting qualities they have to be dressed.
(*b*) The metal being ground—soft materials can clog the spaces between the grits resulting in loading of the wheel. Other metal, such as nimonic alloys, rapidly blunt the cutting edges of the grits and cause glazing.
(*c*) The force on the grits—if cutting forces produce minor grit fracture the blunted cutting points are regenerated. However, if whole grits are forced out of the wheel the surface finish and the geometric accuracy of the part are adversely affected.

The main methods of dressing are described below.

Diamond point. Here a suitably shaped and set diamond is taken across the periphery of the revolving wheel, cutting through both grits and bond. The traverse speed of the diamond is significant. A fast diamond traverse produces an open wheel structure capable of high rates of metal removal, a slow traverse speed results in a close structure and a fine surface finish. It is recommended that a feed rate should be chosen so that each grit is cut twice by the traversing diamond point.[61]

Where a profile has to be generated on the wheel the path of the diamond is controlled by a template, either directly or by means of a pantograph system. A chisel-shaped diamond can be used for simple radii and angles; cone-shaped diamonds are employed for more complicated shapes, such as thread forms.

Diamond clusters. Instead of a single diamond a number of stones are arranged in a holder in a variety of circular or rectangular patterns. These are used to dress large plain wheels. Although lacking the dressing precision of a single point diamond, the use of a diamond cluster does not require such a high level of skill; it is also economical in use.

Crush dressing. A cylindrical roll is brought into contact with the grinding wheel and both rotate without slip at slow speed. The roll, which is of the correct profile, is radially fed into the grinding wheel until the full profile has been formed. The roll material varies from cast iron for small batch quantities, through high speed steel to carbide, as batch quantities increase. Where very large production quantities of the same part justify the cost, diamond coated metal rolls are used. A diamond roll will perform thousands of dressing operations before it is worn out; the dressing is completed in a few seconds and is ideal for automatically redressing profiled wheels.

11.3 HONING

Honing is a finishing operation in which abrasive stones produce surface finishes of the order of 0·05–0·125 μm (2–5 μin). The stones, which are sometimes called sticks, consist of aluminium oxide or silicon carbide abrasive grains (size 30–600) joined by a vitrified, resinoid or shellac bond. Coolants are used to prevent the surface of the stones from being loaded by minute particles of metal and to cool the workpiece.

The main application of honing is to produce internal cylindrical bearing surfaces; usually the surface has already been accurately ground and often only a few ten thousandths of an inch of metal are removed. For internal honing, three or more stones are held in shoes mounted on a metal framework. The shoe positions can be adjusted radially to produce the correct hole diameter. The stone pressure varies between 0·35–0·70 $N\,mm^{-2}$ (50 and 100 lbf/in^2) and the framework is rotated to give the stones a cutting speed of about 1 m s^{-1} (200 ft/min). The framework reciprocates axially at a speed of about 0·25 m s^{-1} (50 ft/min) and the machine is set to give a slight overrun and dwell at the end of each stroke.

11.4 SUPERFINISHING

This process is closely related to honing, similar stones being used to produce very smooth surfaces. The stones are, however, given a small axial oscillation (maximum 5 mm (0·2 in) at about 40 Hz). Superfinish is normally used on external cylindrical surfaces, not internal ones as with honing. Pressures are about 0·07 N m^{-2} (10 lbf/in^2) and the work is rotated at low speed, so there is little risk of thermal damage to the workpiece; in fact superfinishing and honing are often used to remove the thermally damaged surface layer left after grinding. Paraffin and transformer oil are used as cutting fluids.

11.5 LAPPING

Unlike the abrasive processes previously described, lapping employs free abrasive particles which are rubbed against the surface to be machined. Lapping can be used to finish both flat and curved surfaces. The abrasive particles are carried in a liquid or paste and usually applied to a wood or soft metal lap of the appropriate shape. Relative motion between the lap and the work removes minute particles of metal, slowly reducing the size of the workpiece and improving its surface finish. Aluminium oxide is used to lap soft metals, silicon carbide for hardened steels, and diamond powder is used for very hard materials such as tungsten carbide.

Hand lapping is slow and expensive, and where possible is replaced by machine lapping in which a machine is used to produce relative movement between lap and work.

No lap is needed when close conformity is sought between two mating surfaces: in this instance one part acts as a lap on the other, e.g., the pistons and cylinders of fuel pumps for gas turbines and diesel engines. After lapping, the workpiece must be thoroughly cleaned, otherwise lapping will continue in service and premature wear will result.

Although lapping is to be avoided if possible, it is still a necessary process in gauge making and in the production of leakproof metal-to-metal surfaces.

11.6 ABRASIVE BELT MACHINING

This method of machining grinds and polishes. The work is usually hand held against a rotating abrasive belt; it can, however, be mechanically fed and when quantities are large automatic multi-headed machines can be employed. The cutting speeds used are of the same order as those in conventional grinding.

A small backstand machine is shown in Fig. 11.14. The abrasive belt can be tensioned by a simple mechanical adjustment of the idler pulley or by a variety of devices operated by springs or compressed air. Abrasive belts can be obtained in a wide variety of grit sizes, varying typically between extremes of 24 and 320. Small grit sizes produce fine surface finishes; the larger grits give a more rapid rate of metal removal. Aluminium oxide is the most common abrasive medium—the belts are made from woven fabric to which the abrasive is secured by a bonding material. Contact wheels are of two main types. The first is a metal disc rimmed with compressible material, such as polyurethane or rubber of specified hardness. The rim can either be solid or serrated: serrations intensify the grit pressure on the work and produce an aggressive cut. The second

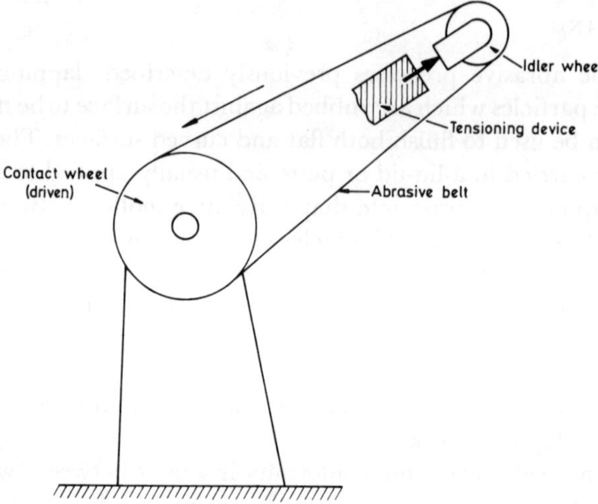

Fig. 11.14 Simple backstand for abrasive belt machining

type of contact wheel is a cloth mop: this provides a much softer support and is more suitable for polishing than for high rates of metal removal. Although flat metal surfaces can be precision ground, abrasive belt machining does not generally produce a high geometric accuracy.

Machines using abrasive belts are relatively inexpensive; belts do not wear rapidly and when worn can be quickly replaced.

12 Recently Developed Techniques of Metal Working

12.1 A range of new techniques has recently been developed, mainly to cope with (*a*) the growing use of materials which are difficult to machine, and (*b*) miniaturization in the electronics industry.

Most of the metal removal techniques to be considered would be uneconomic for machining conventional engineering materials, but the use of new materials has in many cases demanded new shaping processes to supplement grinding, the most usual method of machining hard materials. The techniques described in this chapter are not exhaustive but represent the more important of the new processes.

12.2 ULTRASONICS

Vibrations transmitted through solids or liquids at frequencies in excess of the sonic range can be used for a number of engineering purposes, such as non-destructive testing, cleaning, soldering, welding, machining and electroplating. The frequencies used are generally in the range 16 kHz to 35 kHz.

There are four methods by which vibrations of these frequencies can be produced, using piezoelectric, magnetostrictive or electromagnetic effects, or jet generators. For engineering applications, the first two only are of importance.

12.2.1 Piezoelectric transducers. Some crystals, such as quartz, undergo dimensional changes when subjected to electrostatic fields. The magnitude of these changes for any given crystal is proportional to the applied voltage. If, then, an alternating voltage of the required frequency is applied to the crystal, a vibration will result. The amplitude of the vibration will depend on the resonant frequency of the crystal, so in order to produce a high amplitude the length of the crystal must be matched to the frequency of the generator. To produce resonance the length of the crystal must be half the wavelength of sound in the material.

For quartz, Young's modulus $E = 5{\cdot}2 \times 10^{10}$ N m^{-2} and the density $\rho = 2{\cdot}6 \times 10^{3}$ kg m^{-3}.

$$\text{The velocity of sound } c = \sqrt{\frac{E}{\rho}}$$

$$= 4480 \text{ m s}^{-1} \text{ for quartz.}$$

At a frequency of 20 kHz the wavelength of sound produced $= c/f = 0{\cdot}228$ m. It follows that a crystal used as a transducer at this frequency will be about 110 mm (4·5 in) long. Usually polycrystalline ceramics such as barium titanate are used, for which the required length to give maximum amplitude is 75–100 mm (3–4 in). Practical difficulties preclude the use of large crystals, so sandwich type transducers have been designed (Fig. 12.1).

Conservation of momentum demands that the velocity and hence the amplitude of vibration in the high and low density materials should be inversely proportional to the density, giving a high amplitude of vibration at the radiating face.

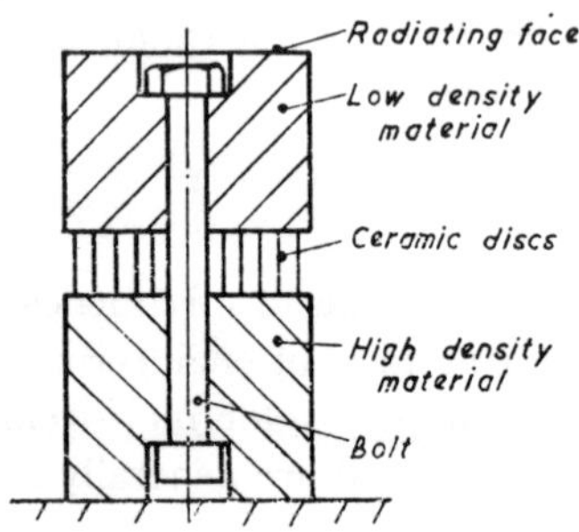

Fig. 12.1 Section through sandwich type piezoelectric transducer

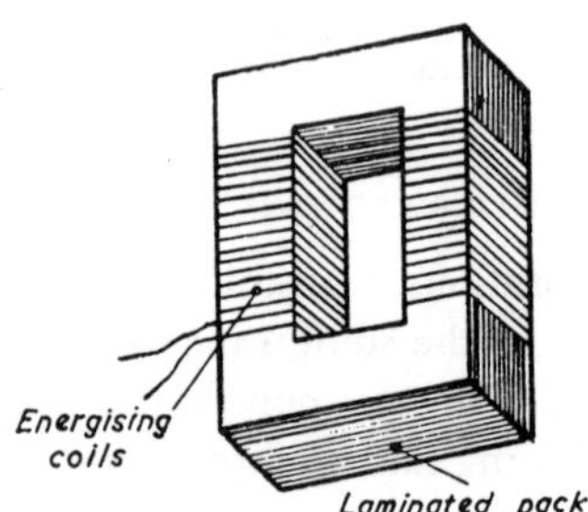

Fig. 12.2 Magnetostrictive transducer

12.2.2 Magnetostrictive transducers. If a piece of ferromagnetic material, such as nickel, is magnetized, a change in dimension occurs. This property is used in magnetostrictive transducers. To reduce eddy current losses, the ferromagnetic material is generally in the form of insulated laminations assembled into a pack (Fig. 12.2).

Electric current of the desired frequency is fed to the energizing coils. Heat is generally dissipated by fitting a water jacket to the transducer.

12.2.3 Ultrasonic machining. Machining is performed by the vibration of a shaped tool tip in an abrasive slurry which forms a cavity of the required shape in a workpiece. A light static load, of about 18 N (4 lbf) holds the tool tip against the workpiece and material is removed by the chipping action of the abrasive particles trapped between the tip and the workpiece. Vibration of the tool tip accelerates the abrasive particles at very high rates and imparts the force necessary for the cutting action. Cavitation occurs in the gap and assists the disposal of chips and the circulation of the abrasive.

Magnetostrictive transducers are usually used, and to increase the amplitude of vibration a velocity transformer is rigidly attached to the

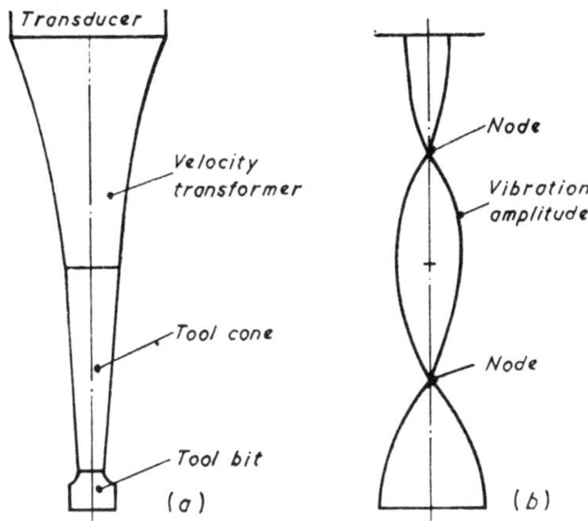

Fig. 12.3 Velocity transformer and tool

radiating face of the transducer. Velocity transformers are made of metals which have high fatigue strength and low energy loss, such as brass, and are usually exponentially tapered (Fig. 12.3), the ratio of the increase in amplitude being inversely proportional to the ratio of the areas of the two ends, as will be apparent from consideration of conservation of momentum.

The tool cone is attached to the end of the transformer and the tool bit is attached to the end of the cone. To give minimum damping the vibrating parts are clamped at the nodes and the various components are acoustically matched. Generally, the best results are obtained using tool bits of alloy steel. Tool wear leads to a shortening in the length of the combined tool cone and tip, so it is usual to make them slightly longer than half a wavelength to increase their effective life.

The abrasive slurry usually consists of water mixed with an abrasive which may be boron carbide, silicon carbide or aluminium oxide. Boron carbide is the hardest and most expensive and is used for machining tungsten carbide, die steels and gems. Silicon carbide is difficult to keep in suspension; aluminium oxide, the softest of the three, wears rapidly and is mainly used for glass and ceramics.

Soft materials are unsuitable for ultrasonic machining because of their damping properties, the process being reserved for hard, brittle materials. Unlike most of the other new metal-removal processes described later, the work material can be non-conducting.

It is difficult to give an accurate idea of the limitations of the process, but under ideal conditions penetration rates of 5 mm min^{-1} (0·2 in/min) are claimed. Power units are usually of 50–2000 watt output although much higher outputs can be used. The specific metal removal rate on brittle materials is 0·018 $mm^3 J^{-1}$ (0·05 in^3/hp min). Amplitudes of vibration of up to 50 μm (0·002 in) can be achieved by suitable design. Normal hole tolerances are about 25 μm (0·001 in), although this can be bettered, and a surface finish of about 0·5–0·7 μm (20–30 μ in) is common.

A recent development in ultrasonic machining by the Ceramics Division of the Atomic Energy Research Establishment combines an ultrasonic axial vibration with a rotation of up to 1500 rev/min. The abrasive grains have been replaced by a diamond-impregnated or electro-deposited tool, vibrated by a piezoelectric transducer. A large variety of forms, including slots, annular grooves and internal and external threads, can be generated by rotation of the tool. Holes from 0·75–12·5 mm (0·030 to 0·500 in) can be sunk to a depth of at least 25 mm (1 in). Cutting speeds are relatively high, and holes 10 mm ($\frac{3}{8}$ in) in diameter and 10 mm ($\frac{3}{8}$ in) long, can be cut in alumina in less than 4 min. This development greatly extends the range of work for which ultrasonic machining can be used.

12.2.4 Ultrasonic cleaning. Ultrasonic vibration in a liquid causes rapid alternating pressures, which produce voids in the liquid. These voids are subject to catastrophic collapse, generating shock waves of pressure amplitude of about 100 N mm^{-2} (1000 atmospheres). When collapse occurs near a soiled surface and the liquid is a suitable cleaning fluid, any oil film on the surface is emulsified and dirt particles are dislodged. Piezoelectric transducers are generally used, fixed in banks to the bottom of the cleaning tank.

Cleaning fluids, must have good wetting properties, and can be trichlorethylene, carbon tetrachloride or water soluble detergents. Detergents are safer, more effective and less expensive.

12.2.5 Other engineering applications of ultrasonics. Cavitation effects produced by ultrasonic vibration can be used to disperse the oxide film on aluminium, enabling it to be soldered. This can be done in a bath of molten solder, or alternatively an ultrasonically vibrated soldering iron can be used.

Welding of dissimilar metals in the form of thin sheet or foil can be performed by lightly clamping the two pieces to be joined between a welding tip and an anvil, and ultrasonically vibrating the welding tip parallel to the metal surface for between 0·5 and 2·0 s. In this way it is possible to join such dissimilar metals as aluminium, copper and stainless steel, although the most important application to date appears to be the joining of thermoplastics. By suitably arranging the tool, lap, butt, or tongued and grooved joints can be made between plastic components of comparatively large cross section, vibrating the tool normal to the surface of the plastic sheet.

It is thought that when applied to metals the welding is brought about by molecular transfer between the contacting surfaces. Certainly, the joint does not exhibit either the cast structure of fusion welds or the deformation pattern of pressure welds. However, when applied to thermoplastics, the hammering and sliding action is sufficient to cause fusion.

Vibrations have been applied to a number of metal cutting operations with interesting results. On lathes, the cutting tool has been experimentally vibrated in the direction of the cutting vector.[62] When cutting creep-resistant alloys it was found that tool life was improved with low intensity vibrations of the order of 1 kW output. Higher intensity vibrations led to a reduction in tool life. Cutting forces were reduced considerably at speeds of up to 0·25 m s^{-1} (50 ft/min), but the effect was less marked at higher speeds. Other advantages claimed were improved surface finish and increased effectiveness of the cutting fluid.

Cutting forces in end milling and tapping have been considerably reduced and tool life has been increased by applying ultrasonic vibrations along the axes of the tool. Apart from some work performed on copper and stainless steel there is little published information about savings resulting from axial vibration of multi-point tools. Colwell[63] showed that by vibrating work at 10–18 kHz while grinding a better surface finish was obtained, although wheel wear was more rapid.

12.3 ELECTROCHEMICAL MACHINING (ECM)

For several years electrolytic baths have been used for de-burring or polishing metal components. The components are immersed in the electrolyte as the anode, and the burrs or surface asperities are dissolved at a rate proportional to the current density and the electrochemical equivalent of the metal.

The successful use of electrolytic methods of controlled metal removal is of much more recent origin. A bath of electrolyte contains the workpiece, connected as the anode to a d.c. electric supply, and a shaped tool which is the cathode. As the cathode is advanced towards the workpiece, erosion of the work surface occurs at a rate which depends on the feed rate of the cathode. Holes of various shapes can be produced, or die cavities can be economically sunk into the surface of metals. There is no

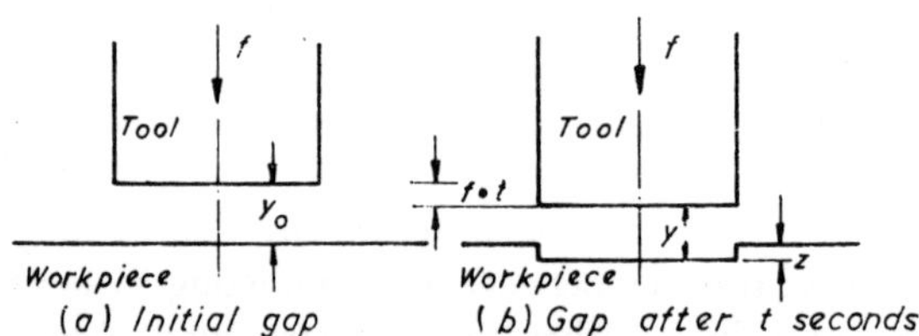

Fig. 12.4 Tool-workpiece gap

tool wear from electrolysis, although the electrolyte may cause some chemical corrosion. Current density in the gap between the workpiece and tool is from 0·4–8 A mm^{-2} (250–5000 amp/in^2), and an applied voltage of not more than 20 volts is used.

Although any conducting liquid could be used as an electrolyte, it has been found that when using neutral salts the eroded metal particles are insoluble and cause scavenging problems. Dilute acids are therefore generally used to dissolve the particles, although in a few cases alkalis are preferred. A great deal of hydrogen is evolved at the cathode, and unless a high-pressure flow of electrolyte is maintained across the gap to dissipate the gas, polarization rapidly occurs.

The following is a simplified approach to the evaluation of the process parameters, assuming a plane parallel gap and constant electrolyte conductivity, (Fig. 12.4). The symbols used are as follows:

y_0 initial gap between tool and workpiece (mm)
z amount of erosion of the workpiece after time t (mm)
f tool feed rate (mm s^{-1})
ρ density of work material (kg mm^{-3})
E electrochemical equivalent (kg s^{-1} A^{-1})
J current density in the gap (A mm^{-2})
y gap after time t (mm)
K conductivity of electrolyte (Ω^{-1} mm^{-1})
J_e current density in the gap under equilibrium conditions (A mm^{-2})

Considering unit area of tool, from the law of electrolysis,

$$\rho \cdot \frac{dz}{dt} = E \cdot J$$

From Ohm's law, with applied voltage V,

$$J = \frac{V \cdot K}{y} \tag{12.1}$$

$$\frac{dz}{dt} = \frac{E \cdot V \cdot K}{\rho \cdot y} = \frac{A}{y}, \text{ where } A = \frac{E \cdot V \cdot K}{\rho}$$

$$\frac{dz}{dt} = f + \frac{dy}{dt} \tag{12.2}$$

$$\frac{dy}{dt} = \frac{A}{y} - f \tag{12.3}$$

Integrating, $$\int_0^t dt = \int_{y_0}^{y} \left(\frac{y}{A - f \cdot y} \right) dy$$

$$t = \int_{y_0}^{y} \left(\frac{A}{f} \cdot \frac{1}{A - f \cdot y} - \frac{1}{f} \right) dy$$

$$t = \frac{1}{f}(y_0 - y) + \frac{A}{f^2} \ln \left(\frac{A - f \cdot y_0}{A - f \cdot y} \right)$$

Figure 12.5 shows how y varies with time, approaching an equilibrium value y_e asymptotically.
Under equilibrium conditions, $dy/dt = 0$.

Hence, from equation (12.3) $$y_e = \frac{A}{f} = \frac{E \cdot V \cdot K}{\rho \cdot f} \tag{12.4}$$

Also, from equation (12.2) $$\frac{dz}{dt} = f \tag{12.5}$$

and from equation (12.1) $$J_e = \frac{V \cdot K}{y_e} = \frac{\rho \cdot f}{E} \tag{12.6}$$

Equations (12.4, 5 and 6) show that the equilibrium gap is proportional to the applied voltage and inversely proportional to the feed rate, and that the equilibrium current density is proportional to the feed. In practice, K increases as the electrolyte temperature increases, so the equilibrium gap will increase as the electrolyte passes through the gap.

Hence to keep the gap substantially constant the electrolyte velocity must be kept as high as possible.

The difficulty of pumping the electrolyte through a small gap eventually results in polarization or in the electrolyte boiling, and this sets a lower limit to the gap size. Usually the gap is about 250 μm (0·010 in) and the electrolyte velocity is between 30–55 m s^{-1} (100–180 ft/s).

An increase in applied voltage causes an increase in current for a specified gap. Therefore if the feed rate is kept constant the gap is increased, or alternatively the feed rate can be increased without decreasing the size of the gap. Unfortunately the larger gaps associated with higher voltages lead to a decrease in accuracy of the form in the workpiece.

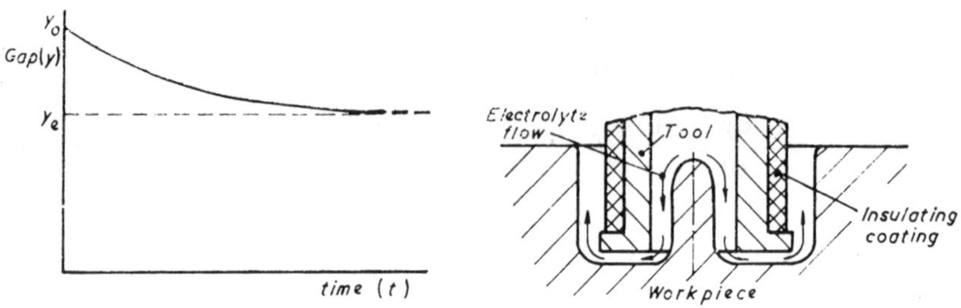

Fig. 12.5 Approach to equilibrium conditions

Fig. 12.6 Hollow tool with insulating coating

When drilling holes in components the tool is hollow so that the electrolyte can pass along the bore (Fig. 12.6). This method also reduces the area at the end of the tool where erosion is occurring. To prevent electrochemical action between the sides of the tool and the hole it is usual to relieve the tool, as shown, and to cover the relieved portion with an insulating coating. Typical feed rates are of the order of 40–130 μm s^{-1} (0·1–0·3 in/min) and a hole accuracy of about 0·8 μm (0·002 in) can be achieved.

Increase in the applied voltage, higher electrolyte conductivity and slower feed rates all tend to cause the hole to be oversize. An elegant analysis of the overcut size in relation to these variables was made by PERA in their report on ECM[64].

When die-sinking or -shaping by electrochemical means it is possible to produce a form on the workpiece which is very nearly the inverse of the form on the tool. Fig. 12.7 shows that the gap at A when using a curved tool will be greater than that at B, since the gap is inversely proportional to the feed rate normal to the surface. It is possible to make allowances for variations of gap if the radius of curvature on the tool is sufficiently large, but small form features or sharp corners cannot be accurately reproduced.

A variant of ECM, known as electrochemical turning, has been developed to permit metal to be removed from solids of revolution as an alternative to grinding. Essentially, the machine has the motions of a lathe, but the metal removal tool is a cathode which is separated from the rotating work surface (anode) by a film of electrolyte. A suitably shaped tool can produce a desired form on a hard metal by this means in a much shorter time than could be achieved by grinding.

Electrolytic grinding is another electrochemical process which is finding popularity for grinding very hard materials such as tungsten carbide. A metal wheel which has abrasive grains embedded in the surface

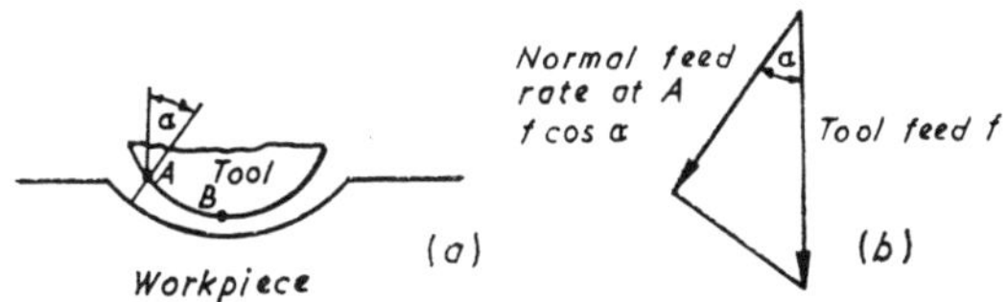

Fig. 12.7 Effect of tool curvature on equilibrium gap

is used. Cutting fluid, which acts as an electrolyte and is usually a cheap non-corrosive alkaline solution, is pumped into the space between wheel and workpiece, and the electric circuit is made by connecting the wheel to the negative terminal and the workpiece to the positive terminal of a d.c. source.

It is estimated that about 90% of the metal is removed electrolytically and 10% abrasively. This considerably increases the rate of metal removal compared with conventional grinding, whilst giving very low wheel wear. In addition, excellent surface finishes of about 0·05 μm (2 μ in) are obtainable.

12.4 ELECTRICAL DISCHARGE MACHINING (EDM)

When a difference of potential is applied between two conductors immersed in a dielectric fluid the fluid will ionize if the potential difference reaches a high enough value. A spark will occur between the conductors which will develop into an arc if the p.d. is maintained. If the p.d. decreases, however, the fluid will de-ionize and the discharge will cease. The temperature of the spark is between 5000 and 10 000°C, and results in local melting of the metals. It is observed that the erosion rate is greater for the conductor connected to the positive terminal. The reason for this is not known, but it is suspected that the electron bombardment of the positive electrode occurs at an earlier moment than the ion bombardment of the negative electrode, causing the rise in temperature at the anode to be greater than at the cathode.

This process is used for producing holes or cavities in hard metal components. It has become an accepted method for producing press tools and dies in hardened metals, thereby obviating distortion due to subsequent heat treatment.

A tool made of the desired form, and the workpiece to be machined, are immersed in a bath containing dielectric fluid and connected as the cathode and anode respectively. The fluid is usually paraffin, transformer oil or white spirit. These all have the essential dielectric properties, they de-ionize rapidly and do not vaporize excessively. The rate of de-ionization determines the maximum rate of sparking, which in turn limits the rate of metal removal.

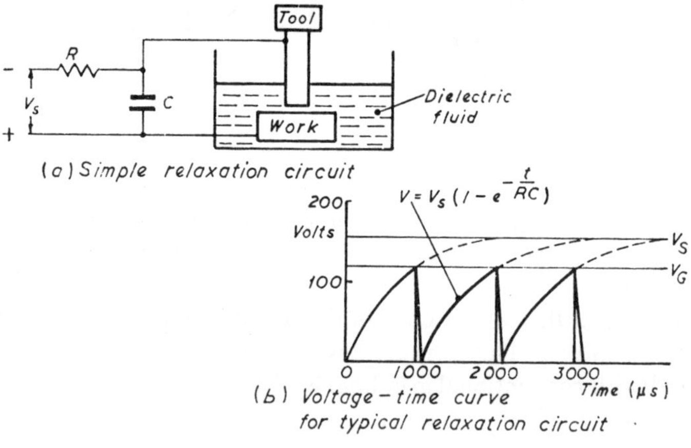

Fig. 12.8

A gap of 25–50 μm (0·001–0·002 in) is maintained between tool and workpiece by a servo-motor, actuated by the difference between a reference voltage and the gap breakdown voltage, which feeds the tool downwards towards the workpiece.

The required gap breakdown voltage can be achieved by using either a d.c. relaxation circuit or a pulse generator. A typical relaxation circuit is shown in Fig. 12.8 (*a*), the power source being either a d.c. generator or a rectified a.c. supply. The capacitor C is charged through the ballast resistance R, giving a charging curve as shown in Fig. 12.8 (*b*). When the voltage across the capacitor reaches the gap breakdown voltage V_G the capacitor discharges across the gap. The sparking rate is limited by the time constant $C \cdot R$, of the circuit, except for very small values of time constant, when the limitation is the de-ionization rate of the dielectric fluid. By suitable arrangement of the parameters, spark rates approaching 10 000/s are possible. Metal-removal rates depend on the power dissipated.

Energy per discharge $= \frac{1}{2}CV_G^2$ joules.

Power consumed $= \frac{1}{2}C \cdot V_G^2 \times$ frequency of sparking $= \dfrac{CV_G^2}{2t}$, where $t =$ charging time.

Substituting for V_G, power consumed, $W = \dfrac{V_s^2 \cdot C}{2t}(1 - e^{(-t/RC)})^2$

$$\frac{dW}{dt} = \left\{\frac{1}{t}(2e^{(-t/RC)} - e^{(-2t/RC)} - 1) + \frac{2}{Rc}(e^{(-t/RC)} - e^{(-2t/RC)}\right\}\frac{V_s^2C}{2t}$$

For maximum power, $dW/dt = 0$. This is satisfied when $t/(RC) = 1{\cdot}2$, and hence $V_G = 0{\cdot}73\ V_s$, where V_s is the supply voltage.

Although the discharge current from a relaxation circuit reaches a high value it is of very short duration, preventing the full erosion effects of the high temperature from being achieved. A better current–time distribution is obtained by using a pulse generator. These are of two types, one having pulse trains independent of machining performance, the other, known as an isopulse generator, having the pulse time adjusted by means of a feedback device which monitors the start of the spark and thereby ensures a constant energy discharge per cycle. When using an independent pulse generator, delay in the onset of ionisation results in variation in the length of the current pulse, Fig. 12.9 (*a*). Isopulse generators enable the voltage pulse time to be varied so that the current pulse is of constant duration, Fig. 12.9 (*b*). Their use ensures better control of electrode wear as well as improved machining rates and well-defined surface finish.

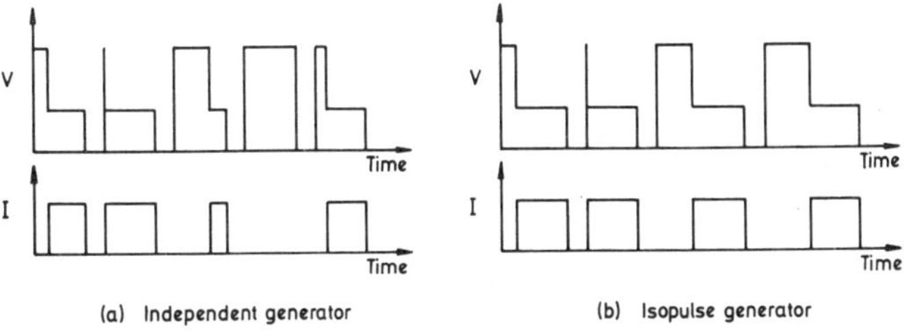

Fig. 12.9

Relaxation generators are now largely superseded by pulse generators for all applications except the production of fine surface finishes and special machining operations with small electrodes.

Electrodes can be manufactured from any conducting material such as brass or copper, but where high accuracy and long electrode life are required higher melting point materials such as copper–tungsten, graphite or tungsten carbide are used. To prevent taper on finishing operations due to sparking at the sides of the electrode when producing holes the sides of the electrode are usually relieved.

The surface finish produced on components has a matt, non-directional appearance, similar to that obtained by shot-blasting, and is particularly suitable for retaining lubricants or for subsequent polishing. The heat from the discharge affects a zone lying between 20–100 μm (0·0008–0·004 in) below the machined surface.

High metal-removal rates lead to poor surface finish, and the selection of parameters is usually a compromise between these conflicting requirements. Maximum metal-removal rates of 80 $mm^3 s^{-1}$ (0·3 in^3/min) can be achieved, or surface finishes of 0·25 μm (10 μ in) can be obtained at very low cutting rates. High rates of metal removal and a good surface finish are possible by taking a roughing and then a finishing cut, using two electrodes. By careful selection of parameters tolerances approaching 0·04 μm (0·0001 in) can be achieved.

A variant of the EDM process makes use of a tensioned travelling wire electrode passing through a workpiece up to 100 mm (4 in) thick. Travelling wire machines are equipped with numerical control, so that contours can be programmed. They also have a facility which enables a suitable off-set to be inserted in a program to allow for the wire radius and overcut due to the spark gap. The wire is tensioned on spools which rotate so that the effect of electrode wear is eliminated. By this means extremely accurate profiles can be machined, suitable for the production of press tools for fine blanking.

12.5 ELECTROFORMING

Electroforming is closely akin to electroplating, and consists of depositing a thick skin of metal on an electrically conducting former. There is no practical reason why any metal should not be electroformed although most applications involve the use of copper or nickel. The low deposition rates, between 25–100 $\mu m\ h^{-1}$ (0·001–0·004 in/h) are a disadvantage, although the basic equipment is not particularly expensive and can be easily duplicated.

Formers may be either metallic or non-metallic. If the former material is non-conducting it is necessary first to metallize the surface by vacuum deposition to render it conducting.

The printing industry has used the technique for many years to form 'electros', as has the recording industry for the manufacture of gramophone record masters. More recent applications are found in toolmaking, and in the electronics and aerospace industries, and many consumer goods such as car accessories and mirrors are now produced by electroforming.

Electroformed components cover a range of sizes, from grids as fine as 30 lines mm^{-1} (750/in) for use in the electronics industry, to nose cones and nozzle liners for the aerospace industry which may be up to 3·6 m (12 ft) long. Nickel moulds for the manufacture of plastics can be economically produced by electroforming. Another important use of electroforms is for electrodes used in ECM and EDM.

12.6 CHEMICAL MILLING

This term embraces a variety of techniques whereby material is selectively removed by etching from the surface of a component. The surface is first coated with a photosensitive resist film; a photographic negative, in which the areas to be removed appear black, is placed in contact with the surface and then exposed to ultraviolet light. When the resist film on the surface is subsequently developed, the portions which have been exposed are resistant to most etching solutions, whilst the unexposed portions are easily etched.

Chemical milling is used extensively for the production of small metal components by etching away the portions not required. A more important application is for printed circuits where the basic connexions of the circuit consist of thin metal strips attached to an insulating base. These circuits are produced from insulating board faced with a thin layer of copper. The copper is coated with photosensitive resist and an image of the required circuit is printed photographically on the surface. Etching removes all the unwanted copper, leaving the circuit standing proud from the insulating base. Line thicknesses can be controlled to less than 0·001 in., and this has led to the development of photo techniques for the manufacture of strain gauges and micro-miniature circuits. Frequently micro-circuits are etched on silicon wafers of about 1 mm^2 area, and without photo-etching these developments would have been impossible.

13 Fabrication by Welding, Brazing or Adhesion

13.1 In this chapter it is not proposed to discuss the methods of temporary or semi-permanent attachment of materials such as bolting or riveting, but to confine attention to the more permanent joints produced by welding, brazing or adhesives.

The majority of high-strength joints between metals are made by welding, which is a term used to cover a wide range of bonding techniques. For convenience the techniques have been grouped under two main headings, fusion and solid-phase, although, as will be seen, some techniques such as flash butt welding do not fit exactly into either group.

Fusion welding techniques involve placing together the parts to be joined and heating them, often with the addition of filler metal, until they melt. The joint is made by the subsequent solidification of the metal. Solid-phase welds are produced by bringing the clean faces of the components into intimate contact to produce a metallic bond. This may be performed with or without the addition of heat, although the application of pressure to induce plastic flow is essential, except for diffusion welding.

Brazing and soldering are processes involving the joining of the components by means of a bridging metal or alloy of lower melting point, melted between the two surfaces to be joined. Almost invariably the strength of the joint is lower than that of the parent metals. Brazing or soldering is frequently used if the parts may subsequently have to be separated, as it is then necessary to melt only the jointing metal.

The widespread use of industrial adhesives has resulted from the development of plastics as engineering materials and the necessity of joining them. Adhesives can also be used to join a metal and a non-metal, or two metals.

13.2 FUSION WELDING AND METAL CUTTING

The processes in this group are distinguished by the method of supplying heat. Often a close study of the advantages and disadvantages of each method is necessary before a decision can be made on the best process for a particular application. Most of the processes are well established, but in recent years new techniques have been developed, mainly to join either very large or very small sections.

13.2.1 Electric arc welding. The heat generated by an electric arc produces temperatures in excess of 5000°C in the immediate area of the arc. The arc may be struck between two carbon or tungsten electrodes positioned close to the surfaces to be joined, or alternatively between the parent metal and an electrode of carbon, tungsten or another metal with a composition similar to that of the parent metal. The last-mentioned method is the most popular, the electrode melting and acting as a filler rod to fill the cavity between the two parts. When carbon electrodes are used, they burn to give a protective shield of gas round the arc, but with tungsten or consumable metal electrodes the weld surface must be protected from corrosion by other means which will be discussed later. Because of their high melting point, tungsten electrodes are virtually indestructible. When using carbon or tungsten electrodes it is necessary in most cases to provide the filler material by means of a separate rod.

A number of methods is available to protect the weld from oxidation and nitrogen embrittlement while at an elevated temperature. With consumable metal electrodes, protection can be obtained by coating or occasionally coring the electrode with a flux. The flux melts and partly vaporizes, forming a blanket of reducing gas around the weld. It also causes the molten metal globules travelling from the electrode to the weld pool to be smaller than when using bare electrodes, and assists in stabilizing the arc. When the flux solidifies on cooling, it forms a brittle slag which covers the weld. Deposition rates of about 1·6 kg h^{-1} (3·5 lb/h) can be achieved by this method. Addition of iron powder to the flux coating enables the weld rate to be increased by up to 80% without increase in current.

Another method of preventing atmospheric attack is to release an inert gas which will blanket the weld around the electrode. The shielding gases are usually argon or carbon dioxide, although in the U.S.A. helium is frequently used. The electrode may be either a tungsten rod, (tungsten–inert gas welding), or a consumable metal wire which is continuously fed from a spool to maintain the arc as the end of the wire melts (metal–inert gas welding). T.I.G. welding is used mainly for magnesium and other light alloys or for welding alloy steels and non-ferrous materials up to 12 mm thick, the gas shield being provided by argon or helium. When used manually the welding rates are lower than those possible with metal arc welding, although the subsequent operation of chipping off the slag is eliminated.

M.I.G. welding usually has CO_2 shielding and enables welding rates up to about 6 $kg\,h^{-1}$ (14 lb/h) to be achieved. Most M.I.G. welding uses the dip transfer technique, where the tip of the wire is allowed to touch the weld pool. The short circuit current causes a droplet to transfer to the weld pool, shortening the wire and re-creating the arc at a frequency of about 60–120/second. With this method it is possible to weld material thicknesses as low as 6 mm. An alternative method of transfer makes use of a

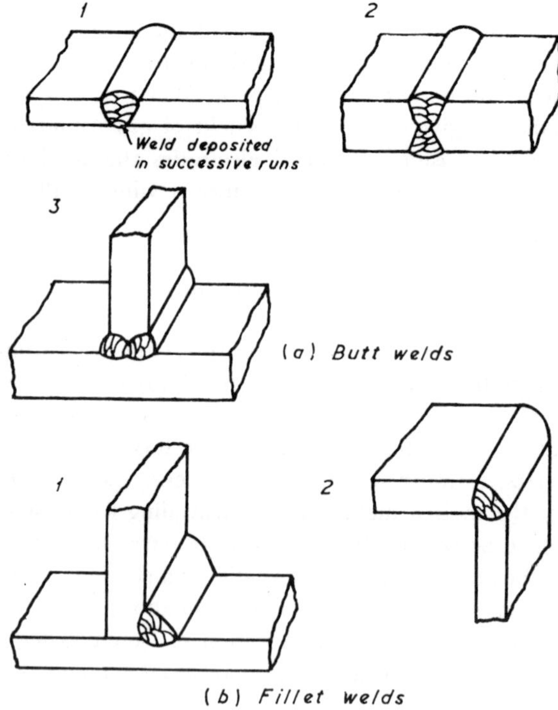

Fig. 13.1 Types of welds

pulsed voltage imposed on a steady voltage. This method gives droplet transfer without the need for short-circuiting. The welding rate is higher than for dip transfer and thinner sections can be welded. Unfortunately it requires an argon shield which is more costly than CO_2. Hence its main use is for low alloy steels, stainless steels and non-ferrous metals, the CO_2 process being more economical for carbon steel. The third technique, spray transfer, projects molten droplets in a continuous stream by electromagnetic forces. This enables a high welding rate to be achieved but results in a large weld pool which renders it unsuitable for vertical or overhead work.

The third method of weld protection is by depositing flux in granular form over the weld area. Heat from the arc causes the flux to melt and the arc is itself submerged in molten flux. This process is used with automatic machines and the high currents which can be employed give correspondingly high deposition rates, of the order of 7 kg h^{-1} (15 lb/h).

Manual arc welding is used mainly for butt and fillet welds (Fig. 13.1), and a skilled operator can produce strong joints over a wide range of work. It can be used successfully on large components since the intense localized heating produces fusion, whereas the heat from other manual welding methods would be dissipated before the metal melted. However, for simple repetitive work automatic arc welding machines, which frequently operate at currents in excess of 2000 amps, are much faster than manual welding, which is limited to a maximum current of about 600 amp.

13.2.2 Gas welding and flame cutting. In gas welding, the heat to produce fusion of the parent metal and filler rod is provided by burning a suitable gas in oxygen or air. A number of gases can be used but acetylene is the most popular, since it burns in oxygen and gives a high flame temperature of 3100–3200°C. Oxygen and acetylene stored in cylinders under pressure are passed through flexible tubes to the torch, which is either hand-operated or mechanically manipulated. By adjusting the proportions of oxygen or acetylene the flame can be neutral, or have either reducing or oxidizing properties. For most materials a neutral flame is used, but for welding high carbon steel, or aluminium and its alloys, where oxidation is a problem, a reducing flame is used. For brass welding, an oxidizing flame is used, as by these means the volatilization of the zinc is suppressed.

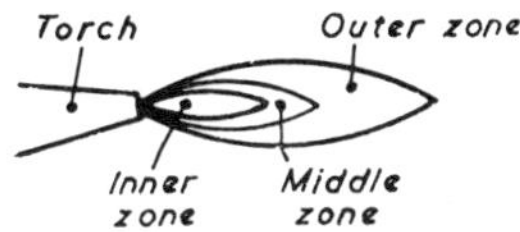

Fig. 13.2 Oxy-acetylene flame

The flame can be divided into three roughly conical zones (Fig. 13.2), the first being where the oxygen combines with acetylene to give a reducing atmosphere of carbon monoxide and hydrogen in the second zone, i.e. $C_2H_2 + O_2 = 2CO + H_2$.

This reaction liberates 492 000 kJ/kg mol, partly due to the exothermic dissociation of acetylene and partly to the partial combustion to carbon monoxide. The third zone is where atmospheric oxygen combines with carbon monoxide and hydrogen, and the following reaction results:

$$4CO + 2H_2 + 3O_2 + 11N_2 = 4CO_2 + 2H_2O + 11N_2$$

The atmospheric nitrogen is included in this reaction to show that a large mass of inert gas is also heated at this stage. A further 807 000 kJ/kg mol of acetylene is liberated by this second reaction.

The capital cost of oxyacetylene equipment is low compared with that for arc welding. The equipment is also easily portable and the process is very versatile. However, its comparative slowness means that it is more expensive than arc welding if there is a considerable amount of welding to be done.

Flame cutting using an oxyacetylene flame is a well-known engineering process. The cutting torch may be manually operated if low accuracy is permissible, or if higher accuracies are required the torch is mechanically mounted and controlled either by tracer or numerically. The gas flame preheats the metal to about 1000°C and is followed by a jet of oxygen which rapidly oxidizes the red-hot metal, enabling thick sections to be cut.

Linde welding is a variant of oxyacetylene welding sometimes used for joining steel. A carburizing flame is used which allows free carbon to be absorbed by the weld pool, lowering the melting temperature. The welding rate is thereby increased and the reducing atmosphere precludes oxygen contamination.

13.2.3 Electroslag welding. Electroslag welding is a recent development which enables plates of thick section to be joined in a single run by the heating effect of an electric current.

The plates are arranged vertically (Fig. 13.3) and the electrodes are lowered into the gap to strike an arc against the bottom damming piece. Flux is poured into the gap, and the arc is eventually shorted. From this point the heat is provided by the passage of the current through the molten flux. The temperature of the slag pool is sufficiently high to melt the parent metal and the electrode, which provides the necessary filler; several electrodes are used for long welds so that the heat is more uniformly spread. To prevent the pool of molten metal from escaping, water-cooled dam plates are fastened to the sides of the workpiece. These plates also assist the solidification process by removing heat, and are

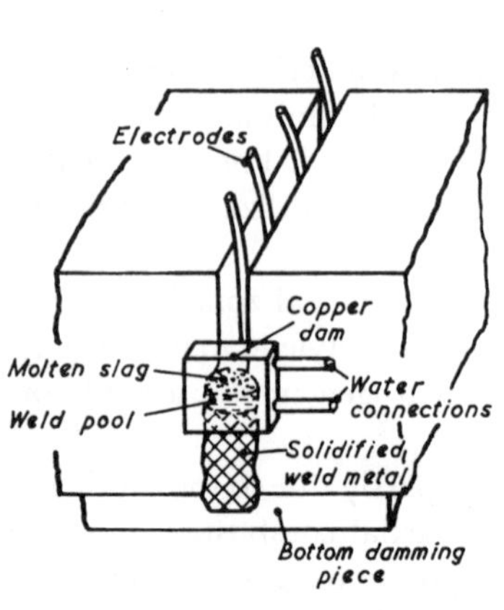

Fig. 13.3 Electroslag weld

moved up the weld as it progresses. Welding speeds of about 1·5 m h^{-1} (5 ft/h) are achieved.

A variant of the electroslag process uses fixed damming pieces, and the electrode is contained in a rigidly positioned steel tube which melts as the molten slag level rises. By these means vertical butt joints up to 1·2 m (4 ft) high have been welded in materials 25–75 mm (1–3 in) thick. The process is thus made more versatile and the cost is considerably reduced.

One limitation is that the process must be performed in a substantially upward direction. However, the seams of boilers can be welded by rotating the boiler with a suitable manipulator.

13.2.4 Welding using exothermic reactions. The heat liberated by an exothermic reaction can be used to weld or cut metal. One such process is Thermit welding, in which the heat to produce fusion is provided by reducing ferric oxide with aluminium, but this is a cumbersome and expensive process and is used only where there is no practical alternative.

A recently developed process for boring holes in metals and non-metals is based on the exothermic effect of burning iron in a current of oxygen. A thermic lance, consisting of a length of steel tube packed with steel wire, is connected at one end to an oxygen cylinder by means of a hose. The free end of the lance is heated to incandescence and the oxygen is then switched on. Once started, the process is self-perpetuating so long as the oxygen and lance last. The heat generated is sufficient for the blast of gas to burn rapidly through concrete or other structural materials. The advantages of this process are its comparative cheapness and high rate of penetration. It is used largely in structural engineering for piercing concrete.

Other exothermic processes in common use do not rely on chemical combination, but on dissociation or ionization.

Figure 13.4 shows how the energy of a gas varies with temperature. Considering first the diatomic gases, nitrogen and hydrogen, the curves steepen in the region of 4000°C and 2000°C respectively. At around these temperatures the molecules start to dissociate to atomic form. The curves then flatten out until, at about 10 000°C, ionization begins. Ionization is also strongly endothermic, causing the curves to become steeper again.

The monatomic gases, argon and helium, have curves which rise steadily until ionization begins at about 10 000°C and 14 000°C respectively.

Rise in enthalpy due to dissociation is used in atomic hydrogen welding. Hydrogen is passed through an arc struck between two tungsten electrodes, and the intense heat of the arc causes dissociation of the gas. The atomic hydrogen plays on the weld metal and recombination occurs, accompanied by the liberation of heat which raises the metal temperature to about 3700°C. A strongly reducing atmosphere envelops the weld owing to the presence of nascent hydrogen. Although atomic hydrogen welding is expensive, it is frequently used to weld difficult metals such as stainless steel and heat-resisting alloys.

Plasma torches make use of ionization to transfer heat to the workpiece. The gas is ionized by passing it through a constricted arc, where the

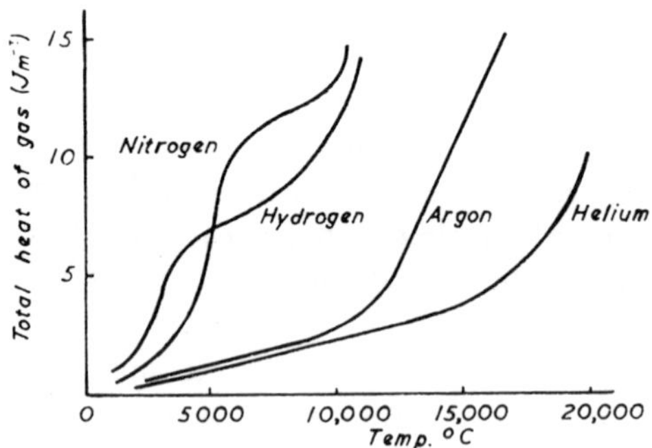

Fig. 13.4 Enthalpy–temperature relationships

high current density causes a spectacular rise in temperature. The arc can be struck either between the casing of the torch and a centre electrode (Fig. 13.5 (*a*)) when it is known as a non-transferred arc, or between the centre electrode and the workpiece (Fig. 13.5 (*b*)), when it is known as a transferred arc. The power supply is from a d.c. unit of 20–30 kW capacity. Flame temperatures are in the order of 20000°C. An auxiliary gas jet, not shown in Fig. 13.5, passes through a chamber surrounding the nozzle and supplies a gas shield to the weld.

Non-transferred arcs are used mainly for spraying plastics, refractory metals and ceramics, while transferred arcs, which operate at higher temperatures, are used mainly for cutting metals, particularly stainless steel and non-ferrous alloys, and for welding. When used for metal cutting, plasma torches are about four times as fast as oxyacetylene, cutting 40 mm ($1\frac{1}{2}$ in) thick mild steel plate at about 20 mm s^{-1} (50 in/min). Another advantage is that the thermally disturbed zone is much reduced.

Although the high velocity of the plasma jet, about 1200 m s^{-1} (4000 ft/s), creates difficulties if used in welding, a number of processes have been developed. The usual application is in butt welding, where deposits 6 mm ($\frac{1}{4}$ in) thick can be laid in a single run. A transferred arc torch with three gas passages is used. These passages are set in a line along the weld pool and leaves an oval hole or keyhole; as the jet is moved the hole is closed behind it by the surface tension of the molten metal. This technique gives good penetration, with welding rates up to 100% faster than other methods.

'Keyhole' welding torches of this sort are essentially for automatic operation, but a low-powered 'micro-plasma' torch, operating between 1–10 A, has been developed for manual operation. This does not produce a 'keyhole' and can be used also for brazing and soldering.

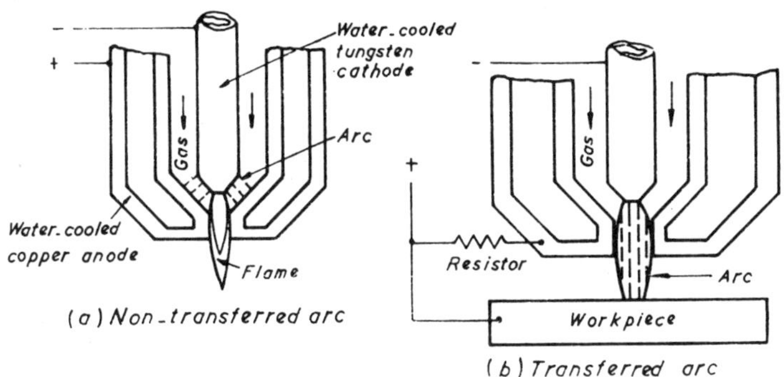

Fig. 13.5 Plasma torches

Weld surfacing can also be performed with a transferred arc plasma torch. This process involves the deposition of a surfacing metal, and is used in place of metal spraying where the component is subjected to stress which might cause stripping of the sprayed metal. The essential requirement is that the parent metal can be metallurgically fused to the deposit, which has a lower melting point.

A variant of the conventional spraying technique, using a plasma arc, consists of spraying a molten refractory metal or ceramic onto a former. When the deposit has hardened the former is removed, leaving a refractory shell. This process is known as spray forming.

Other applications of plasma torches are in sintering refractory metals, in gouging grooves in stainless steels and other materials which are difficult to machine, and in turning. Turning is performed by pointing the torch tangentially towards the workpiece which is then rotated. When turning materials which are difficult to machine with conventional tools, some success has been claimed for preceding the tool by a plasma flame which thermally softens the workpiece.

13.2.5 Welding using high energy beams. If a beam of high energy is focused on a small area it is possible to use the energy of the beam to heat a workpiece. These beams may consist of elementary particles such as electrons, or alternatively may be in the form of light rays.

Electron beam welding machines consist essentially of an electron beam gun and a magnetic focusing lens. In the electron gun, a cathode of refractory material such as tungsten or tantalum is heated to incandescence and emits electrons which are accelerated through a ring-shaped anode to about 160 000 km s^{-1} (100 000 miles/s) by a pulsed voltage of 20–200 kV. The electron beam is focused by means of a coil on a spot a few μm in diameter. The higher the beam velocity, the more accurate the focusing, but the high anode potentials necessary to achieve the velocity give rise to X-rays, and protective shielding has to be fitted.

When the electrons strike the workpiece, their kinetic energy is converted to heat, causing temperature rises up to as high as 6000°C at the surface of the workpiece.

When welding at atmospheric pressure, the beam efficiency is reduced due to collisions between electrons and gas molecules. For this reason, many applications require a hard vacuum which reduces output rates due to manipulation problems and the need to evacuate the welding chamber. Partial or non-vacuum systems are also used, but these require a short beam path, usually less than 120 mm (5 in) and an inert atmosphere is desirable to avoid contamination.

Although any material can be heated by these means, the best results are obtained when welding metals have low thermal conductivities and high melting points. Electron beam welding is particularly useful for joining metals with dissimilar thermal characteristics, and where distortion is unacceptable. Applications vary from joining components in miniature circuits to welding steel parts 100 mm (4 in) thick using a machine of 25 kW capacity.

Other uses are for cutting or drilling. Thin films of evaporated metal on glass can be slit to make miniature resistors, and holes can be drilled in synthetic sapphire watch jewels at a rate of 600/h.

Lasers are another source of high energy beams, in the form of coherent monochromatic light. Early attempts at laser welding used ruby lasers which were pulsed, and had a low energy conversion. More recently CO_2 lasers, having an efficiency of 15–25%, have been used, with an output of several kilowatts. These are continuous in operation and can be focused to give a power intensity comparable with that of electron beams, around 10^9 W cm^{-2}. Their application to welding has so far been restricted to joining polythene sheet, metal foil, wire and thin sheet up to about 2·5 mm (0·1 in) thick. Unlike the electron beam process, a vacuum is not necessary for laser welding. Lasers can also be used for drilling holes and for cutting.

In drilling applications, holes with a length/diameter ratio exceeding 25:1 have been successfully pierced. For cutting, a CO_2 laser is mounted coaxially with an oxygen jet; the laser pre-heats the metal and the oxygen jet promotes the exothermic reaction necessary to cause melting. By this means a narrow molten zone is created, and the thermally disturbed region is small. Steel sheet 5 mm thick can be cut at speeds up to 10 mm s^{-1}; ceramic materials can also be cut.

13.2.6 Resistance fusion welding. These welding processes all use the heating effect caused by a large current passing across the interface

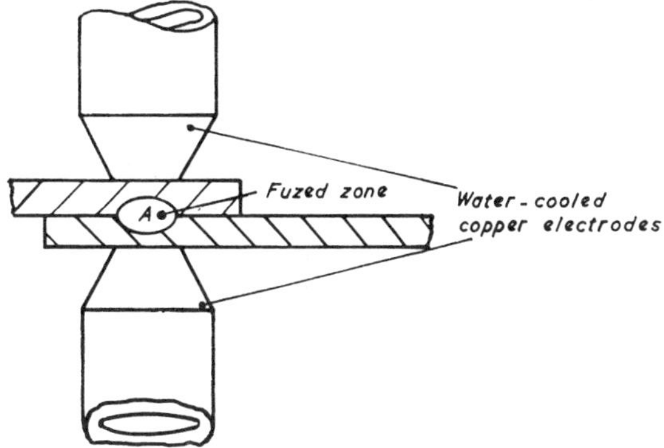

Fig. 13.6 Section through spot weld

of the two metals to be joined. A high pressure is applied to the metals at the interface to localize the passage of current. This causes some plastic deformation, although the fact that melting occurs makes these processes fusion welding rather than solid-phase welding. The processes considered are known as spot, projection and seam welding.

Spot welding has found many applications where the rapid joining of sheet metal is required. The car industry, in particular, makes extensive use of the technique. The two sheets of metal to be welded are cleaned, lapped and squeezed at pressures of 70–100 N mm^{-2} (10 000–15 000 lbf/in^2) between the tips of two electrodes (Fig. 13.6). A current, the intensity of which is 120–300 A mm^{-2} (80 000–200 000 A/in^2) at the electrode tips, is passed at low voltage for a specified time, and the high resistance at interface A leads to a localized rise in temperature, causing fusion. Efficient welds depend on the correct combination of pressure, current and cycle time; experience has shown that materials with high thermal conductivities, such as aluminium, require higher currents for shorter periods than materials of lower conductivity. Most ductile metals, including many dissimilar metals, can be spot welded provided there is not too great a difference in the conductivities and melting points.

Until recently it was impossible to predict whether spot welds were good or bad. Even a sampling technique, where a proportion of the welds is tested to destruction, does not ensure that all the welds are of good quality, since current may vary from one weld to the next due to:

(*a*) oxide films on the surfaces to be welded altering resistance;
(*b*) shunting of current through adjacent welds;
(*c*) electrode wear altering the current density.

The BWRA discovered a correlation between the thermal expansion across the electrodes during welding and the quality of the weld. Figure 13.7 shows the effect of thermal expansion across the electrodes when the

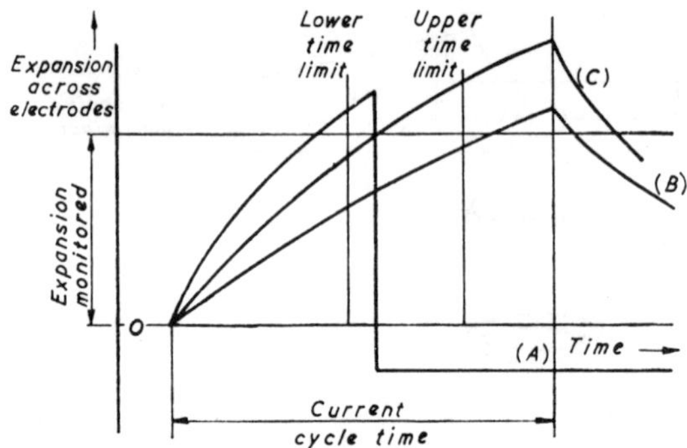

Fig. 13.7 Expansion–time graph for spot welds

current is too high (*A*), too low (*B*) and correct (*C*). With too high a current density the weld splashes, and the pressure on the electrodes causes the weld to collapse at an early stage, producing a poor weld. Too low a current density results in a slow rate of expansion and the weld is insufficiently fused.

A spot-welding monitor developed by the BWRA[65] is based on the rate at which expansion occurs. The monitor, which is fitted to the electrodes, measures the time taken to achieve a given expansion during welding, and top and bottom time limits are set. Welds which take longer than the maximum time allowed or shorter than the minimum are rejected.

To increase productivity multiple spot-welding machines can be used, with which a large number of welds can be made simultaneously. A limitation on the number of simultaneous welds is imposed by the heavy surge of current drawn from the mains.

A scaled down version of conventional spot welding techniques, known as microresistance welding, makes use of two parallel mounted electrodes separated by a layer of insulating material. The bottom surface of the lower component to be welded rests on an insulating substrate and the current passes down one electrode and up the other, so that two small series welds are simultaneously produced. Microresistance welding is particularly suited to small electrical components as an alternative to soldering.

Projection welding is similar to spot welding except that projections are raised by coining or embossing one of the parts to be joined. Projections can be cheaply produced by adding a part-shear to an existing press operation. The main advantage of projection welding is that the areas of the ends of the electrodes are independent of the size of the weld and the heat generated is almost completely confined to the weld area. When the welding temperature is attained, the projection collapses under the applied pressure, and the two parts fuse together.

Projection welding permits a greater variety of metals and sheet thicknesses to be used than in spot welding. It is customary to form the projections on the thicker component or on the component having the higher thermal conductivity, to help equalize the heat losses.

Seam welding is another variant of the spot-welding technique, in which current is fed either continuously or intermittently to two thin copper alloy wheel electrodes which replace the stick electrodes used in spot welding. This process is suitable for producing fluid-tight joints in cans, but if the intermittent method is used the welds must be at close enough intervals to overlap.

13.3 SOLID-PHASE WELDING

Solid-phase welds are made by the creation of a metallic bond between the two surfaces being joined. A bond is created when the surfaces are brought into such close contact that the atoms are separated by less than the relaxation distance over which the inter-atomic forces act. Good metallic bonds are not easily produced, particularly in cold metal, because of the difficulty of bringing a sufficiently large area into intimate contact. This is due to surface asperities on the metals and the oxide films and impurities on the surfaces, which act as a barrier between the metals. However, with reasonable care a weld of strength comparable to the parent metals can often be achieved.

In solid-phase welding it is essential to apply sufficient pressure to cause plastic flow of the two surfaces, thereby forcing the maximum area into contact and fracturing any contaminant surface films. Pressure may also be applied as a means of friction heating.

Most solid-phase welding processes are performed hot, the application of heat having the following beneficial effects:

(*a*) plastic flow occurs at a lower pressure due to the reduction in yield strength;
(*b*) surface contaminants are melted or evaporated;
(*c*) recrystallization occurs, causing the growth and coalescence of grains across the interface;
(*d*) surface diffusion at the interface causes a modification to the shape and size of voids produced, and volume diffusion can transfer metal across the interface.

13.3.1 Butt welding. This method, used extensively for butt-joining parts, is heat-assisted pressure welding. The heat source may be an oxy-acetylene flame or, more usually, an electric current. Resistance butt welding is performed by butting together under pressure the ends to be joined and passing current across the interface. The interface resistance causes a rise in temperature, and when the welding temperature is reached the applied pressure causes plastic flow, resulting in a shortening of the component and thickening in the weld area.

A more popular method is flash butt welding, where the current is passed before the parts are forced together. An arc is formed across the contact points on the joint faces, and when the required temperature is attained the current is switched off and the parts forced together. This results in a more efficient joint than plain butt welding since the arc cleans the surface of impurities. Also the thermally disturbed zone is smaller and the upsetting effect is reduced. Since melting occurs at the surface from the heat of the arc, this process should more properly be considered as one of fusion welding.

The correct selection of upsetting pressure is important, as a low pressure produces a porous weld whereas a high pressure causes brittleness. It is difficult to butt weld metals with widely differing melting points and thermal conductivities, although a wide range of metals can be joined if heating time is reduced and the pressure rapidly applied.

Cold butt welding is possible, although the forces involved are so high as to render it unattractive except for joining small sections such as electrical conductors of aluminium and copper.

13.3.2 Friction welding. Friction welding makes use of the heat generated by rubbing the two surfaces to be joined against each other. To date the only convenient method of producing relative motion between the surfaces has been by rotating one piece of metal against the other.

One component is rotated at constant speed whilst the other is held rigidly against it. Without making a detailed analysis it will be obvious that the higher rotational speeds at the periphery will cause a temperature gradient across the surfaces. When the required welding temperature has been attained the parts are pressed hard together and rotation is stopped. The parameters which can be varied to suit the requirements of the metals are the initial pressure, the heating-up time and the final upsetting pressure.

Bars of dissimilar metals and of varying cross section can be joined by friction welding, the essential requirement being that the ratio of thermal conductivities should be in approximately inverse proportion to the ratio of the melting points.

13.3.3 Indent lap welding. This is a cold process, in which the two non-ferrous metals to be welded are cleaned, lapped and pressed between two platens to reduce the thickness. The process is essentially one of plastic deformation, and it is found that below a certain threshold deformation, the metals do not join.

As the amount of indentation is gradually increased, the failure of the joint is initially due to shear across the interface; beyond a certain value of area reduction, failure occurs due to simple tension across the reduced section. If the platen thickness b is increased (Fig. 13.8), the area resisting shear also increases and the optimum deformation decreases, giving an improvement in weld strength (Fig. 13.9).

Indent welding is used for joining foil and thin strip materials; dissimilar metals may be joined, a typical example being the joining of the metal foil in capacitors to the aluminium or copper containers.

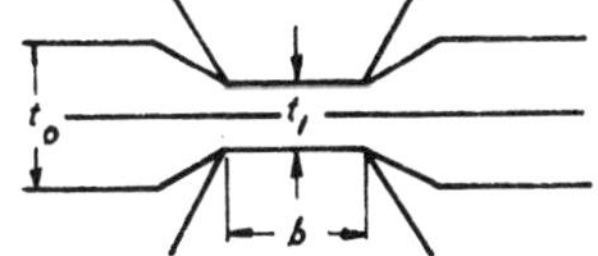

Fig. 13.8 Indent welding

13.3.4 Diffusion welding. As a process, diffusion welding has little practical use. The two parts to be joined are finished to a high standard of flatness and held, or preferably wrung together, under pressure. Heat is applied to raise the temperature and increase the diffusion rate, which approximately doubles for a temperature rise of 20°C. The process is usually performed in an inert atmosphere and takes from about 15 min to several days at temperatures which seldom exceed 500°C.

13.3.5 High frequency welding. High frequency eddy currents can be used to raise the temperature of joint surfaces before applying a deformation force. The current can either be induced by means of induction coils or can be fed to the components by means of sliding contacts. Since high frequency currents travel preferentially along metal surfaces the heating can be localized near the joining faces. The most usual application of high-frequency welding is in the manufacture of axially or helically welded tube.

13.3.6 Explosive welding. This process has been developed mainly for cladding large plates before fabrication. The surfaces to be joined are inclined to each other at a small angle, and are in contact along one edge. One plate rests on an anvil and the other, known as the flyer plate, is backed by a rubber spacer, on the outside of which is a layer of explosive. When the

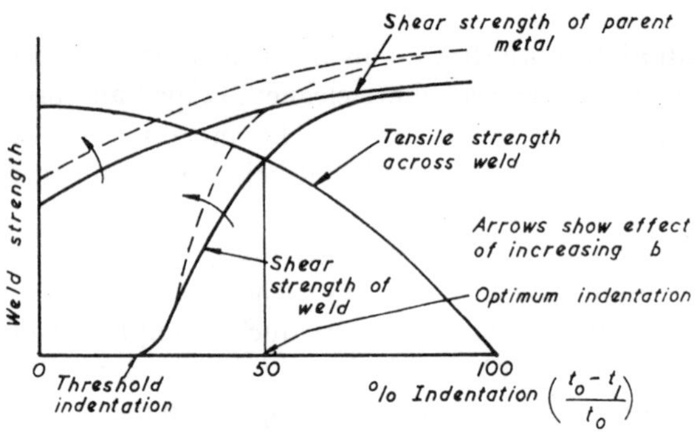

Fig. 13.9 Strength of indent weld

explosive is fired, the flyer plate comes progressively into contact with the bottom plate causing rapid deformation. A jet of molten metal travels ahead of the contact line scrubbing the surfaces of both plates and removing the oxide layers, resulting in a wave-like weld interface.

13.4 BRAZING AND SOLDERING

These processes, unlike fusion welding operations, involve small clearances between the mating surfaces. The filler is drawn into the gap by capillary action, and the bond is produced either by the formation of solid solutions or intermetallic compounds of the parent metal and one of the metals in the filler. Essentially the strength of the joint is provided by metallic bonding, but where solid solutions or intermetallic compounds are formed there is inevitably some inter-diffusion, which creates a small transition zone between the filler and the parent metal. This improves the chance of effecting a good bond and may allow oxide films to be penetrated.

Table 13.1 shows the variations in melting point and shear strength of various solders and brazing alloys. It will be seen that the strength of the filler material increases with increase in melting point.

TABLE 13.1

	Melting point (°C)	*Joint clearance* (mm)	(in.)	*Shear strength* (N mm^{-2})	(tonf/in^2)
Soft solders	70–305	0·05–0·20	0·002–0·008	30–45	2–3
Silver solders	620–870	0·05–0·13	0·002–0·005	150–185	10–12
Brazing bronze	850–900	0·08–0·25	0·003–0·010	250–310	16–20
Copper	1083	0–0·05	0·0 –0·002	310–380	20–25

A suitable flux is always necessary to remove oxide films on the unjoined metal, to protect the surface of the finished joint from oxidation, and to reduce the surface tension of the filler and thereby assist its penetration. For soft soldering, the fluxes used are generally zinc chloride or resin; for brazing operations, borax is generally used.

Heat may be applied in a number of ways. Soft solders can be melted by the heated copper tip on a soldering iron, or a molten solder bath into which the articles are dipped can be used. When soldering the connections on printed circuit boards it is essential that the board itself is not immersed in solder. For this purpose wave soldering and cascade soldering have been developed, where standing waves of molten solder are produced and the boards are passed over them so that only the connections are immersed.

Gas torches may be used for any of the fillers, or alternatively the components can be assembled with the filler metal and heated in a furnace. A convenient method of heating large quantities of parts is by high frequency induction coil. Copper brazing, being a higher temperature process than the others, is best performed in a furnace with a controlled atmosphere to prevent oxidation.

A mis-named process in this group is bronze welding, which is used for joining metals with high melting points, such as mild steel, cast iron and copper. As in welding, large clearances are used; these are filled with bronze or brass which is melted by an oxyacetylene torch or electric arc. However, unlike welding, fusion of the parent metal does not occur and the process is therefore essentially a brazing operation.

13.5 ADHESIVES

The use of adhesives as a method of fastening metal parts together has gained considerable popularity during recent years. Reasons for this popularity are listed below:

(1) Temperatures necessary to cure or solidify the adhesive are comparatively low, and hence unlikely to affect the metal.

(2) Two dissimilar materials or two widely dissimilar sections can be joined without difficulty and galvanic corrosion is inhibited.

(3) The load imposed on the structure can be distributed over a large area compared with the load carrying area of bolted, riveted or spot welded joints.

(4) A wide range of electrical conductivities can be obtained by choosing a suitable adhesive.

(5) Adhesive joints can be made to resist water or gas pressure.

(6) Adhesive bonds have high damping capacity which can improve fatigue resistance.

There are, however, disadvantages or limitations to the use of adhesives, and these are:

(1) The restricted range of temperatures between which they are neither prone to brittle behaviour or thermal softening; typically, the strength of epoxy/phenolic adhesives at 200–300°C is only about 25% of their strength at room temperature.

(2) The need for careful preparation of the surfaces to be joined, and the long curing times needed for some adhesives.

The strongest joints are normally obtained if only sufficient adhesive is used to just fill the voids and irregularities of the surfaces to be joined plus sufficient to allow for shrinkage during solidification. To achieve the intimate contact with the joint surfaces which is necessary if an efficient joint is to be formed, the adhesive must have good wetting properties.

The type of bond created by adhesives is different from those previously considered because there is no metallic bonding. It appears that the bond may be effected in one of three ways. In certain cases chemical action may occur between the adhesive and the material being joined, for instance the formation of copper sulphide when bonding brass with rubber. Usually, however, when bonding metals the strength is due to either homopolar bonds being formed between the adhesive and an oxide film on the metal surface, or to secondary bonding, the effect of forces of molecular attraction between the adhesive and the metal.

Adhesive bonds are of comparatively low strength, usually being limited by the bulk strength of the adhesive. The strength of a lap joint is normally from 30% to 60% of the bulk shear strength of the adhesive, but if the adhesive film is sufficiently thin a strength greater than that of the bulk adhesive can sometimes be achieved. The reason for this is not known with certainty.

Adhesives used are generally synthetic thermosetting resins which polymerize on solidification. It is usually essential that the surface of the material to be joined is degreased and suitably roughened, although epoxy adhesives which will bond to oily surfaces with little loss of strength have been developed. Oxide films to promote homopolar bonds may be produced on the surface by anodizing, or other suitable surface treatments. The adhesive is applied, usually in a liquid solvent and the parts are brought together under light pressure and cured. Curing is usually performed by heating to about 150°C for 30 min, but more rapid curing times can be achieved by increasing the temperature.

Some adhesives can be cured by chemical reaction without heating, using a suitable activator. Others solidify at room temperature due to evaporation of the solvent, whilst yet another type is applied at a temperature above its melting point but solidifies when cooled to room temperature.

14 Casting and Sintering of Metals

PART 1. CASTING

14.1 CASTING PROCESSES

Casting is a method of producing a large variety of components in a single operation by pouring liquid metal into moulds and allowing it to solidify. Ingots, which are afterwards shaped by further working into a large range of finished products, are originally shaped by casting. Subsequent working improves the mechanical properties of the metal and because of the superior properties of most worked materials, cast parts are often limited to low stress applications.

Some of the more important casting operations are described below.

14.1.1 Sand casting. Sand moulds are the most commonly used method of producing ferrous and non-ferrous castings. This method of casting is used chiefly when small batches are required, but if necessary it can be used for large quantity production, such as for internal combustion engine parts. Patterns which correspond to the external shape of the part are made from wood, or, if the quantity of castings warrants the extra cost, from metal. To enable the pattern to be removed, the mould is split longitudinally into two or occasionally more sections. Small and medium-sized moulds are normally made in two boxes or flasks, open at the top and bottom. The upper and lower boxes accurately register on each other and are called the 'cope' and 'drag' respectively.

If a hollow casting is required, an appropriately shaped sand core is placed in the mould cavity to exclude metal from that part of the mould (Fig. 14.1). The sand used for mould making is normally moist so that the grains cohere, and in this condition it is called green-sand. Green-sand is not, however, strong enough to make cores, so dry-sand cores or cores produced by the CO_2 process are used. Dry-sand cores are prepared from sand mixed with a bonding agent such as linseed oil;

after moulding, the cores are hardened by baking in an oven. The CO_2 process makes strong cores without the need for baking. Sand is mixed with a solution of sodium silicate, and when the core has been moulded, CO_2 is passed through it. The silica gel which is formed strongly bonds the sand grains together. The CO_2 process can also be used to produce moulds.

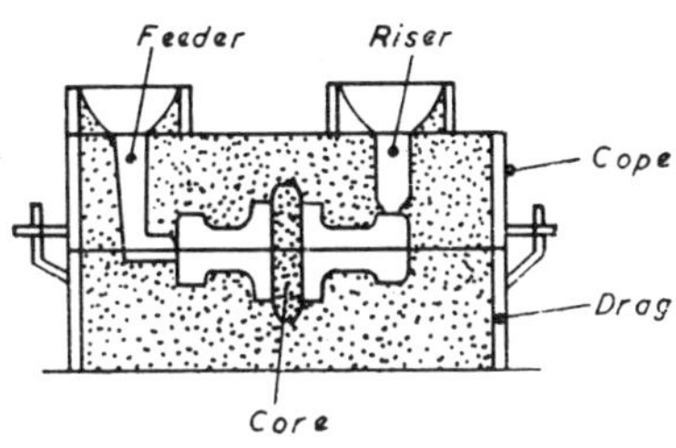

Fig. 14.1 Sand mould ready for pouring

A channel called a feeder must be provided through the sand to allow the metal to be poured into the mould. A cavity above the casting, referred to as a riser, is made to act as a reservoir for molten metal which feeds the casting during solidification. In large castings more than one feeder and riser may be required. Moulds should be strong enough to withstand the pouring of the metal, but should collapse when subjected to shrinkage stresses and be sufficiently permeable to allow gases to escape.

The use of medium and high-pressure moulding machines in the mass production of castings has facilitated highly automated production of moulds. High-pressure mould-making machines produce castings having closer tolerances and better surface finishes than those obtainable with conventional sand casting. Small castings can be produced without moulding boxes.

Although synthetic and natural sands are used to cast most metals, steel requires a more refractory material and 'compo' a coarse fireclay mixture, is used. The moulds are lined with a fine grain refractory material and dried out before the metal is poured.

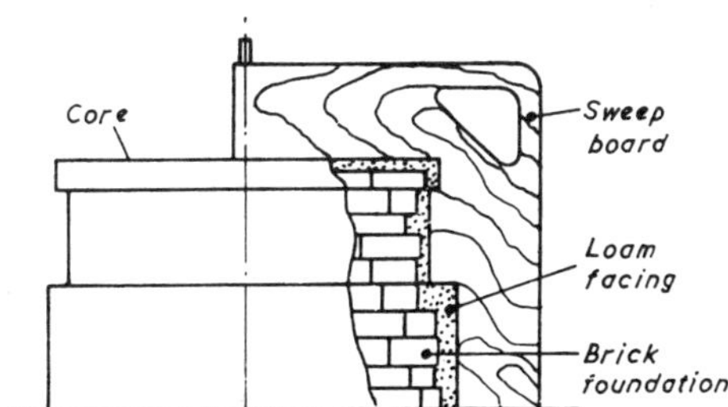

Fig. 14.2 Loam moulding

Large castings are made in loam moulds, which are built up from bricks reinforced with iron plates. The bricks are surfaced by a loam which contains sand mixed to a mortar-like consistency. Solids of revolution can be generated by rotating a sweep board of the appropriate shape about a vertical axis (Fig. 14.2).

14.1.2 V-process. This is a method of casting in which the mould is made from dry sand vacuum sealed in plastic film. The pattern is mounted on a pattern plate, beneath which is a vacuum chamber; the vacuum chamber is connected to the pattern. A heated plastic film is placed over the pattern and drawn down on to it by the vacuum. A moulding box is placed over the pattern, filled with dry sand and vibrated. After the feeder and the risers have been formed the top of the mould is sealed with a second plastic sheet. A vacuum is applied to the box to compact the dry sand. The other half of the mould is produced in a similar manner and after assembly of the two halves the mould is filled with molten metal. During pouring a vacuum is applied to both upper and lower boxes. The vacuum holds the mould rigid during the pouring and solidification of the casting; it also prevents the plastic film from burning on contact with molten metal.

The process uses easily reclaimable dry sand and the castings have an excellent surface finish.

14.1.3 Freeze moulding. This method of casting is also called the Effset process and uses sand, clay and water to make the moulds from permanent patterns. Liquid nitrogen is used to freeze the moulds, which have good permeability and are rigid enough to be used without boxes. As in the V-process the castings have an excellent surface finish and the sand is easily reclaimable.

14.1.4 Shell moulding. This development of sand casting uses a mixture of sand and thermosetting resin which is formed into thin shells by contact with heated metal patterns (Fig. 14.3). After removal from the patterns, the shells are cured for a few minutes and then brought together to make a mould. This process enables moulds to be produced quickly by less skilled workers, and the castings have a relatively smooth surface finish and better dimensional accuracy than normal sand castings.

Zircon sand is sometimes used in place of the conventional silica sand. Although more expensive, this produces a better surface finish, because of its small round grain, and a dense mould of high thermal conductivity which rapidly and evenly cools the casting.

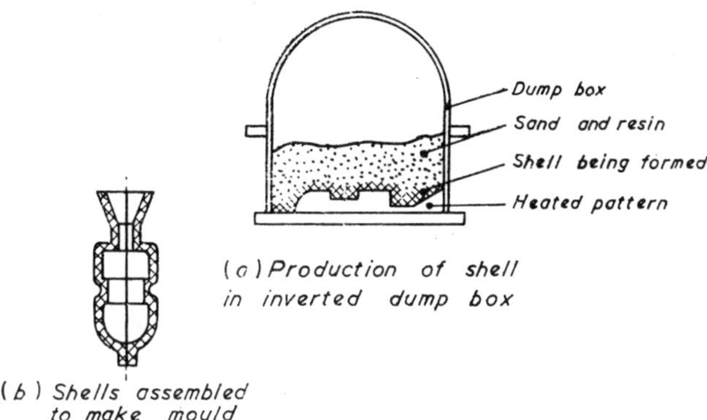

Fig. 14.3 Shell moulding

Cold rather than hot-set processes have been developed; these have the advantage over shell moulding of lower energy requirements and less expensive equipment costs. A number of techniques are available to set the sand and resin mixture, some of which require gassing.

14.1.5 Centrifugal casting. Centrifugal casting can be used to produce large hollow cylindrical castings, and avoids the use of cores, feeders and risers (Fig. 14.4 (*a*)). Small parts can also be cast centrifugally and Fig. 14.4 (*b*) shows a centrifuge being used to produce a number of small castings from a multi-impression mould. Rotational speeds used in centrifugal casting cause centripetal accelerations of the order of 60 g and the resulting casting has a dense structure, with non-metallic inclusions segregated at the inner surface.

14.1.6 Investment casting. Small accurate shapes in almost any metal can be produced by investment casting which is sometimes termed the 'lost wax' process. The process is costly and involved, but justified when a good surface finish and close tolerances are required. Firstly a pattern is made by injecting wax into a split mould; this is usually metallic but can also be made from rubber, plastic or plaster of Paris for short-run work. As the patterns are usually small, a number are frequently assembled on a wax feeder using a heated spatula, and in this way several parts can be cast in a single mould.

The next step is the investment of the feeder and its cluster of patterns by dipping into refractory slurries. If a solid mould is required the invested pattern is placed, feeder downward, in an open-ended can into which a slurry of refractory material or cement is poured. Shell moulds can be used in place of solid ones; these are produced by dipping the pattern a number of times into refractory slurries until the requisite thickness has been built up.

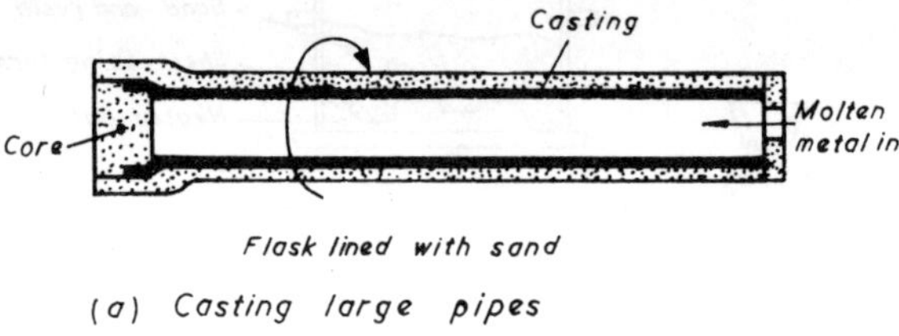

(a) Casting large pipes

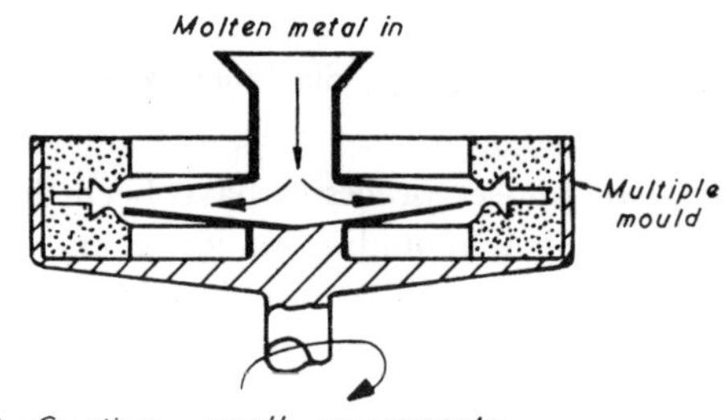

(b) Casting small components

Fig. 14.4 Centrifugal casting

To complete the mould the wax pattern must be removed; this can be done by melting out in a furnace. The metal may be poured under gravity, but if a denser casting is required it can be injected under pressure or cast centrifugally.

Sometimes included under the heading of investment casting is full-mould casting. Here a pattern is made of foamed polyurethane or polystyrene by cutting and assembling stock material. Once the pattern is made it is placed in a mould box and sand packed round it. When the casting is poured the liquid metal melts and vaporizes the foamed plastic pattern. The casting will have similar properties to a green-sand casting. The process is particularly suitable for one-off production since the pattern costs only about a third that of a wooden pattern.

An alternative to dry sand in full mould casting is dry iron shot. This material fills the moulding box around the foamed plastic pattern and is then held rigid during the pouring and solidification of the casting by a strong magnetic field. This method of casting is suitable for non-ferrous metals.

14.1.7 Shaw casting. This recently introduced method of casting is somewhat similar to investment casting, in that the pattern is invested with slurry to produce the mould. The pattern is, however, a permanent one, usually made from metal. The mould is made in two halves by pouring round the pattern a slurry made from graded refractory materials, an ethyl silicate binder, water, alcohol and an accelerator. The mould is flexible before solidifying, and in this state the pattern is removed. Because of its rubber-like nature, the mould will spring back when distorted, rendering tapered patterns unnecessary and even allowing small undercuts to be incorporated. When the mould has set, it is heated in a furnace which causes the mould surface to be completely crazed with minute cracks. These counteract the contraction which takes place on the setting of the investment mould, and the mould cavity remains the same size as when it was in contact with the pattern.

Although Shaw casting cannot compete with investment casting for very small components, it is becoming widely used for the precision casting of medium-sized components.

14.1.8 Gravity diecasting. This process, sometimes called permanent mould casting, is used to produce non-ferrous castings with a better surface finish and higher dimensional accuracy than sand castings. The usual mould material is close grained cast iron, with alloy steel or sand cores. Gates and risers are still required with this method of casting. Gravity diecasting is often used in preference to sand casting when batch quantities justify the higher die costs.

14.1.9 Low-pressure diecasting. This process is used largely for aluminium alloy castings, e.g. beer barrel halves, although using graphite dies, steel railway wheels have been cast. The die is usually made from cast iron and is connected by a tube to a container of liquid metal (Fig. 14.5). Air pressure is used to lift the metal and fill the mould cavity. To avoid turbulence the filling of the die is slow, but there is less oxidization of the metal than with gravity diecasting. The cost of a low-pressure diecasting machine is about $\frac{1}{3}$ of the cost of a (high) pressure diecasting machine.

An alternative to low-pressure diecasting is the counter-pressure process, where the pressure used is approximately ten times greater than the die filling pressure of 0·1 N mm^{-2} (7 lbf/in^2) employed in the low-pressure process. At the start of the casting cycle pressures are equal, both in the die cavity and over the melt; liquid metal is then gradually admitted to the die by slowly reducing the die cavity pressure. During solidification full pressure is maintained over the melt.

14.1.10 Pressure diecasting. Pressure diecasting is an important process used to produce castings of high dimensional accuracy and excellent surface finish from zinc, magnesium and aluminium alloys. The dies are manufactured from alloy steels and molten metal is forced into them at pressure of 25–200 N mm^{-2} ($1\frac{1}{2}$–12 tonf/in^2). The rapid delivery of the casting metal minimizes heat losses to the die and enables thin sections to be successfully cast. Shrinkage cavities can also be avoided as pressurized metal will continue to be fed into the die during solidification. The cost of pressure diecasting dies and the machines in which they are mounted is high, and in consequence this process is confined to large quantity production.

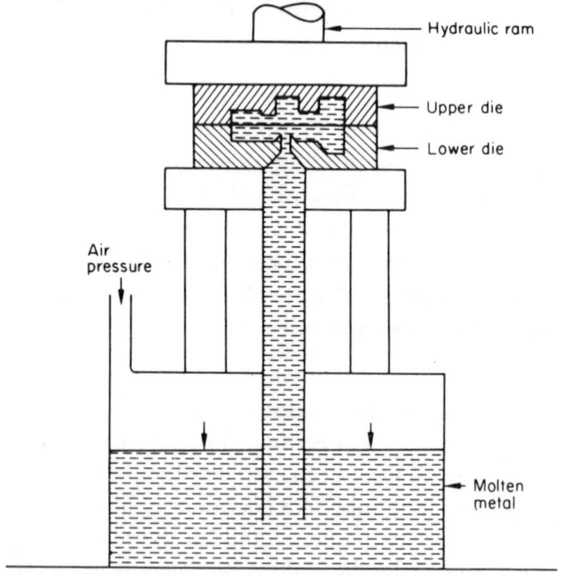

Fig. 14.5 Low-pressure diecasting (filling the mould)

Diecasting machines are classified as either hot or cold chamber types depending on the method of transporting the molten metal to the dies (Figs. 14.6 (*a*) and (*b*)).

Diecast parts which have similar properties to forged components can be produced by the squeeze casting process. Squeeze casting is described in section 14.1.11.

Brass can be pressure diecast but the relatively high temperatures of casting cause 'heat checking' or 'crazing' of the dies and in consequence relatively short die lives. However, when the product application rules out other diecasting alloys, as in certain electrical and plumbing components, brass pressure diecasting is employed.

Much work is being done to make possible the pressure diecasting of ferrous materials. Here the casting temperatures are an even greater problem than with brass and die lives are correspondingly shorter; sintered tungsten or a titanium/zirconium/molybdenum alloy are being used as the die materials. The most likely application of ferrous diecasting is to parts of up to 0·25 kg ($\frac{1}{2}$ lb) in weight, produced from special steels such as stainless.

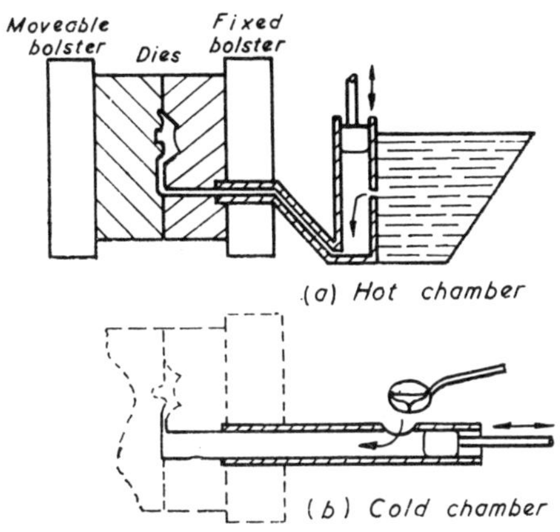

Fig. 14.6 Pressure diecasting

14.1.11 Squeeze casting. This process was developed in the U.S.S.R. and produces non-ferrous castings having mechanical properties comparable with forgings. After liquid metal has been metered into the open lower half of the die, a closely fitting upper die moves down and compresses the liquid at high pressure, and this pressure is maintained during solidification. Much greater component complexity is possible than with closed die forging, and the yield is almost 100% of the poured metal.

14.1.12 Continuous casting. This method of casting has been used to produce ingots on a continuous basis, the molten metal being continuously poured into one end of a vertical open-ended mould of the appropriate cross section (Fig. 14.7 (*a*)). While the metal is passing down the mould it cools and solidifies. The walls of the mould are water cooled, and there is further water cooling of the solid metal when it emerges from the bottom of the mould. Ingots are produced by cutting off the cast metal with a flying saw which moves down at the same speed as the metal. The downward movement of the ingot is controlled by a piston, and the flow of metal into the mould is adjusted by a float in the pool of liquid metal at the top of the mould.

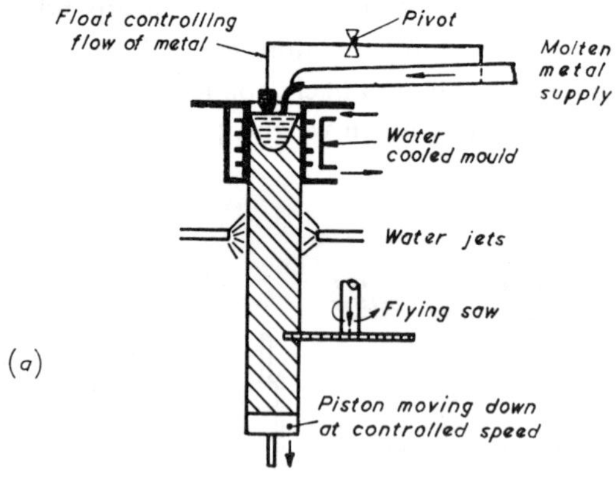

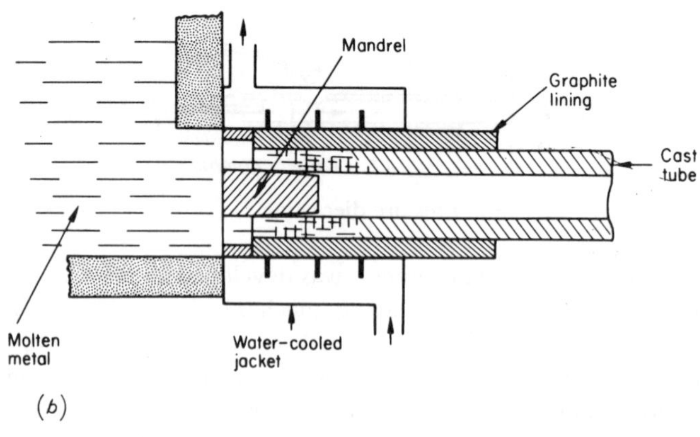

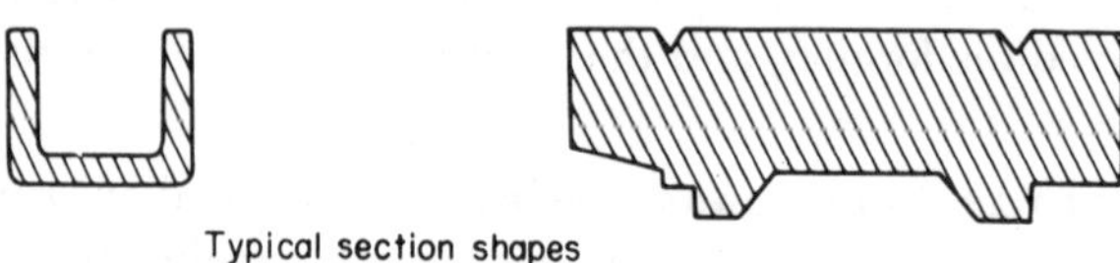

Fig. 14.7 Continuous casting

As well as producing ingots, which are subsequently hot worked, continuous casting can be used to cast sections from which components are manufactured, e.g. milling machine tables. To improve the quality of the product a closed system is used in which the liquid metal is fed direct to the die. This is in contrast to the open system used in billet casting, where the molten metal supply is open to the atmosphere and the product is likely to suffer from oxide inclusions and surface imperfections. The continuous casting of grey-iron hollow section using a closed system is shown in Fig. 14.7 (*b*).

The mechanical properties of continuously cast grey-iron are superior to those obtained by sandcasting; this is due to its fine grained dense structure, the surface finish is also improved and there are no sand inclusions.

14.1.13 Summary of casting processes. The casting processes already described are summarized in Table 14.1. The figures should be taken as a rough guide only.

TABLE 14.1

Type of casting	*Labour cost/ casting*	*Equipment cost*	*Surface finish (μ CLA)*	*Usual accuracy (mm)*	*Minimum section thickness (mm)*
Sand (green)	medium	low	500–1000	±2·50	5·0
Shell	low	medium	100–300	±0·25	2·5
Centrifugal	low	medium	100–500	±0·70	8·0
Investment	high	medium	25–125	±0·06 (25 mm part)	0·6
Shaw	medium	medium	80–180	±0·08 (25 mm part)	3·0
Gravity diecasting	low	medium	100–250	±0·40 plus 0·05 per 25 mm	2·5
Low-pressure diecasting	low	high	40–100	±0·05 plus 0·05 per 25 mm	1·2
Pressure diecasting	very low	very high	40–100	±0·05 plus 0·05 per 25 mm	0·5
Continuous casting	low	high	100–200	±0·12 per 25 mm	8·0

14.2 SOLIDIFICATION OF METALS

In the comparatively short time taken for a casting to solidify its original crystal structure is formed, and faults such as seams (defects at the junctions of two streams of metal), gas porosity (entrapment of gas during pouring and cooling), and hot tears (cracks due to tensile stresses during solidification) may be caused. A knowledge of the mechanism of solidification and the rate of heat loss from the metal to the mould facilitates prediction of the way in which the casting will solidify.

14.2.1 Solidification of pure metals. Solidification requires energy to produce a crystalline structure, and for this reason cooling below the freezing point, i.e. supercooling, is necessary before the liquid starts to solidify. Solidification can be induced with comparatively little supercooling if particles of foreign matter are available to provide sites around which crystals can grow. An initial source is provided by the walls of the mould and subsequently by the solidified particles of the metal itself.

When metal is poured into a mould, it can be assumed initially that there is no temperature gradient in the molten metal. Heat is then extracted by the mould walls, and the metal in their vicinity is cooled more rapidly than that elsewhere. The result is a temperature gradient in the liquid and, just below the freezing point of the metal, crystals start to grow from the mould walls. The process continues as more heat is lost, with crystals growing inwards until the whole of the metal has solidified. Provided the mould walls have high conductivity the crystals near them will be small, with their axes randomly orientated. As solidification proceeds, crystals with their axes perpendicular to the mould grow more rapidly and their shape changes from equiaxed to columnar (Fig. 14.8).

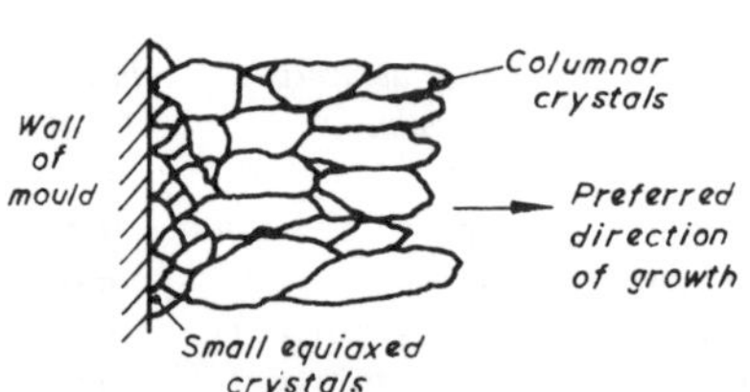

Fig. 14.8 Solidification of pure metal near mould wall

14.2.2 Solidification of alloys. In this section the solidification of solid solution alloys will be discussed; these are alloys in which one metal is dissolved in the other to form a single-phase alloy. A major difference between the solidification of alloys and of pure metals is that alloys normally freeze over a range of temperature, whereas pure metals have a single freezing point. When the alloy has cooled to a given temperature it starts to solidify, but remains in a mushy, part-liquid part-solid state, until further cooling renders it completely solid. The temperatures at which alloy solidification starts and finishes will vary with its composition, and are indicated by the liquidus and solidus lines respectively in Fig. 14.9.

If metal B is added to metal A to produce an alloy, the usual effect is to depress the melting point below that of metal A (Fig. 14.9). When the alloy is poured into a mould, the metal near its walls will cool more rapidly, and solidification will start on the walls at a temperature depending on the alloy composition. The frozen metal adhering to the mould walls will have a different composition from that of the original alloy. This can be seen by following the dotted line in Fig. 14.9, which shows that an alloy with an initial composition of 80% metal A and 20% metal B

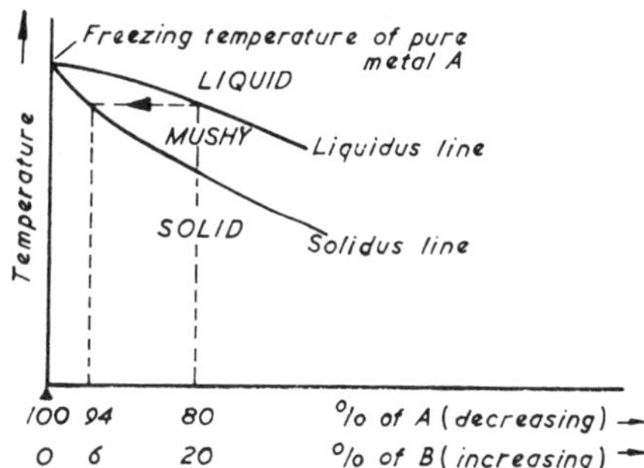

Fig. 14.9 Effect on solidification of alloying metals *B* with metal *A*

will first precipitate crystals of 94% A and 6% B. Assuming a low rate of diffusion, the concentration of rejected metal B in front of the solidification front rises above that of the parent alloy, and depresses the temperature at which solidification starts to below that of the parent alloy. This effect is referred to as constitutional supercooling, and will cause spikes of solidified metal to be pushed out perpendicular to the mould wall into the liquid metal. The solidification of the metal at the sides of the spikes will be further delayed as the local concentration of the rejected metal is progressively increased. If the rate of constitutional supercooling is high, the spikes will grow arms perpendicular to their main growth direction to produce a dendritic structure. When supercooling is extreme, equiaxed randomly oriented crystals will form in the liquid metal ahead of the solidifying interface.

14.3 SOLIDIFICATION OF CASTINGS

When a pure metal is poured into a mould, the temperature at a particular point in the liquid falls steadily until freezing commences. While solidification is occurring, the temperature at that point remains constant due to a release of latent heat; in fact, there will be a slight increase in temperature if supercooling has occurred (Fig. 14.10 (*a*)). With a solid solution alloy, there is a similar check to the temperature drop at the commencement of crystallization, followed by a period of less steep temperature reduction while the metal is passing through the mushy state (Fig. 14.10 (*b*)).

By using a number of thermocouples it is possible to find the distribution of temperature as a casting cools and hence to plot the time when

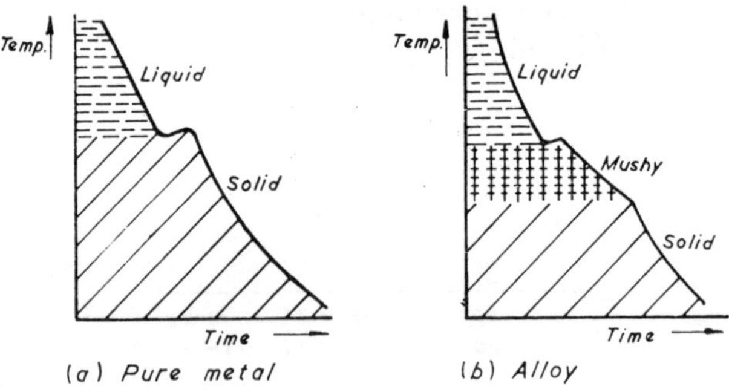

Fig. 14.10 Cooling of pure metals and alloys

freezing begins and ends at various distances from the centre line. In a sand mould, the extraction of heat will be considerably slower than in a metal mould. This can be seen by comparing Figs. 14.11 (*a*) and (*b*),

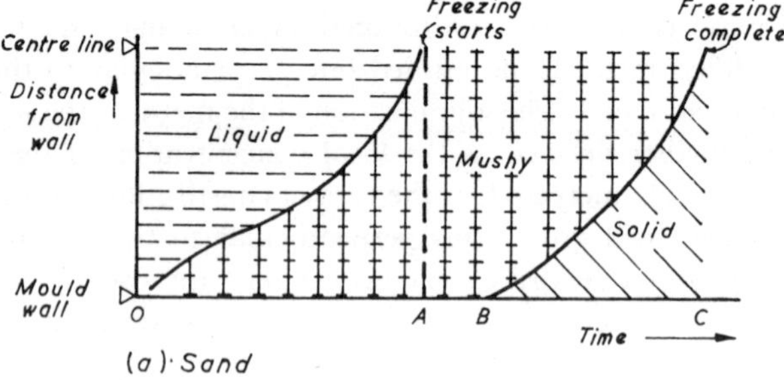

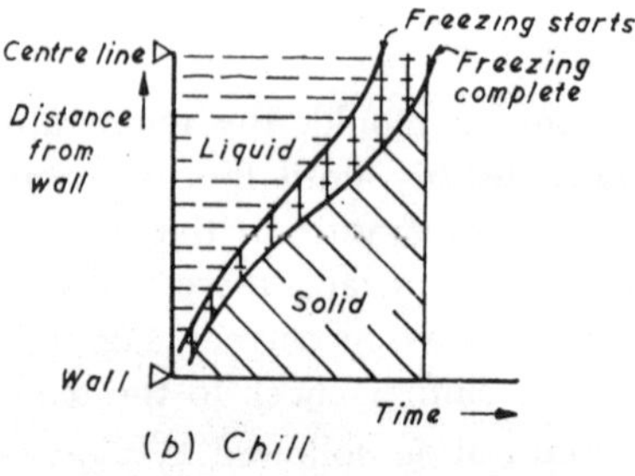

Fig. 14.11 Solidification of castings

which are typical of thick uniform sand and chill castings respectively. It should be noted that when heat is rapidly extracted from the casting, a narrow mushy zone quickly sweeps through the cooling metal (Fig. 14.11 (*b*)), whereas if heat is slowly extracted, the mushy zone may extend throughout the casting (e.g. AB in Fig. 14.11 (*a*)).

The presence of free crystals ahead of the solidification front makes the feeding of metal more difficult and the risk of voids greater. In general, alloys having the smallest temperature difference between the start and finish of solidification, i.e. the narrowest mushy zone, are the easiest to feed. An expression indicating the ease of feeding is the centre line feeding resistance (CFR),

where CFR =

$$\frac{\text{time during which crystals are forming at centre line}}{\text{solidification time of casting}} \times 100 \qquad (14.1)$$

Referring to Fig. 14.11 (*a*)

$$\text{CFR} = \frac{\text{AC}}{\text{OC}} \times 100 \simeq 42\%$$

In practice it is found that alloys with CFR $> 70\%$ are difficult to feed.

14.4 HEAT LOSS FROM CASTINGS

Risers, which act as a reservoir of liquid metal above the casting, must be able to supply liquid metal to the casting throughout its solidification period. So that a riser of adequate size can be provided, it is necessary to calculate the rate of heat removal through the wall of the mould.

Starting from the diffusion equation, derived in Appendix I, and assuming the mould wall to be a semi-infinite plane the diffusion equation reduces to one of conduction in the x direction only.

i.e.
$$\frac{\partial^2\theta}{\partial x^2} - \frac{1}{\kappa}\frac{\partial\theta}{\delta t} = 0$$

Carslaw and Jaeger,[66] by assuming that there was a sudden temperature rise at the mould face of $\theta_1 - \theta_0$, due to the metal being poured, and that κ, the thermal diffusivity of the mould, was independent of temperature, derived an expression for the temperature at distance x from the mould metal interface.

$$\theta_x = \theta_0 + (\theta_1 - \theta_0)\{1 - \operatorname{erf} x/(2\sqrt{\kappa t})\} \qquad (14.2)$$

where θ_0 is the initial mould temperature,
θ_1 is the mould/metal interface temperature, assumed constant during solidification,
t is the time after pouring.

Values of the error functions $\{\operatorname{erf} x/(2\sqrt{(\kappa t)})\}$ can be read from mathematical tables or the graph in Appendix 1.

The heat flow through the mould wall can be obtained by differentiating (14.2) with respect to x.

$$\frac{\partial\theta_x}{\partial x} = (\theta_0 - \theta_1)\frac{\mathrm{d}}{\mathrm{d}x}\left(\operatorname{erf}\frac{x}{2\sqrt{(\kappa t)}}\right)$$

expanding, $$\operatorname{erf} x/\{2\sqrt{(\kappa t)}\} = \frac{2}{\sqrt{\pi}}\int_0^{\frac{x}{2\sqrt{(\kappa t)}}}\left(1 - a^2 + \frac{a^4}{2!} - \frac{a^6}{3!} + \ldots\right)\mathrm{d}a$$

$$\operatorname{erf} x/\{2\sqrt{(\kappa t)}\} = \frac{2}{\sqrt{\pi}}\left[a - \frac{a^3}{3} + \frac{a^5}{5 \times 2!} - \frac{a^7}{7 \times 3!} + \ldots\right]_0^{\frac{x}{2\sqrt{(\kappa t)}}}$$

$$\operatorname{erf} x/\{2\sqrt{(\kappa t)}\} = \frac{2}{\sqrt{\pi}}\left[\frac{x}{2\kappa^{\frac{1}{2}}t^{\frac{1}{2}}} - \frac{x^3}{3 \times 8\kappa^{\frac{3}{2}}t^{\frac{3}{2}}} + \frac{x^5}{5 \times 2! \times 32\kappa^{\frac{5}{2}}t^{\frac{5}{2}}} - \frac{x^7}{7 \times 3! \times 128\kappa^{\frac{7}{2}}t^{\frac{7}{2}}} + \ldots\right]$$

$$\frac{\partial\theta_x}{\partial x} = \frac{2}{\sqrt{\pi}}(\theta_0 - \theta_1)\left[\frac{1}{2\kappa^{\frac{1}{2}}t^{\frac{1}{2}}} - \frac{x^2}{8\kappa^{\frac{3}{2}}t^{\frac{3}{2}}} + \frac{x^4}{64\kappa^{\frac{5}{2}}t^{\frac{5}{2}}} - \frac{x^6}{768\kappa^{\frac{7}{2}}t^{\frac{7}{2}}} + \ldots\right]$$

The thermal gradient at the mould face, i.e. at $x = 0$, can be obtained by substituting 0 for x in the above equation.

$$\frac{\partial\theta}{\partial x} = \frac{\theta_0 - \theta_1}{\sqrt{(\pi\kappa t)}}$$

The rate of heat flow/unit area J is given by

$$J = -K\frac{\partial\theta}{\partial x}$$

where K is the thermal conductivity of the mould material.

$$\therefore \qquad J = \frac{K(\theta_1 - \theta_0)}{\sqrt{(\pi\kappa t)}} \tag{14.3}$$

If we consider the time t_s required to solidify a plate of surface area A.

$$Q = A\int_0^{t_s} J\mathrm{d}t = \frac{2AK(\theta_1 - \theta_0)\sqrt{t_s}}{\sqrt{(\pi\kappa)}}$$

where Q = quantity of heat passing mould/metal interface.
An expression for Q can also be obtained by considering the amount of heat which must be lost to produce solidification.

$$Q = \rho V\{L + C(\theta_p - \theta_1)\}$$

where ρ is density of metal,
V is volume of metal,
L is latent heat of fusion of metal,
C is specific heat of metal,

θ_p and θ_1 are pouring and mould surface (solidification) temperatures respectively.
Equating the two expression for Q

$$\frac{2AK_{\text{mould}}(\theta_1 - \theta_0)\sqrt{t_s}}{\sqrt{(\pi\kappa_{\text{mould}})}} = \rho_{\text{metal}}V\{L + C(\theta_p - \theta_1)\}$$

$$t_s = \left[\frac{\rho^{\text{metal}}\{L + C(\theta_p - \theta_1)\}}{2K_{\text{mould}}(\theta_1 - \theta_0)}\right]^2 \pi\kappa_{\text{mould}}\left(\frac{V}{A}\right)^2$$

$$t_s = B\left(\frac{V}{A}\right)^2 \tag{14.4}$$

where B is the mould constant.
Equation (14.4) shows that the solidification time for a plate is proportional to (volume/area)^2. Chvorinov[67] has shown that this relationship is true for a large range of casting and riser shapes.

It is possible to use a similar method to find how time varies with depth of solidification. Considering one side of a large plate, area A, and half its volume, the quantity of heat Q needed to solidify it to depth d can be found.

$$\frac{2AK_{\text{mould}}(\theta_1 - \theta_0)\sqrt{t}}{\sqrt{(\pi\kappa_{\text{mould}})}} = \rho_{\text{metal}}Ad\{L + C(\theta_p - \theta_1)\}$$

and $$t = Bd^2 \qquad (14.5)$$

Equation (14.5) shows that the time needed to solidify to a given depth is proportional to the square of the distance from the mould/metal interface.

14.5 RISER DESIGN

If risers are to act as sources of molten metal for a solidifying casting they must be large enough to remain liquid after the casting has solidified, as well as contain sufficient metal to make good contraction losses. Risers should be positioned so that they can continue to supply metal throughout the solidification period.

If the solidification of a casting without risers is considered, the liquid metal will soon be completely surrounded by a solidified shell of fixed outer dimensions and the contracting liquid will inevitably produce voids towards the centre of the casting. Further contraction occurs when cooling in the solid state; this does not produce shrinkage voids, although it can set up undesirable stresses in the casting.

Caine[68] has produced a relationship for steel castings which enables the adequacy of riser size to be checked by taking account of the shape

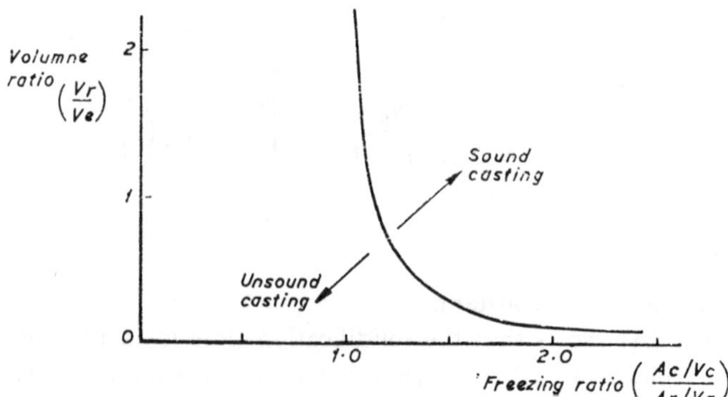

Fig. 14.12 Typical risering curve (*after Caine*)

of both riser and casting. It was assumed that the heat loss from a casting or riser is proportional to its surface area A and the heat content is proportional to its volume V: hence cooling rate $= \propto (A/V)$

Consider first a casting which, because of its chunky shape, has a very slow freezing rate (A/V is low). In this instance the accompanying riser must be large so that it remains liquid after the casting has solidified, i.e. $(A_{casting}/V_{casting}) > (A_{riser}/V_{riser})$. At the other extreme, the size of the riser for a fast freezing casting (A/V is high) need not be so large, but must at least be large enough to make up the contraction as the casting is cooling and solidifying.

A risering curve which enables the minimum riser size to be determined for a given material is shown in Fig. 14.12. The x axis represents the freezing ratio $(A_c/V_c)/(A_r/V_r)$ and the y axis the volume ratio V_r/V_c. The relationship between the x and y co-ordinates is a hyperbolic function expressed by the equation

$$x = \frac{a}{y - b} + C \qquad (14.6)$$

where x is freezing ratio $(A_c/V_c)/(A_r/V_r)$

a is the freezing characteristic constant for the metal,

y is volume ratio V_r/V_c

b is contraction ratio from liquid to solid,

C is relative freezing rate of riser and casting (unity if same mould material around casting and riser).

A typical value of equation (14.6) for steel when the casting and riser are surrounded by the same mould material is

$$x = 0{\cdot}1/(y - 0{\cdot}03) + 1{\cdot}0$$

Risers are connected to the casting by a neck of metal called a gate, which enables the riser to be removed easily from the casting after solidification. Gates do not freeze before their risers as the continual passage of liquid metal locally heats the mould material to a higher temperature than that of the rest of the mould surface.

14.6 RISER PLACEMENT

Although the graph in Fig. 14.12 indicates a suitable riser size, it assumes that the riser is able to feed the solidifying casting effectively. Where the casting has a chunky shape, approximating to a cube or sphere, (A_c/V_c is low) there is usually little difficulty in feeding it from a single riser. Bar and plate shaped castings (A_c/V_c is high) may however require more than one riser, otherwise the slushy state just prior to solidification may restrict metal flow from a single riser and cause centre-line shrinkage. Bishop and Pellini[69] found for steel plates 12–100 mm ($\frac{1}{2}$–4 in) thick, one central riser could be used, provided that the maximum feeding length $< 4{\cdot}5 \times$ plate thickness. It was also found[70] that bars of square cross section in thickness from 50–200 mm (2–8 in) could be satisfactorily fed from a central riser for distances $< 6 \times \sqrt{\text{bar thickness}}$. If chills (blocks of metal) are built into the mould, they will increase the cooling rate and reduce the centre-line feeding resistance, thereby permitting longer feeding distances. Chills can be placed at the ends of bars and plates with central feeding; when more than one feeder is used, a chill can be positioned midway between two feeders (Fig. 14.13).

14.7 APPLICATION OF STEEL RISERING DATA TO OTHER METALS

Most of the quantitative risering data which has been published applies to steel. The same data can however be used for the sandcasting of alloys with a lower centre-line feeding resistance than steel. If it is applied to metals with a centre-line feeding resistance higher than steel, chills will be needed to ensure soundness of those parts of the casting requiring the greatest strength. Grey cast iron is especially interesting since solidification occurs in two stages. The first is similar to that of solid solution alloys, where metal from the riser is required to make good initial contraction; in the second stage, expansion occurs and liquid metal is returned to the riser. With certain grey irons the contraction is balanced by the expansion and these require no risers.

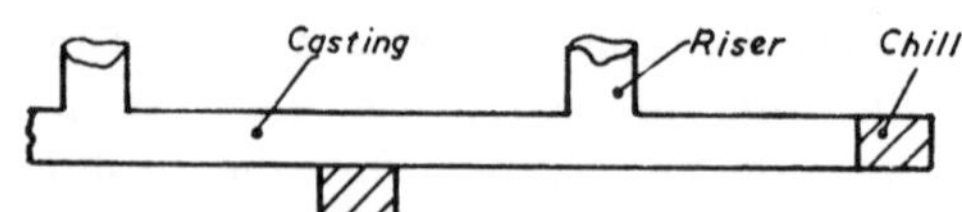

(a) Chills used to extend feeding distance

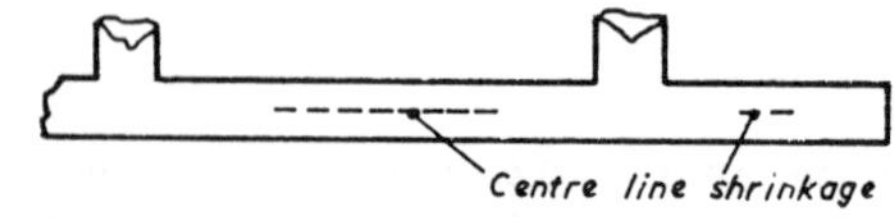

(b) Same casting without chills

Fig. 14.13 Effect of chills

14.8 POURING OF CASTINGS

When metal is being poured into a mould it should be done with minimum loss of temperature; the liquid flow should not damage the mould, nor should it carry sand, dross (oxide) or air into the casting. Much can be done by good gating design to achieve these ends; gating systems will vary with the metal being poured, and metals such as aluminium which oxidize easily demand special attention. Some general principles of gating design are discussed below.

The shape of the vertical passage down which the metal is poured must be designed so that no air is absorbed by the liquid metal on its downward passage. A gate feeding directly into the mould with a cylindrical shape similar to that shown in Fig. 14.14 (*a*) will be considered.

By applying Bernoulli's equation for the total energy of a liquid, it is possible to find pressures at various levels in the gate. Neglecting friction and assuming an impermeable mould where the pressure is atmospheric, i.e. $p_2 = p_0$, we have at points x and 2.

$$h_x + \frac{v_x^2}{2g} + \frac{p_x}{w} = h_2 + \frac{v_2^2}{2g} + \frac{p_0}{w}$$

$$h_x + \frac{v_x^2}{2g} + \frac{p_x}{w} = 0 + \frac{v_2^2}{2g} + \frac{p_0}{w}$$

$A_x = A_2$ (cylindrical gate)

$\therefore\ v_x = v_2$ (as $A_x v_x = A_2 v_2$ for continuity of flow) hence

$$p_x = p_0 - h_x w \tag{14.7}$$

The pressure of the liquid at point x will therefore be below atmospheric, by an amount depending on its distance up the gate and the density of

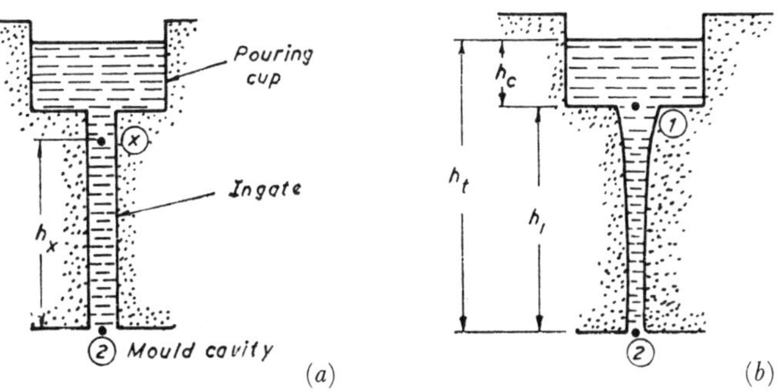

Fig. 14.14 Cylindrical and tapered ingates

the metal.

A sand mould is, however, permeable and the mould gases will be in contact with the metal at the mould/metal interface. Gas can therefore pass into the metal during its journey down the gate, the amount depending on the pressure reducing term $h_x w$ (equation 14.7), and on the pressure to which the mould gases rise as the metal is being poured.

To reduce the aspiration of gases into the stream of molten metal the pressure in the gate should not be allowed to fall below atmospheric. This can be achieved by using an appropriately tapered gate (Fig. 14.14 (*b*)). Considering the entrance and exit of this gate and applying Bernoulli's equation to points 1 and 2:

$$h_1 + \frac{v_1^2}{2g} + \frac{p_1}{w} = 0 + \frac{v_2^2}{2g} + \frac{p_0}{w}$$

To avoid aspiration, p_1 should not fall below p_0

$$\therefore \qquad h_1 + \frac{v_1^2}{2g} = \frac{v_2^2}{2g}$$

For continuity of flow $A_1 v_1 = A_2 v_2$

Substituting $v_2\,(A_2/A_1)$ for v_1

$$v_2^2\left\{\left(\frac{A_2}{A_1}\right)^2 - 1\right\} = -2gh_1$$

$$\frac{A_2}{A_1} = \sqrt{\left(1 - \frac{2gh_1}{v_2^2}\right)} \qquad (14.8)$$

Assuming that the pouring cup is kept filled to height h_c then the velocity v_2 of the metal at the exit of the gate can be found $v_2 = \sqrt{(2gh_t)}$ where h_t is the total head.

From Fig. 14.14, $h_1 = h_t - h_c$

Substituting from v_2 and h_1 in equation (14.8)

$$\frac{A_2}{A_1} = \sqrt{\left[1 - \frac{2g(h_t - h_c)}{2gh_t}\right]}$$

$$\frac{A_2}{A_1} = \sqrt{\left(\frac{h_c}{h_t}\right)} \tag{14.9}$$

Therefore, to avoid aspiration A_1 should be greater than A_2 as h_c is always less than h_t. The sides of the downgate should be hyperbolic in section to satisfy equation (14.9), although a straight-sided taper is normally used as it is easier to produce.

Most mould cavities are not filled directly from the simple vertical gates just described but via a down sprue and a short horizontal gate, as shown in Figs. 14.15 (*a*) and (*b*). This arrangement minimizes oxidization and reduces the damage to the mould cavity because the force of the incoming metal is reduced. Some cavities are filled from the bottom (Fig. 14.15 (*a*)); this method is likely to minimize mould damage and oxidization but the time to fill the mould is considerably increased.

A number of methods of preventing impurities from going into the casting can be used in the pouring system (Figs. 14.15 (*a*) and (*b*)).

(*a*) *Pouring cups.* These break down the eroding force of the stream of molten metal being poured from a ladle. They also help to maintain a constant pouring head.

(*b*) *Strainers.* Ceramic strainers can be set in down sprues to prevent dross from the ladle entering the casting.

(*c*) *Splash cores.* Splash cores are made from ceramic material and are placed at the bottom of down sprues. As they are not eroded by the stream of molten metal, they minimize the amount of sand entering the mould.

(*d*) *Skim bobs.* These are traps placed in horizontal gates to catch both the heavier and lighter impurities flowing towards the casting.

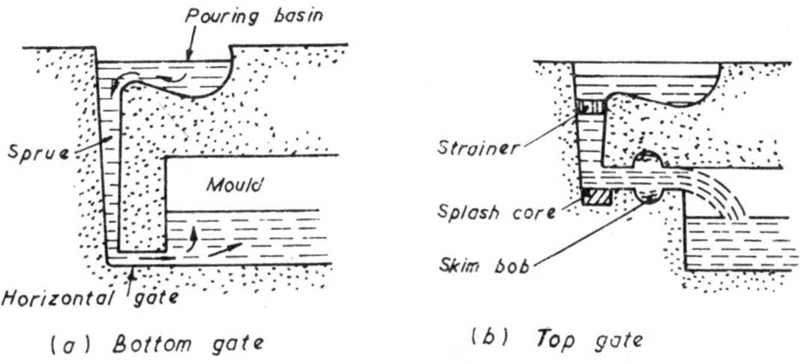

Fig. 14.15 Sprues with horizontal gates

14.9 GASES IN CASTINGS

Faulty castings may result from the presence of gases in the metal. Gases can appear as gas holes, pin holes and porosity, depending on their size and distribution (Fig. 14.16). Not all gases appear as voids in the solidified casting, for instance nitrogen can embrittle steels by forming nitrides.

Aspiration, the mechanical entrapment of gases during pouring, has been described in section 14.8. It can be prevented by proper riser design and by the use of adequately vented and permeable moulds.

A second cause of gases in castings is that many common gases may already be dissolved in the metal before it is poured. The solubility of a gas in a metal is a function of its temperature and a typical solubility curve for a pure metal and an alloy is shown in Fig. 14.17. It will be seen that the solubility drops sharply as the metal transforms from liquid to

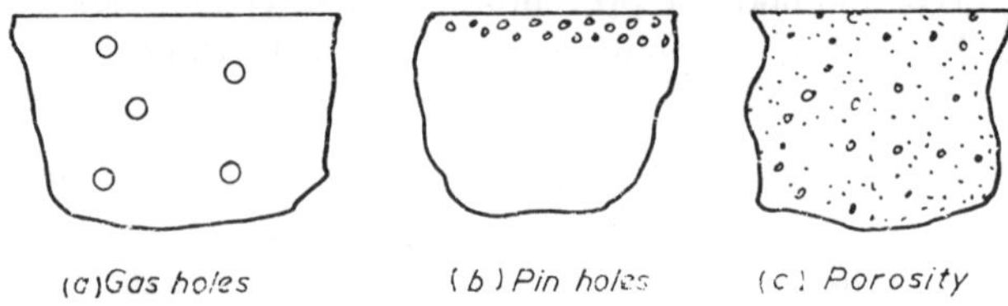

Fig. 14.16 Casting defects caused by gases

solid. For instance, the solubility of hydrogen in iron when the partial pressure of hydrogen surrounding the melt is 0·1 N mm^{-2} (1 atm) is $2{\cdot}7 \times 10^5$ mm^3 kg^{-1} of iron just prior to solidification, but falls to $0{\cdot}7 \times 10^5$ on solidification. Although large amounts of superheat above the liquids will produce high fluidity assisting mould filling, this must be balanced against the larger quantities of dissolved gases that the molten metal can hold. Superheat should therefore be kept to a minimum, but up to 250°C of superheat may be necessary for intricate castings with thin sections.

The pressure of the gas in contact with the liquid metal also affects its solubility. Sievert's law states that the solubility of gas in the melt varies as the square root of the partial pressure of the gas over the melt ($S = K\sqrt{p}$). For instance, at 0·025 N mm^{-2} (0·25 atm) the solubility will be half that at 0·1 N mm^{-2}. Vacuum melting and vacuum degassing can be used to reduce gas in melts. Almost all the gas can be removed by vacuum melting and pouring at pressures of about 10^{-3} mm of mercury. Vacuum degassing is a little less effective and involves placing the liquid metal in a low-pressure chamber for a few minutes to remove the dissolved gases, and then removing and pouring.

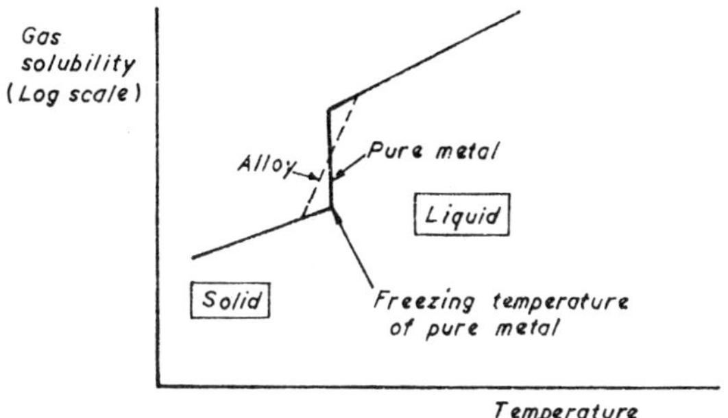

Fig. 14.17 Variation of hydrogen solubility with temperature

Purging is frequently used to remove hydrogen from brass and aluminium. The process consists of bubbling a dry gas, such as chlorine, nitrogen or argon, through the melt. The partial pressure of the hydrogen in the bubbles is initially zero, and the hydrogen from the metal diffuses into them, being released to atmosphere when the bubbles reach the surface. Nitrogen is an unsuitable purging gas for steel as it dissolves in iron, but carbon monoxide is satisfactory.

Apart from simple gases such as hydrogen and nitrogen, which are present in almost all cast metals, complex gases can also occur in castings. They are particularly troublesome in copper alloys where hydrogen, oxygen and sulphur in the metal may react chemically during solidification to produce SO_2 and H_2O.

Gas inclusions in high-pressure diecastings are difficult to eliminate as the inrush of liquid metal into the die entraps air. The oxygen in the air dissolves in the metal, but the nitrogen remains as a gas, weakening the casting. The Japanese have developed a process which purges the air in the die with oxygen prior to metal injection; this eliminates the nitrogen and the strength and ductility of the castings are significantly improved.

PART 2. SINTERING

14.10 APPLICATIONS OF SINTERED METALS

The fabrication of shapes from small particles of metal by sintering has been practised for several thousands of years and a six-ton iron pillar at Delhi was produced in this way about 300 AD. The large-scale use of sintered metal components has, however, been a post-war development, although cemented carbide tool tips, porous oil-impregnated bearings, porous metal filters and powder magnetic cores were all available in the 1920s. The sintering process is also used in the ceramic industry and in recent years there has been increased use of cermets, a sintered combination of metals and non-metals.

Sintering has a number of advantages:

(*a*) the parts produced have an excellent surface finish and high dimensional accuracy; this often avoids machining operations;

(*b*) the porosity inherent in sintered components is useful in specialized applications such as filters and bearings;

(*c*) refractory materials which are impossible to shape by other processes can be shaped by sintering with a metal of lower melting point, e.g. sintering tungsten carbide with cobalt;

(*d*) a wide range of parts with special electrical and magnetic properties can be produced.

Often sintering is the only satisfactory method of production, but sometimes there are alternative processes. In these instances the limitations of sintered parts must be considered. Sintered components are appreciably more brittle than those made from solid metal and certain shapes cannot be produced. Because of tooling costs, small quantities of sintered parts are expensive and a figure of 20 000 parts is often quoted as the break-even quantity between sintering and machining from solid metal.

Manufacture normally takes place in three stages: production of the powder, shaping the powder, and strengthening by sintering.

14.11 PRODUCTION OF METAL POWDERS

The manufacturing methods normally fall broadly into two categories, chemical and mechanical. The former includes reduction, precipitation, chemical reaction and electrolysis, and the latter atomization and disintegration. Very fine particles, down to a few microns in size, can be produced if required. Sintered parts are normally made up of particles having a range of sizes; a typical specification (using BS mesh sizes) might be:

50% between 100 and 150 mesh
25% between 150 and 200 mesh
25% between 200 and 300 mesh

The mesh number refers to the number of holes in the mesh per linear inch.

The particles vary not only in size but in shape, the different manufacturing processes producing characteristic shapes. Particle shapes are difficult to describe but vary from spherical (those with the smallest surface area per unit volume) to irregular shapes with large surface areas.

Typical methods of powder production are outlined below.

14.11.1 Reduction of the metallic oxide. This method is widely used to produce powders from oxides of the metals with higher melting point, the most common reducing agents being hydrogen and carbon. At high reduction temperatures the metal particles sinter together into a spongelike mass which has subsequently to be crushed into a powder; at lower reduction temperatures a lightly caked powder is formed. The carbon reduction method is economical and large tonnages of iron powders are produced in this way. A disadvantage of this process is that impurities in the oxide also appear in the powder.

14.11.2 Precipitation from vapours. Zinc powders may be produced by condensation from metal vapour. Iron and nickel powders can be obtained by the precipitation of carbonyl vapour, the carbonyls being obtained by passing carbon monoxide at high pressure over the heated metal. Although powders of high purity are obtained from the carbonyl method, it is an expensive one.

14.11.3 Electrolysis. This process again uses a salt of the metal. Metal is either deposited as a sludge at the bottom of the tank or as a spongy mass on the electrode.

14.11.4 Atomization. Fine powders can be produced by the mechanical disintegration of molten metal caused by a jet of gas or water at high pressure.

About 80% of powders produced are iron powders. The main methods of manufacturing these in the U.K. are by atomisation (Domitar process) and the reduction of iron ore (Hoganas process).

14.12 FURTHER TREATMENTS

The products of the above processes require further treatments. Crushing or grinding may be needed to reduce the product to a powder. Annealing may be required to remove work hardening; if annealing is performed in a reducing atmosphere it will also remove oxides from the powder. Spheroidizing of particles for use in filters is achieved by allowing the powder to fall through heated space. Finally the particles are segregated into their size ranges by screening.

14.13 SHAPING OF POWDERS

Before sintering, the powder is shaped into the form of the part. The processes used fall into three main categories: pressureless forming, cold pressing, and hot pressing, cold pressing being by far the most widely used.

14.13.1 Pressureless forming. When highly porous parts are required they can be produced by loose sintering. The powder is poured or vibrated into a mould, which is then heated to sintering temperature. Considerable shrinkage occurs during heating, and only those shapes which can be withdrawn from the mould after sintering are suitable for manufacture by this method.

Other methods of pressureless forming are slip and slurry casting but these are not widely used.

14.13.2 Cold pressing. In this process the powder is formed into shapes suitable for sintering by mechanical or hydraulic pressure. The pressures used are comparatively low, from about 78 N mm^{-2} (5 tonf/in^2) for soft bearing alloys to 780 N mm^{-2} (50 tonf/in^2) for steels. The pressures used are however, high enough to produce cold welding of the powder; this occurs because the contact areas between granules are small and produce correspondingly high welding pressure. Cold welding imparts a green strength, which holds the parts together and allows them to be handled;

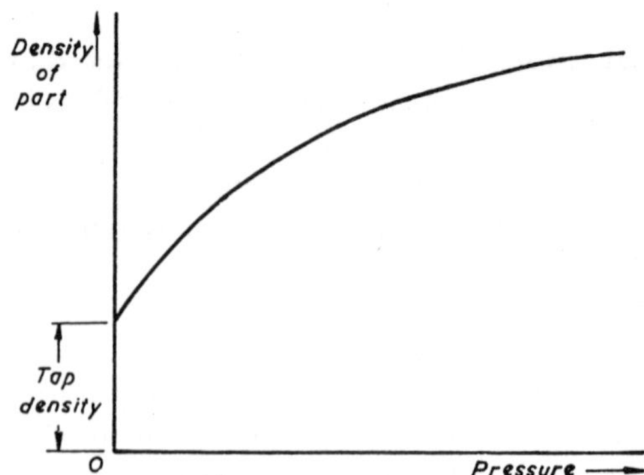

Fig. 14.18 Effect of compacting pressure on density

it also increases the density of the powder beyond its tap density (the density of the powder obtained by tapping it in a graduated cylinder). A typical pressure/density curve is shown in Fig. 14.18. When under pressure the powder does not behave as a fluid, with pressure equally distributed throughout its volume: because of friction between the granules themselves and between the granules and the tool, the compacting pressure varies. Typical pressure variations in a cylindrical part produced by single and double ended pressing are shown in Figs. 14.19 (*a*) and (*b*). It will be seen that if pressure is applied at both top and bottom a much more even distribution is obtained, resulting in a stronger and denser part. Waxes and soaps are used to lubricate powders and to produce a more even pressure distribution. They are particularly valuable in reducing friction at the die walls, but inhibit cold welding and thereby reduce green strength. Ideally only the die wall should be lubricated, but this is difficult to achieve under production conditions.

Other methods of cold pressing include isostatic pressing, where the powder is placed in a flexible mould which is then subjected to a high hydrostatic pressure, about 70 N mm^{-2} (10 000 lbf/in^2). Using this method, green densities of up to 90% of the solid metal density can be obtained with ductile powders, but close tolerances cannot be held. Another method of achieving high green densities is to compact powders by an explosive charge; parts weighing 12 kg (26 lb) have been compacted by this method.

14.13.3 Hot pressing. If pressing is carried out above the recrystallization temperature of the metal, work hardening is eliminated and accurately shaped high density parts are produced. In particular, very hard powders which cannot be satisfactorily cold pressed may be compacted by

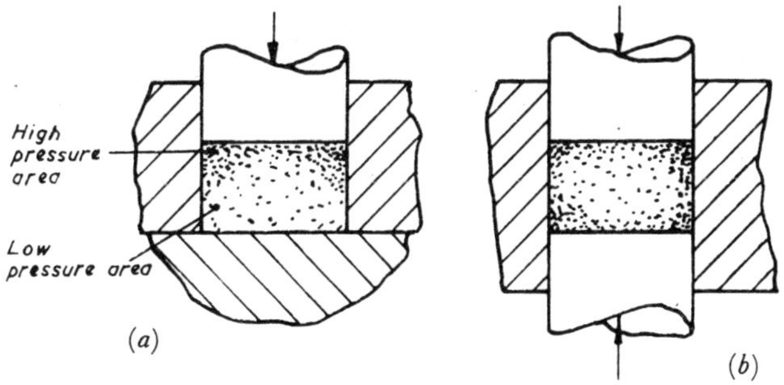

Fig. 14.19 Pressure variation with single and double ended pressure

this process; for instance, diamonds used in cutting tools can be hot pressed in a metal matrix.

A disadvantage of hot pressing is that unless it is carried out in a vacuum, a neutral or a reducing atmosphere, the metal will oxidize. Die materials must be appropriate to the pressing temperature. Below 1000°C suitable metallic dies can be employed but above this limit graphite is almost universally used. Graphite provides its own reducing atmosphere, and is cheap, easily machined and resistant to thermal shock. It is, however, weak, and pressing loads must be below 30 N mm^{-2} (2 tonf/in^2); this is not a serious disadvantage as the loads needed to produce a dense part at high temperatures, even with very hard materials, need not exceed this limit.

Other methods of hot-forming powders are to roll, forge or extrude them. In most of these processes the powder is 'canned' in a container from which air can be evacuated to prevent oxidation. Experimental work has shown that hot isostatic pressing is possible, although it is difficult to find a suitable envelope for the higher ranges of temperatures envisaged.[71]

14.14 SINTERING SYSTEMS

Prior to sintering, the components are heated to a temperature of around 400°C to evaporate any volatile lubricants.

Sintering is the process normally used to strengthen the fragile green compacts produced by the pressing operation and to convert them into engineering components. It may be defined as heating a powder to below the melting point of at least one of the major constituents to produce bonding. Many parts consist of a single metal (a single-phase system) and in these instances, and in some multi-phase systems, the sintering temperature is such that no melting occurs. Apart from strengthening the component, sintering increases electrical conductivity, density and ductility. The variation of these properties with sintering temperature for a

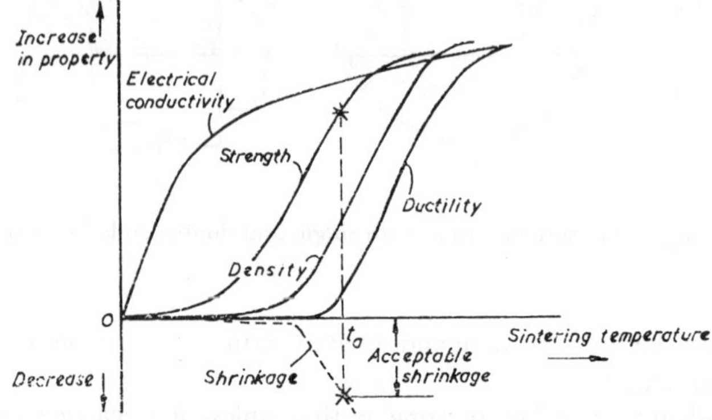

Fig. 14.20 Effect of sintering temperature on properties

typical single-phase sintered part is shown in Fig. 14.20. Most materials have a range of possible sintering temperatures and the one selected for a particular part is usually a compromise. In most instances the choice is between strength and dimensional stability, and the highest temperature which provides an acceptably small dimensional change is chosen, e.g. t_a (Fig. 14.20). Due to the high sintering temperatures and to the large surface area of the particles, severe oxidation will occur during sintering unless precautions are taken. These normally consist of using a reducing atmosphere in the furnace or of sintering in a vacuum.

14.14.1 Stages in single-phase sintering. Although sintering is an operation in which the changes are gradual, three stages can be identified. The first is the growth of necks at the points of contact between individual particles (Fig. 14.21). As the necks enlarge, the mid-points of the particles approach each other and there is a small amount of shrinkage. In the second stage, the grain growth which started when sintering began is accelerated. Individual particles start to lose their identity and the interconnecting network of pores changes to isolated vacancies. Most of the shrinkage occurs at this stage, and the density begins to approach that of the solid metal. The third stage, which may not be reached in some sintering operations, is identified by a spheroidization and reduction in the number and size of isolated pores, and a slow increase in density.

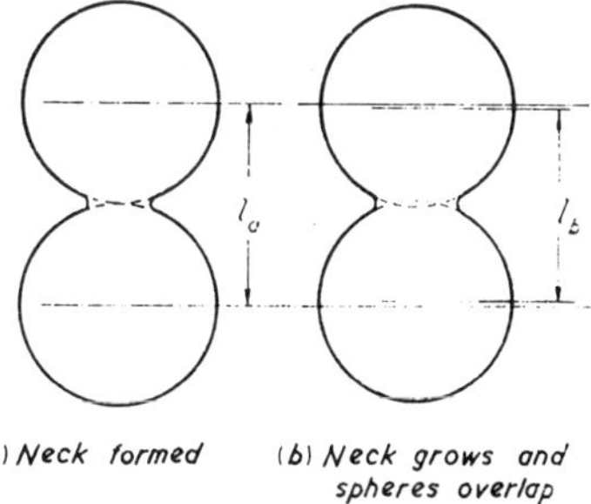

Fig. 14.21 Neck growth in sintering

14.14.2 Sintering in multi-component systems. Many sintering powders consist of two or more materials; for instance, a typical carbide tool might be made from tungsten carbide, titanium carbide and cobalt. In multi-component systems diffusion is no longer limited to self-diffusion; heterogeneous diffusion can occur, i.e. different materials can diffuse into each other across the particle boundaries. In some multi-component sintering, one of the materials may melt and liquid-phase sintering will result. This is an important process, in which particles of high melting point material such as tungsten carbide can be bonded together, and under certain conditions porosity can be eliminated.

Three stages may be identified in liquid-phase sintering. The first is the re-arrangement of the residual solid by viscous flow in the liquid. This gives a closer packing of the solid particles and is accompanied by rapid densification. Provided there is limited solubility of the solid phase in the liquid, a second stage can be distinguished. This is the disappearance of some of the smaller particles and a rounding-off in shape of the larger ones. If wetting by the liquid phase is complete, the particles are separated from each other by a film of liquid, and there is no third stage. A third stage, however, occurs at unwetted particle boundaries. This is similar to the sintering which would occur if there were no liquid phase present, and is accompanied by a gradual densification.

No analytical treatment of sintering has been attempted in this chapter, but those seeking further information can find a comprehensive list of references in an article by Thümmler and Thomma,[72] reviewing the sintering process.

14.15 SINTER FORGING

The sintered part is coated with graphite, reheated and forged in a die. The product has a density approaching 100% of the theoretical maximum, with a strength equal to that of a conventional forging; it has a good surface finish and close dimensional tolerances. Sinter forgings are in competition with conventional closed die forgings where large quantities of parts are required, as in the automobile industry.

15 Polymer Processing

It is estimated that by the 1980s the volume of plastics production will commence to outstrip that of all metals.

The reason for this lies in the particular merits of high-polymer materials which are:

low relative density that leads to high specific strength and stiffness,
corrosion resistance,
mostly both tough and resilient, with good vibration damping capacity,
good electrical and thermal insulators,
low coefficients of friction,
ease of fabrication.
cheaper than metals on a volume basis.

The disadvantages are:

low permissible operating temperatures,
low strength in comparison with metals,
high thermal expansion,
degradation due to sunlight, in some instances,
inflammability.

A peculiar characteristic of thermoplastic polymers is their ability to recover visco-elastic strains during a period of time after unloading. This is an advantage in providing shrink-fit packaging, but not when it leads to post-mould shrinkage of components in service. This phenomenon is illustrated in Fig. 15.1.

Most of the processing techniques have similarities to those developed for metal forming, but variations occur between the methods for thermoplastic and thermosetting polymers. Fig. 15.2 shows the processes applicable to the two groups of materials. Thermoplastics consist of long molecular chains entangled with one another, but not actually bonded together. Heating reduces the viscosity of such polymers, and cooling increases the viscosity in a reversible process.

In thermosets long molecular chains are not only intertwined but also bonded together by additional covalent bonds, i.e. 'cross-linked'. The curing of thermosetting polymers by cross-linking is an irreversible change and they cannot subsequently be softened by heating.

The molecular structures can be modified and 'alloyed' by forming copolymers to give desirable characteristics, even as metals are tailored to specific requirements. Properties are further modified by other chemical additives and reinforcing materials.

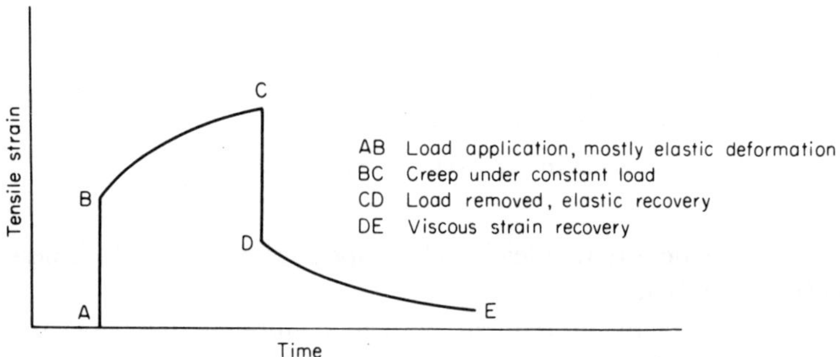

Fig. 15.1 Time-dependent strain of a thermoplastic polymer

Detailed information on the processing of particular polymers for specific applications is readily available on request to the principal suppliers of the raw material, whose expertise is invaluable.

The techniques normally employed for a range of commercially available polymers are shown in Fig. 15.3.

15.1 THE GENERAL REQUIREMENTS OF POLYMER PROCESSING

An understanding of the most important aspects of polymer processing can be achieved by considering the extrusion process, which lies behind many production systems.

The essential parts of an extruder are shown in Fig. 15.4. Polymer in a granular form, together with any necessary additives, enters the feed zone of the barrel from a hopper. This region is water-cooled to prevent premature softening of the feed and a blockage in the supply system. A rotating screw transfers the mixture along the barrel during which it is homogenized by the shearing action. The resulting frictional heat, together with that from the external barrel heaters, lowers the viscosity of the melt ready for its passage through the breaker plate and die. The latter gives a profile to the extrudate.

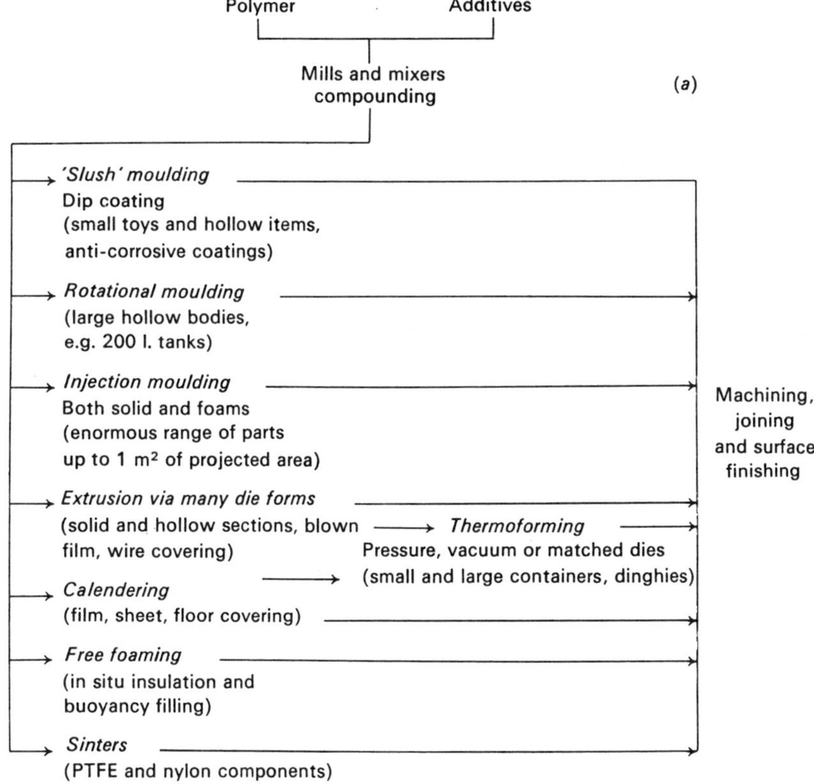

Fig. 15.2 Plastics processing

The stages involved in most polymer processing are then;

(1) mixing, homogenizing and melting,
(2) melt transfer,
(3) shaping,
(4) removal.

15.2 THE MAIN METHODS OF POLYMER PROCESSING

15.2.1 Mixing and compounding. Compounding is mixing during which a change of state occurs. Plastics are formulated from several components such as; polymer, stabilizers, plasticizers, colourants, lubricants, fillers, anti-statics. These may be solids or liquids of varying viscosity. If the finished product is to be homogeneous the mixing must be complete, and at a uniform temperature before passing to the next production stage.

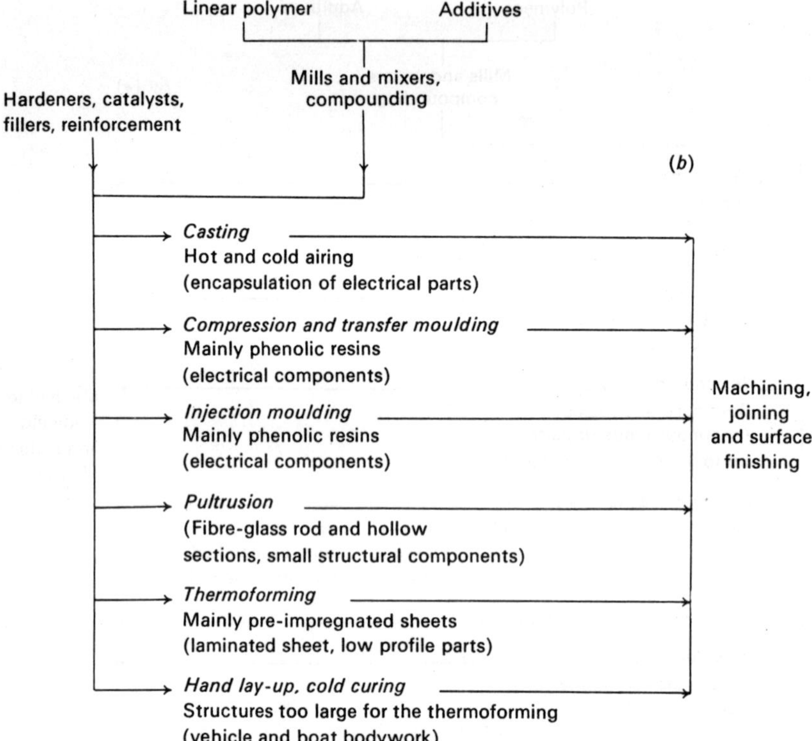

Fig. 15.2 Plastics processing

The simple mixing of powders and low viscosity pastes is carried out in rotating drums, or 'tumblers', and with rotary paddles of the simple stirrer type. Larger capacity mixers are often of the 'ribbon blender' and 'Banbury' types, although there are many other patented systems. These are normally fitted with a means of heating or cooling the charge as necessary. Where large production rates exist such machines may have a capacity of several thousand kilograms per charge.

The thorough mixing of a formulation normally requires its passage through a roll mill. Fig. 15.26 shows the pre-calendering preparation of a polymer, and it will be noticed that an extruder which discharges through a strainer is also included in the line.

Several types of mixing and compounding devices are shown diagrammatically in Figs. 15.5–7.

TYPE OF PLASTICS	Compression moulding	Transfer moulding	Injection moulding	Extrusion	Blow moulding	Thermoforming (vacuum forming)	Calendering	Rotational moulding	Dip moulding	Powder coating	Sheet for forming	Casting	Laminating	Expanding (foaming)	Encapsulation	Sintering	Glass fibre reinforcing
THERMOSETS																	
Alkyds (DAP and DAIP)	X	X	X									X			X		X
Aminos (urea and melamine formaldehyde)	X	X	X[1]										X	X[7]			
Epoxides	X									X		X	X		X		X
Phenolics	X	X	X	X								X	X	X	X		X
Polyesters	X										X	X	X	X	X		X
Polyimides	X															X	X
Polyurethanes	X			X								X	X	X			
Silicones	X	X										X	X	X	X		X
THERMOPLASTICS																	
Acrylics			X	X	X	X		X[3]			X	X	X	X[5]			
Acrylonitrile butadiene styrene (ABS)			X	X	X	X	X				X			X[3]			X
Cellulose acetate	X		X	X		X					X						
Cellulose acetate butyrate (CAB)	X		X	X		X	X	X		X	X						
Cellulose propionate	X		X	X		X				X	X						
Chlorinated polyether			X	X	X	X				X							
Chlorinated polyethylene			X	X	X	X		X			X			X[3]			
Ethylene vinyl acetate (EVA)			X	X	X												
Fluorocarbons: PTFE				X[2]							X		X			X	
PCTFE	X	X	X[2]	X[2]													
FEP			X	X		X					X						X
Ionomers			X	X	X	X											
Nylons			X	X	X[1]			X[4]		X		X				X	X
Polyacetal			X	X	X												X
Polycarbonate			X	X	X	X		X[3]									X
Polyethylenes			X	X	X	X		X			X			X			X
Polyethylene terephthalate			X	X													X
Polyphenylene oxide (PPO)			X	X	X												
Polyphenylene sulphide	X		X		X					X							
Polypropylene			X	X	X	X				X	X			X[6]			X
Polystyrenes			X	X	X	X		X			X			X			X
Polysulphones			X	X	X	X					X						X
Polyvinyl chloride (PVC)	X	X	X	X	X	X	X	X	X	X	X		X	X			X
Styrene acrilonitrile (SAN)			X	X	X												X
TPX (methylpentene polymer)			X	X	X	X											
Polyurethane (thermoplastic)			X	X			X										
Phenoxy			X	X	X												

NOTES: 1. Some grades. 2. With difficulty. 3. Not known to have been exploited commercially at this time. 4. Nylons 11 and 12. 5. Produced in Japan. 6. Produced in USA. 7. Urea formaldehyde.

Fig. 15.3 Principal types of plastics and their normal methods of forming

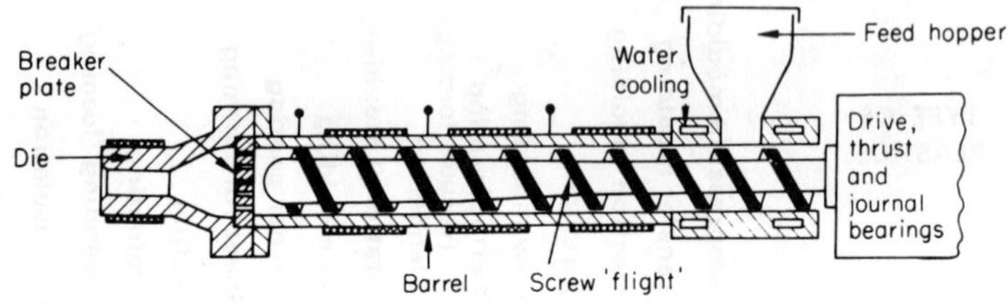

Fig. 15.4 Elements of a single screw extruder

Fig. 15.5 Typical arrangement of ribbon blender

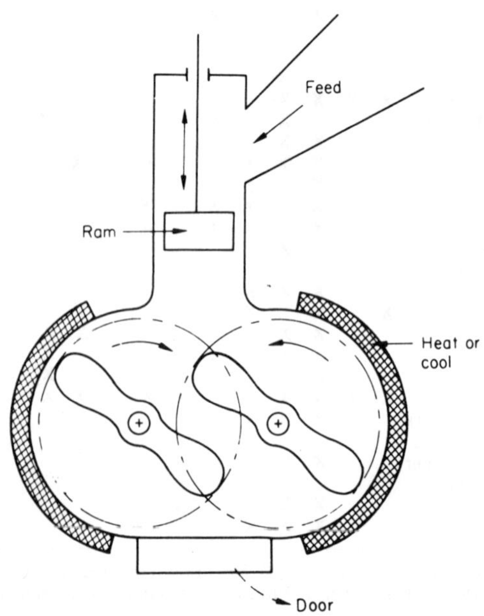

Fig. 15.6 Section through Banbury type mixer

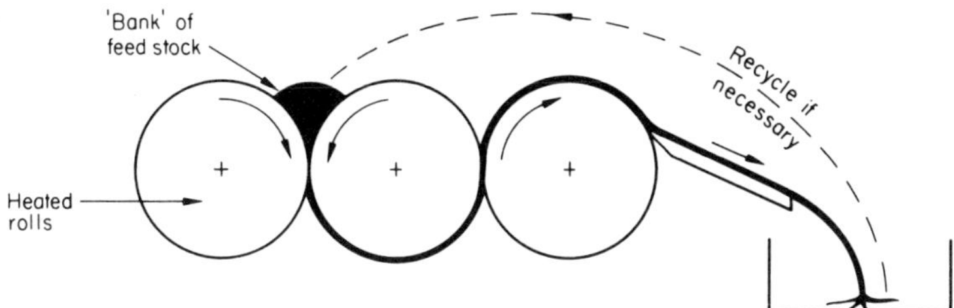

Fig. 15.7 Three-roll mill

15.2.2 Extruding. This is normally only used for thermoplastics, except when used as part of the system for injection moulding thermosetting materials. The single-screw extruder shown in Fig. 15.4 represents a continuous process of feeding, compressing and metering the polymer being forced through the shaping die. Although most extruders are single screw, twin-screw machines are available with the screws side by side in a common barrel. These can respond to die changes more readily than single-screw machines, and are especially efficient as compounders.

Apart from size, screw extruders vary most in the form of the screw which is designed to suit the various possible extrudates. Generally the screw has feed, compression and metering zones as shown in Fig. 15.8 (*a*), compression being achieved by one of the methods shown in Fig. 15.8 (*b*). If

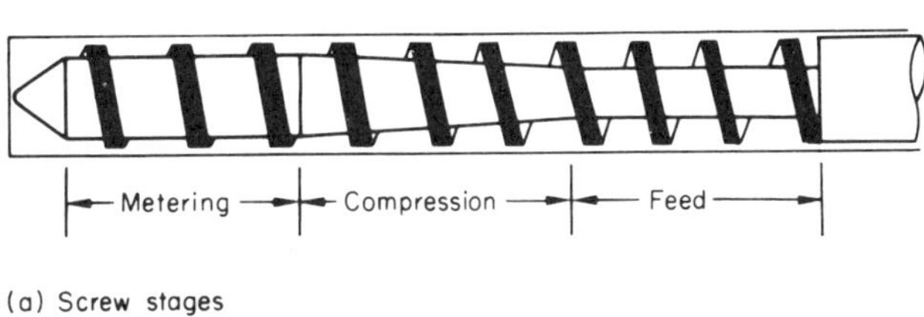

Fig. 15.8 Extruder screw details

volatiles arise during the compression and heating zones, a vented barrel is used as shown in Fig. 15.9.

The function of the screen and breaker plate (Fig. 15.4) is to create the necessary back pressure required for adequate mixing and shear heating. They also filter off any extraneous solids which might damage the die. Pressures across the die are normally in the range 15 to 50 $\mathrm{N\,mm^{-2}}$.

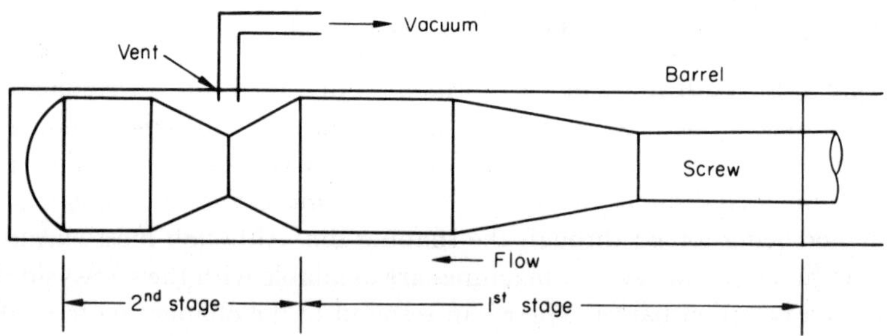

Fig. 15.9 Vented barrel and two-stage screw
N.B. The pumping capacity of the second stage must be greater than the first stage to prevent polymer passing out of vent.

Dies. By interchanging dies the extruder exhibits great versatility. The following extrudates are commonly produced; sheet to about 1·5 m width (both plain and coated), solid and hollow sections of many forms (both solid and foamed), blown film to about 2 m diameter, wire covering and blow mouldings. Some typical die forms are shown in Figs. 15.10–12. After passing the die the extrudate passes to a related take-off system. The material is often modified to some extent at this stage while cooling, as for example in the draw down of polypropylene for tapes where the high degree of orientation gives very high tensile strength. Several take-off systems are shown in Figs. 15.13–16.

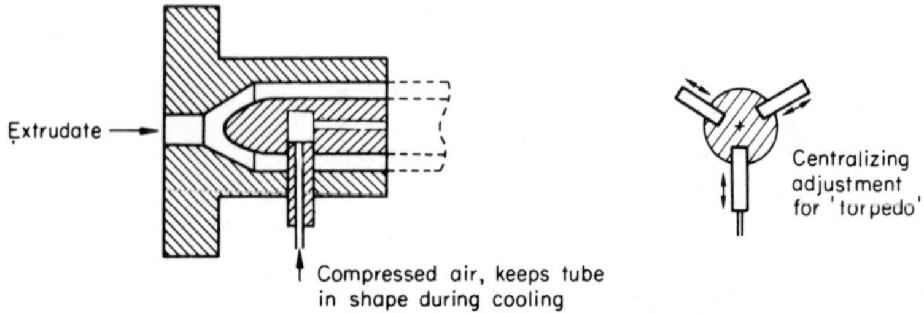

Fig. 15.10 Principle of tube-forming die

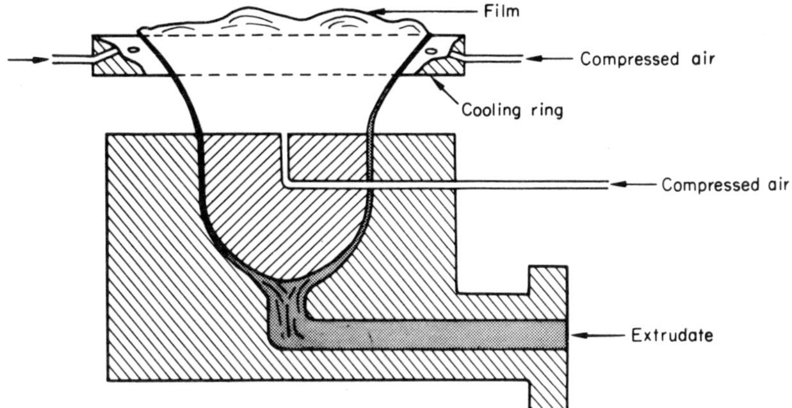

Fig. 15.11 Die for film blowing

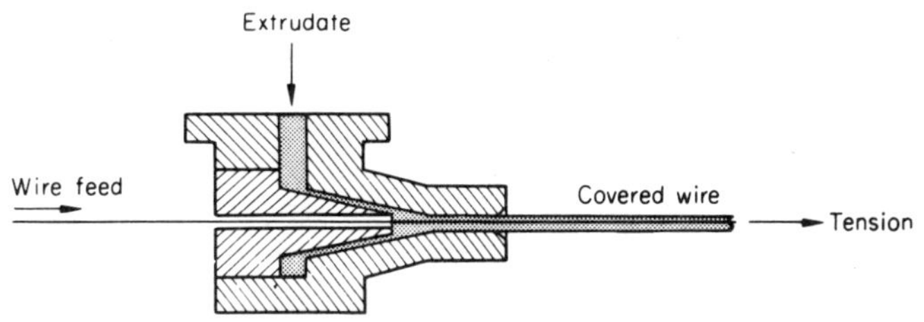

Fig. 15.12 Die for wire covering

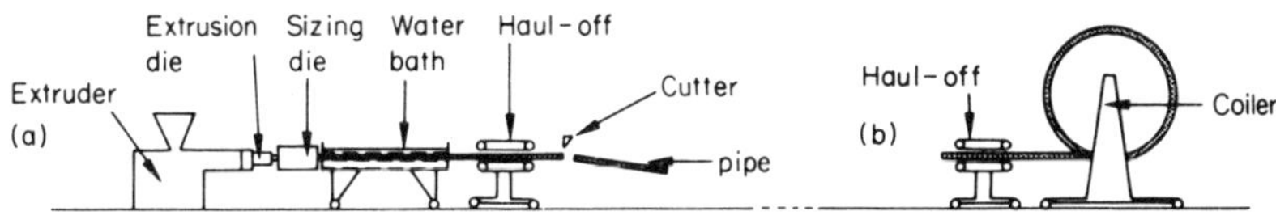

Fig. 15.13 Pipe-extrusion process

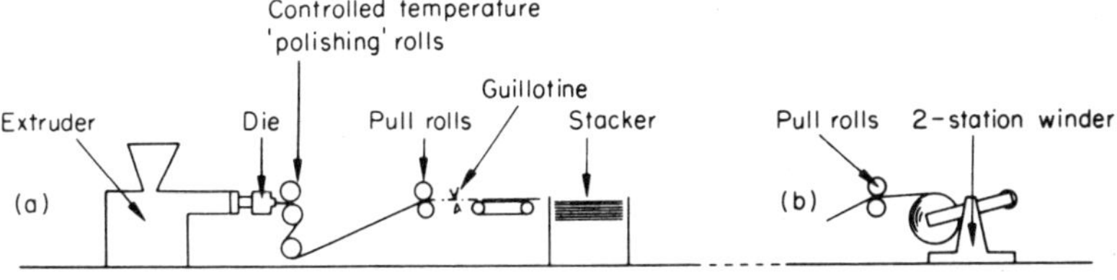

Fig. 15.14 Sheet-extrusion process

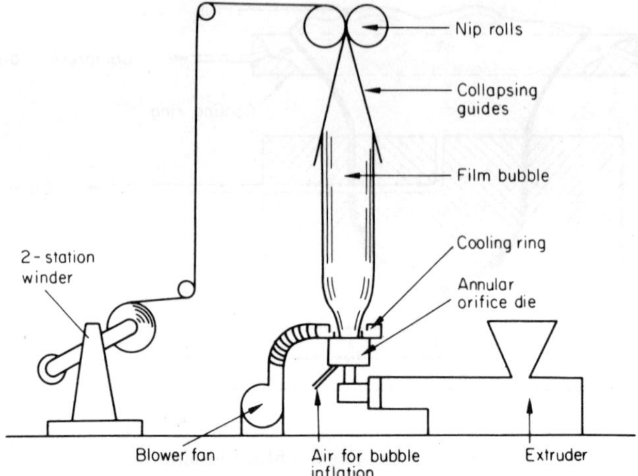

Fig. 15.15 Film-blowing process

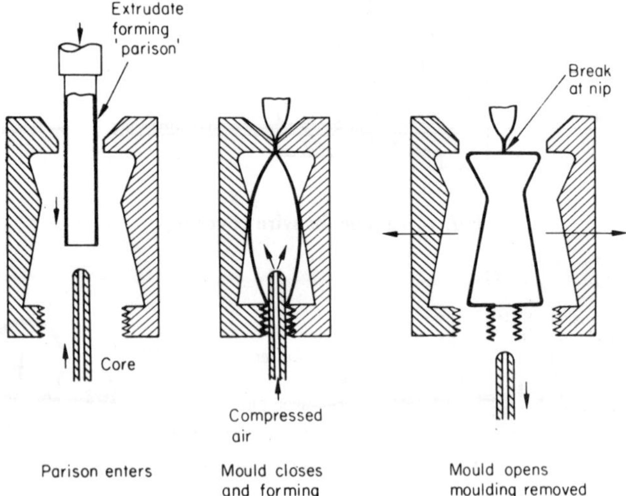

Fig. 15.16 Stages in blow moulding

Pultrusion. This is the polymer equivalent of metal wire drawing, and is largely used for the production of long sections of glass reinforced resins. The glass tow passes through a bath containing a resin solution, the composite material then proceeds to a sizing and shaping die. Finally curing of the resin occurs as the composite traverses a long oven. Sections are then sawn off at appropriate lengths. Fig. 15.17 illustrates the principle of pultrusion.

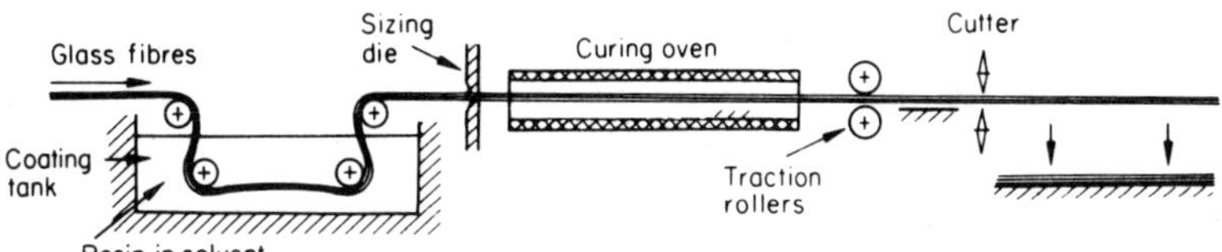

Fig. 15.17 Principle of pultrusion

Advantages of extrusion

(1) Die costs are low relative to those for injection moulding.
(2) Continuous production is especially valuable for long lengths of rod, pipe, film and covered wire.
(3) Anisotropy can be introduced if required to give high uniaxial strength.

Disadvantages of extrusion

(1) Take-off and post extrusion accessories may be expensive.
(2) Further work may be needed in assembling components.

15.2.3 Injection moulding. This technique was developed from the die casting methods used for low melting point metals. It is fundamentally simple and consists of heating the polymer to the required viscosity, metering the 'shot' size (volume of polymer required to fill the mould cavity completely), and transferring it through a sprue and gates into the closed mould. Where thermosetting materials are used much of the heat required for cross-linking is produced by shear heating during transfer through the injector nozzle; the mould is also kept hot to ensure completion of the curing. Moulds for thermoplastics are kept cool to speed the stiffening of the moulding to the stage when it can be ejected without deformation.

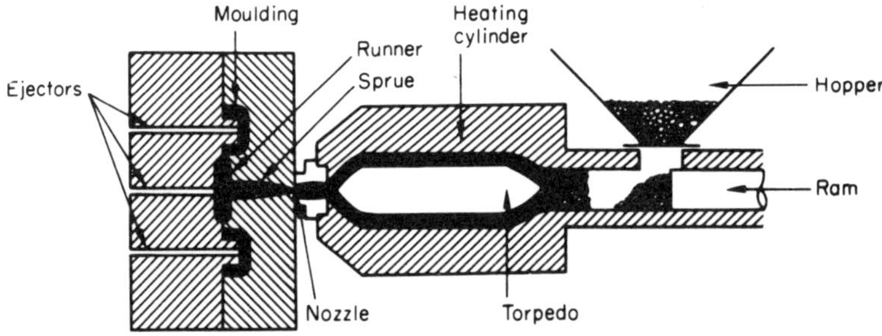

Fig. 15.18 Injection moulding, ram type

While the process is theoretically simple the design and manufacture of moulds to produce both large and intricate components is a very complex task. In consequence injection moulding is only likely to be economical when large quantity production is required. Undercut and screwed mouldings cause particular difficulty, especially in automatic ejection systems. As in all polymer processing close temperature control is essential for optimum results.

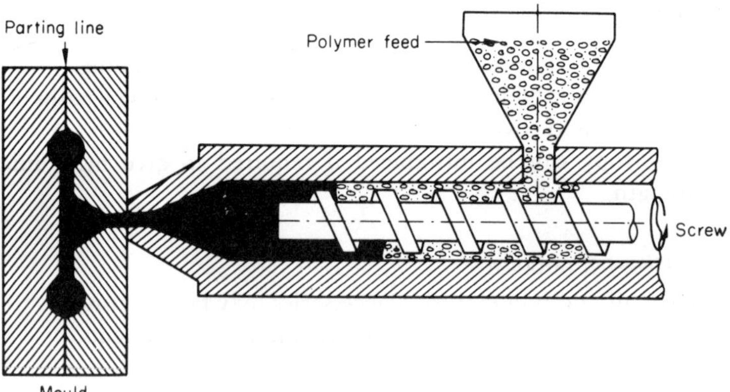

Fig. 15.19 Injection moulding, screw type

The injector system can be either a simple ram as shown in Fig. 15.18, or a particular form of extruder Fig. 15.19 in which the screw itself is used as the injector plunger at the appropriate stage in the moulding cycle. This axial screw motion is normally produced hydraulically, and controls the shot size.

As injection pressure may be as high as 120 Nmm^{-2}, components of large projected area, say 1 m^2, require a very high force, greater than 120 MN (12 000 tonf), to lock the mould. Hence injection moulding machines tend to be very large and expensive.

A typical injection moulding cycle is shown in Fig. 15.20.

Advantages of injection moulding

(1) The moulding normally requires no further machining.
(2) High output rates are possible, even with thermosetting polymers where the time saving in comparison with compression or transfer moulding is very great.

Disadvantages of injection moulding

(1) The capital cost of both the basic machine and its mould is high, particularly when automatic systems are involved.

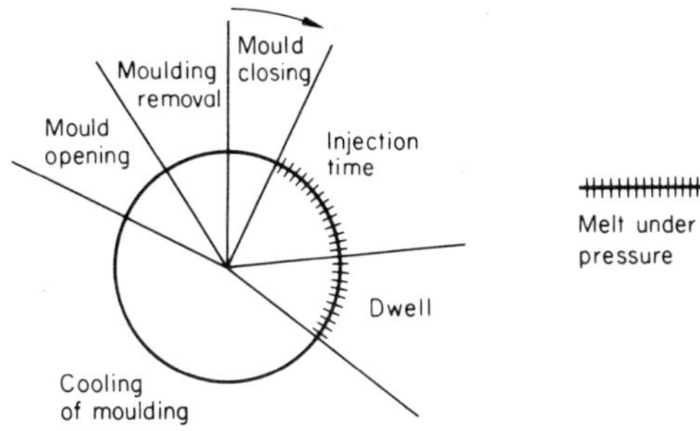

Fig. 15.20 Typical injection-moulding cycle

(2) It is difficult to mould parts with large variations of wall thickness, owing to the resulting warping that appears after moulding.
(3) With thermosetting materials the sprue and runner volume is wasted as recycling is not possible (see Section 15.5).

15.2.4 Compression and transfer moulding. These methods were traditionally used for the cheaper phenol–formaldehyde resins, mixed with fillers such as wood flour, chopped cotton, asbestos and other materials, giving specific physical properties. They require heat for curing, or cross-linking, of the polymer chains. Such curing must occur in the mould otherwise transfer becomes impossible. With compression moulding the metered weight of the powder or granules is placed in the female half of the mould, often compacted into pellets that have been warmed. Closure of

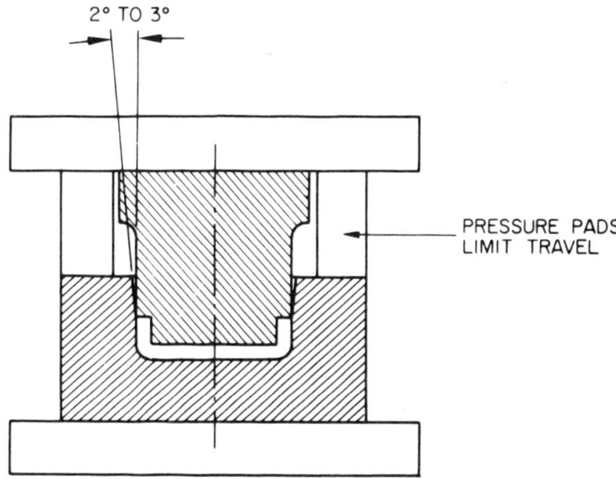

Fig. 15.21 Simple compression mould

the mould distributes the material, and subsequent heating leads to the necessary cure; after the required time a 'rigid' component is ejected from the opening mould.

Transfer moulding was introduced to give a lower viscosity melt in the mould cavity without the risk of curing in the transfer pot. Shear heating during transfer raises the temperature so that a reduced cure time is needed in the mould. In addition the lower viscosity of the transferred polymer ensures a completely filled mould.

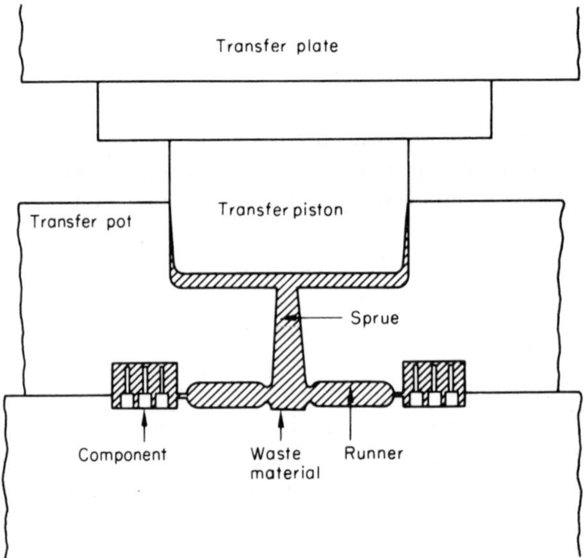

Fig. 15.22 Simple transfer mould

The injection moulding of thermosetting polymers, to which reference has already been made, is a development from the transfer moulding system, and makes use of the same shear heating principle.

Figs. 15.21 and 15.22 show the methods of compression and transfer moulding, and Fig. 15.23 shows a typical press.

15.2.5 Thermoforming. Compression moulding is also used for laminated thermosetting materials, often called 'pre-pregs' when the paper or cloth reinforcement is impregnated with an uncured resin. Flat or low contour shapes are most common, and the forming may be carried out between matched moulds (see Fig. 15.24). Curing due to cross-linking occurs during the heated pressing operation.

Single part moulds are also used for the moulding of thermoplastic sheet.

Either compressed air or vacuum systems are used to force the hot, ductile sheet against a mould of the required form (see Fig. 15.25).

Very large components (to 10 m^2 in area) are possible with comparatively simple machines.

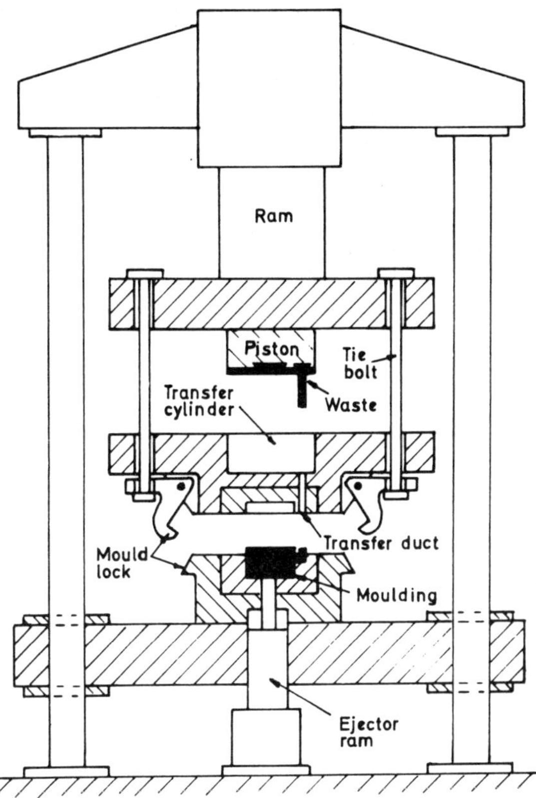

Fig. 15.23 Typical downstroking press

15.2.6 Cold-forming of thermosetting sheet. The technique of thermoforming can be modified to a cold process if the cross-linking is initiated by the introduction of a chemical accelerator prior to pressing. This means that the thermal effect is replaced by a chemical one.

15.2.7 Calendering. This is a rolling process for the production of sheet and film. As an alternative to extrusion through a sheet die, it permits better surface finish and thickness control. Also for materials like polyvinyl chloride that are subject to thermal degradation it is an easier process. Fundamentally the process is simple. The compound formulation is passed

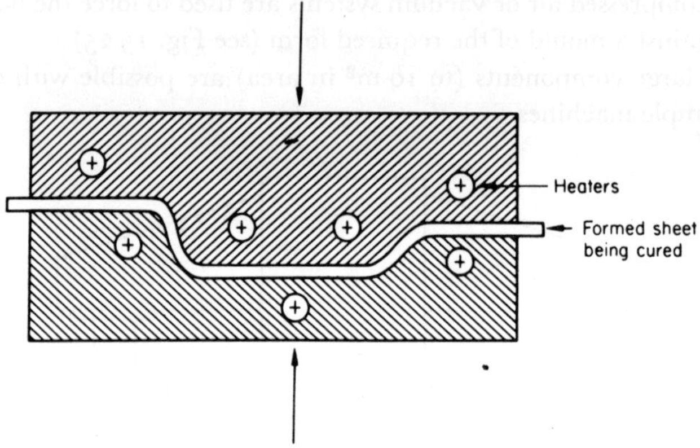

Fig. 15.24 Matched mould forming of 'Pre-Preg' sheet

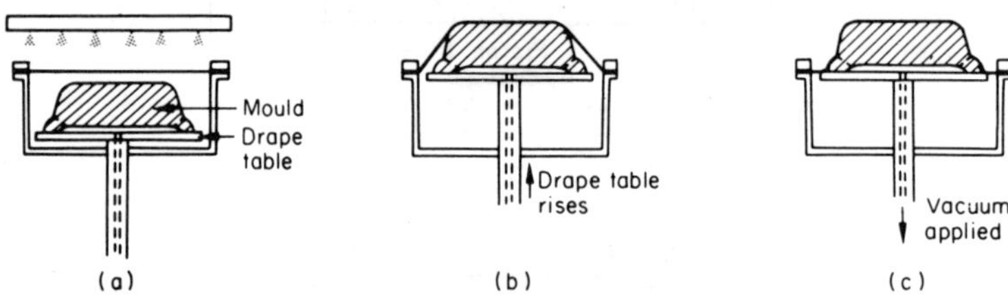

Fig. 15.25 Vacuum forming

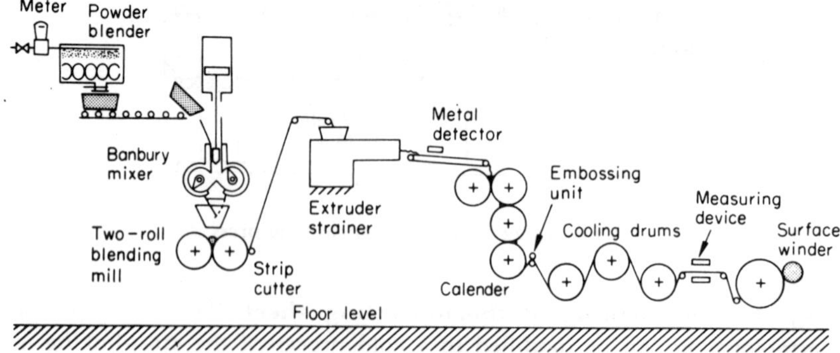

Fig. 15.26 Layout of typical calendering system

through several heated rolls until it is formed at the required thickness. Reference to Fig. 15.26 shows that in practice a more sophisticated system is necessary. A major problem is the roll arrangements to provide roll gaps that produce a sheet of constant thickness across a span of two metres or more.

15.2.8 Dip coating. The simplest way of depositing a surface coating either internally, externally or both is to dip the article in a container of polymer that is molten or dissolved in a volatile solvent. If the coated article is then allowed to cool, or kept warm until the solvent has evaporated a coating is formed. Rubber coated gloves and gum boots can be produced in this way.

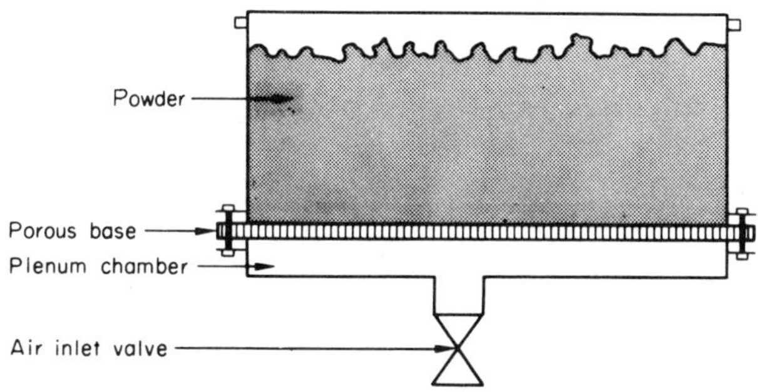

Fig. 15.27 Fluidized bed for dip coating

Alternatively a fluidized bed of polymer powder (see Fig. 15.27) can be used to coat the surface of heated items that are immersed in the powder. This is restricted to metals that can be heated above the melting point of the polymer without any resultant damage.

15.2.9 Rotational moulding. This is an extension of the principle of dip moulding, and an important alternative to blow moulding in the production of hollow containers. When really large containers, of say 200 litres capacity, are required it supplants blow moulding. Although symmetrical forms are most commonly produced, asymmetrical shapes can also be moulded.

The required amount of powdered polymer is introduced into a female mould which has the same form as the component. Whilst being externally heated in an oven the mould is slowly rotated about two orthogonal axes. As the polymer tumbles and melts it adheres to the mould, forming a uniform lining which becomes the wall of the finished component. When the moulding has been formed the mould is moved to another machine station where rotation continues in a cooling air flow. Subsequently the rotation is stopped and the rigid moulding removed. Arrangements of a rotational moulding machine are shown in Fig. 15.28.

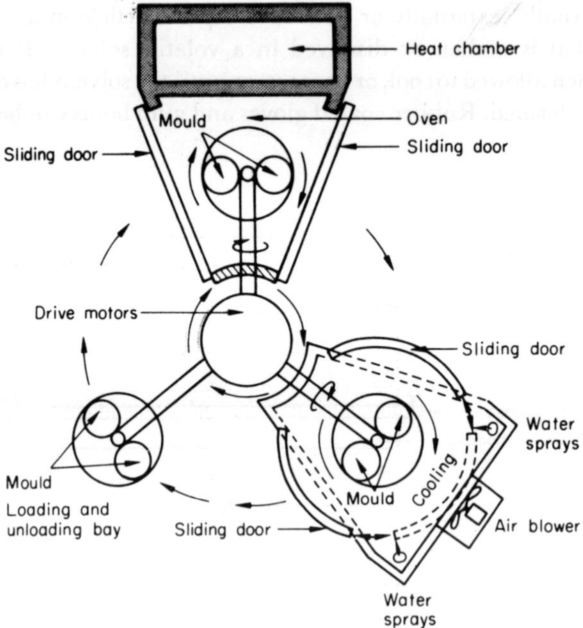

Fig. 15.28 Three-station rotational moulding machine

15.2.10 Cold lay-up. Two factors limit the size of mouldings, the size and the cost of the machine. If cold lay-up methods are used economy can be improved, even in labour intensive operations. For really large items like ships' hulls cold lay-up is the only possible production technique. It is important to use a fibre-reinforced resin that will cure rapidly at ambient

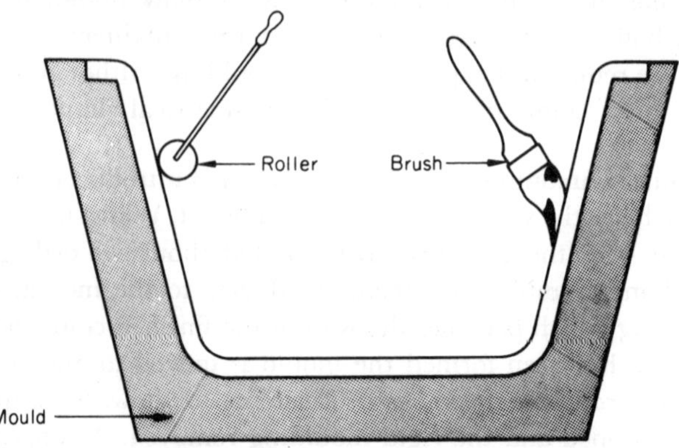

Fig. 15.29 Hand lay-up. Resin brush applied to reinforcing fabric within a female mould

temperatures. Mechanization reduces labour costs; some systems of production are shown in Figs. 15.29 to 15.31.

An alternative is cold-press moulding, to which reference has already been made. This gives mouldings with a smooth surface on both inside and outside.

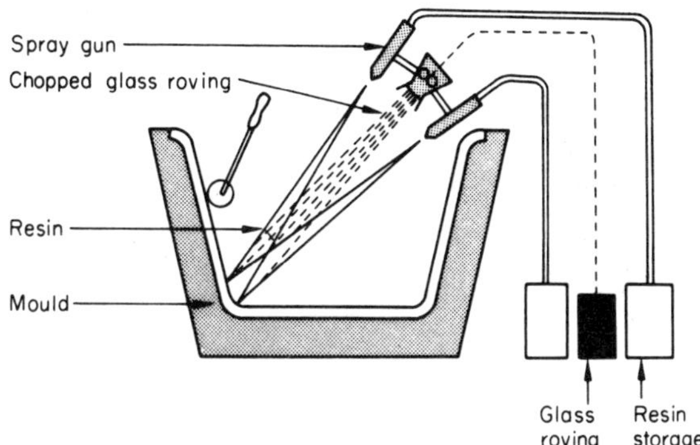

Fig. 15.30 Spray application within female mould

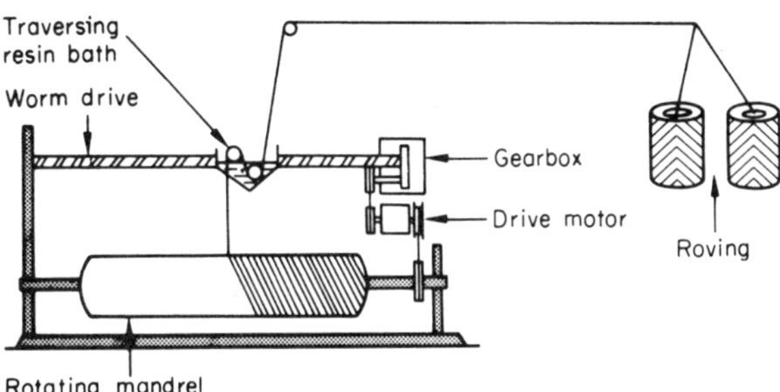

Fig. 15.31 Filament winding of impregnated reinforcement on to mandrel

15.2.11 Forming of expanded polymers. Most polymers can be made cellular and reduced in density by foaming actions. Controlled density then gives controlled strength, stiffness and insulation properties. The lowest densities occur when the polymer is allowed to expand freely. By restricting expansion, as in a closed mould, the average density can be increased and the cold mould surface leads to a higher density 'skin', which is often a structural advantage.

Typical methods are as follows.

(*i*) *Polyurethanes.* A urethane is produced by a reaction between an isocyanate and an alcohol. If a suitable catalyst and water are also present the reaction is accelerated and the carbon dioxide produced acts as a cell former. These components are mixed in a machine supply nozzle which discharges onto a moving belt. This forms a continuous slab that can be cut to size (Fig. 15.32). Alternatively the mixture can be pumped into hollow sections which become filled by a buoyant, insulating cellular structure.

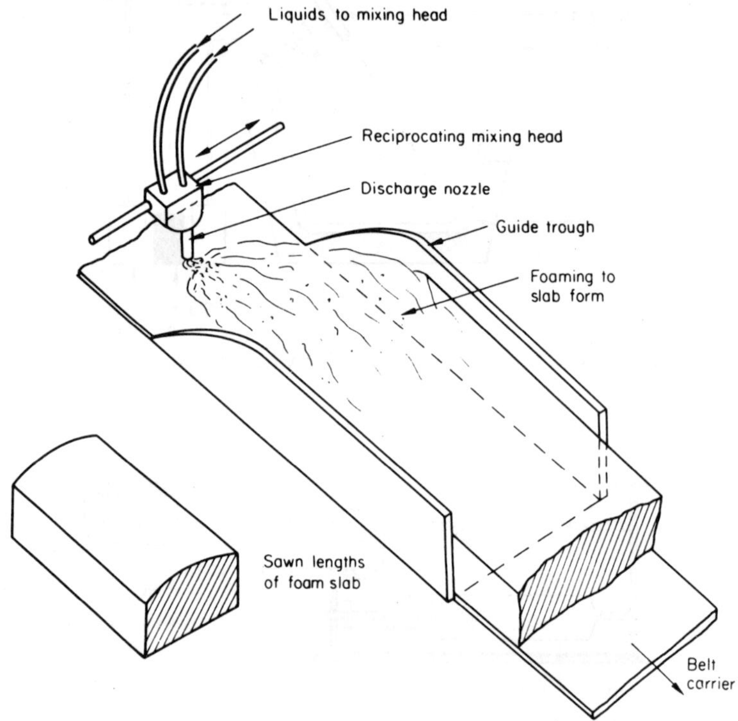

Fig. 15.32 Production of foam slab from two-part mix

(*ii*) *Polystyrene*. The raw material is hollow spheres or 'beads' of polystyrene about 1 mm diameter, each of which is filled with a gas such as pentane. These are prepared by a suitable polymerization process. A low-density polymer is obtained by steam heating the beads. This raises the pressure of the gas and expands the thermally softened polymer. Such beads may then be used as a 'loose fill' type of thermal insulator.

Of greater importance is expanded polystyrene moulding. By expanding the beads in a closed mould they are pressure welded into a rigid mass. Fundamentally, a strong mould with small holes allowing the access of steam at about 2 $N\,mm^{-2}$ is the item of equipment used in production.

By the use of alternative 'blowing agents' it is possible to produce most polymers in cellular form. If foams are injection moulded the higher densities approximate to that of timber for which such mouldings are a common substitute, e.g. in furniture.

15.2.12 Encapsulation. This involves the casting of a polymer around a part as a permanent enclosure and is often referred to as 'potting'. The main engineering application is the total enclosure of electronic assemblies to give complete protection. Glass-clear formulations are available, but these are mostly used for preserving zoological and botanical specimens. Cold curing epoxy or polyester resins are most suitable as they avoid the risk of thermal damage to the embedded parts.

15.2.13 Sintering. Some polymers have both a high melting point and high melt viscosity that make the normal processes of extrusion and injection moulding difficult. Polytetrafluoroethylene (PTFE) is the main example. In such cases sintering is necessary. The process is basically similar to that described for metals.

Improved dimensional accuracy is obtained and it is also possible to include higher filler contents by sintering. Sintered nylon parts containing a lubricating filler are valuable as small bearing type components. It is also possible to produce porous structures that can absorb lubricating oils or printing inks.

15.2.14 Machining. Most plastics machine extremely well using conventional wood or metal working tools. Two important factors will call for modified techniques, namely the high thermal expansion and low thermal conductivity compared with metals. In consequence reference should be made to the plastics suppliers' literature for information about cutting angles, speeds and feeds. If post-machining cooling is appreciable, dimensions must provide for contraction effects.

15.3 FIXING, FASTENING AND ASSEMBLY

15.3.1 Welding. This process is limited to thermoplastics since melting is required as in the fusion welding of metals.

(*i*) *Ultrasonic.* The higher damping capacity of polymers facilitates the generation of heat at the joint faces, but machine requirements restrict application to smaller components.

(*ii*) *Hot plate welding*. The two halves of the joint are heated by being placed in contact with a hot metal surface. As soon as the surfaces have softened sufficiently they are brought together under pressure and a weld forms. The nature of this method restricts its application to flat joint areas.

(*iii*) *Friction welding*. This follows the pattern adopted for metals, and is normally restricted to circular sections.

(*iv*) *Resistance heating*. A length of resistance wire is included at the joint face. After assembly a current is passed through the wire, heating and melting the local area where welding then occurs.

(*v*) *Induction heating*. This involves the introduction of a metallic particle filled sheet of polymer similar to the polymers being welded. By sandwiching this sheet between the joint faces and inserting the unit in a high frequency electromagnetic field, local heating provides the weld energy. Whilst this is similar to (*iv*) it offers a shorter cycle time, but is limited to surfaces of high radii of curvature.

(*vi*) *Hot-gas welding*. Nitrogen is mainly used as it prevents oxidation of the hot polymers. As in metallic welding chamfered edges are prepared and filler rod used. This technique is generally used in the fabrication of sheet and tubular structures of large size.

(*vii*) *Solvent welding*. In this method chemical softening of joint faces is produced by dipping them in the appropriate solvent. After the joint has been made the solvent evaporates leaving a gradually strengthening weld.

15.3.2 Adhesives. Many types of both 'hot' and 'cold' curing systems are available and designed to suit particular pairs of joint materials. Where dissimilar materials with widely differing melting temperatures are involved adhesives must replace welded joints. It is found that polyethylene and polytetrafluoroethylene are very difficult to join with adhesives.

15.3.3 Fasteners. As plastics replace metals an increasing number of components that require frequent assembly and dismantling are appearing. In general the techniques developed for metals are applicable, but certain factors are unusual.

(1) Concentrated loads tend to produce local deformations instead of the more rapid load dispersal that exists with metals. This means

that several small fasteners are preferable to a few large ones. 'Bigheads' as in Fig. 15.33 can be embedded in a plastic, particularly a cold lay-up resin to spread the load.

(2) Screwed inserts can be moulded into compression mouldings, or forced into injection mouldings. Ultrasonic loading is a help to the fitting in the latter case.

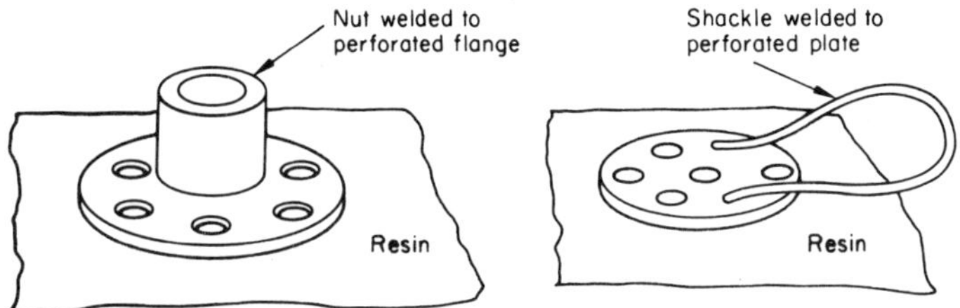

Fig. 15.33 'Bigheads' embedded in fibre-reinforced plastic

(3) Snap fasteners are simple to provide due to the resilience of thermoplastics.

(4) Moulded screw forms are available, but a special 'unscrewing' mould will be needed for most components and such moulds are expensive.

15.4 PRINTING AND SURFACE COATING

The largest demand for these techniques is from the packaging industry where film and blow-moulded containers are widely used. Polyethylene is a popular plastic but its waxy surface is difficult to print with simple methods. Other glossy surfaces are also likely to give printing problems.

One method that overcomes these difficulties is to roll a printed paper into the hot and soft surface of the freshly moulded part. Other printing methods, viz. screen printing and dry off-set printing, require surface pretreatments of which three types are available; these are chemical oxidation, electrical discharge and 'flaming' which produce a wettable surface. After printing, some drying arrangement is necessary to prevent smudging of the work.

Surface coating of metals and fabrics has been mentioned as part of other techniques, but one form of spread coating is shown in Fig. 15.34.

15.4.1 Metallizing

(*i*) *Vacuum deposition.* Desirable brilliance and attractive colours are mainly achieved with aluminium, but gold alloys may be also used. As films are typically 0·025 μm (1 μin) thick the material cost is low.

To give the required surface polish the surface is lacquered, this also seals the moulding and prevents any loss of volatiles that would break vacuum and affect the coating bond. The general scheme for vacuum deposition is shown in Fig. 15.35. There are two ways of protecting the

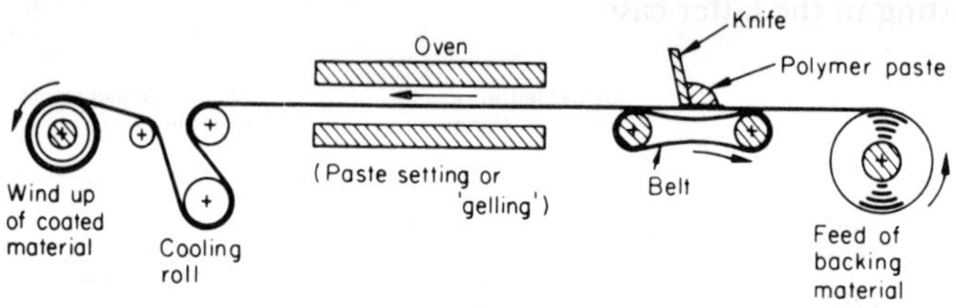

Fig. 15.34 Spread coating, e.g. PVC on woven fabric

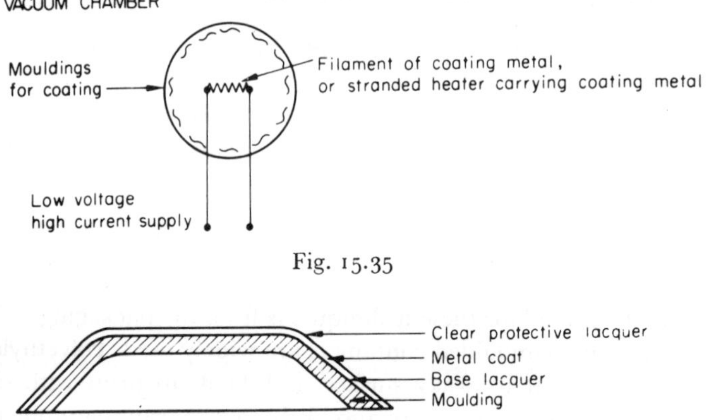

Fig. 15.35

Fig. 15.36 Plating on working surface

metal coating after deposition, depending upon whether the film is on or beneath the working surface of the moulding.

(1) On the working surface of the moulding. In this case protection of the film is dependent on a top coating lacquer as seen in Fig. 15.36.
(2) Beneath the working face of the moulding. Protection is given by the moulding itself, which clearly must be transparent (see Fig. 15.37).

(*ii*) *Electrodeposition.* Unless a polymer has a conductive filler introduced during moulding it will be an insulator incapable of responding to electrodeposition. Normally a plastic requires:

(1) A surface etch in acid reagent to give a good 'key' for metal layers.

(2) Surface preparation by metal deposition from a solution, often copper sulphate.
(3) Electrodeposition in required electrolyte.

Acrylobutadienestyrene (ABS) is the polymer most suitable for electroplating, and as it is one of the 'engineering' thermoplastics the metallic appearance is an advantage. Such mouldings are replacing metallic die-castings in many lightly loaded components.

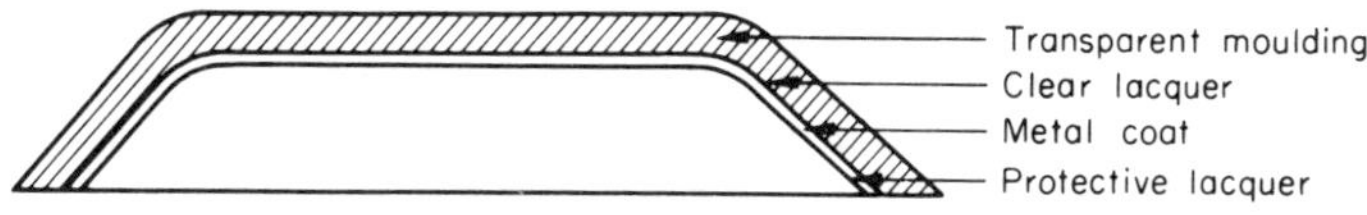

Fig. 15.37 Plating beneath working surface

15.5 PLASTICS WASTE RECOVERY

Appreciable waste material arises in plastics processes as sprues, runners and flash. Also during the setting-up procedures defective mouldings and unwanted extrudate occur. In consequence recycling is most desirable.

The nature of thermosetting polymers precludes their re-use, except after regrinding, when they may be used as fillers in other polymers. Thermoplastics are generally quite suitable for regranulating and recycling. However the recycling of unplasticized polyvinyl chloride is not recommended owing to appreciable degradation during reheating. In all cases some slight reduction of desirable characteristics is likely, so blending of new and recycled polymer is adopted. Such methods require a knowledge of the scrap material, and can be undertaken within a factory. Salvaged plastics of unknown origin are less useful as uncontrolled compositions with unpredictable properties would arise. However, for some applications an *ad hoc* mixture may offer average properties that are acceptable. Types of block board formed by the compression moulding of mixed granulated scrap and waste paper are commercially available.

Various granulating machines are used for preparing plastic waste for recycling.

16 Control of Machine Tools

16.1 The main headings under which the control of machine tools can be considered are shown in Fig. 16.1. It is not intended to describe the technical details of the various systems, which are amply covered in publications on machine tool design. Rather, it is intended to show the pattern into which the various systems of control fit, and to dwell at

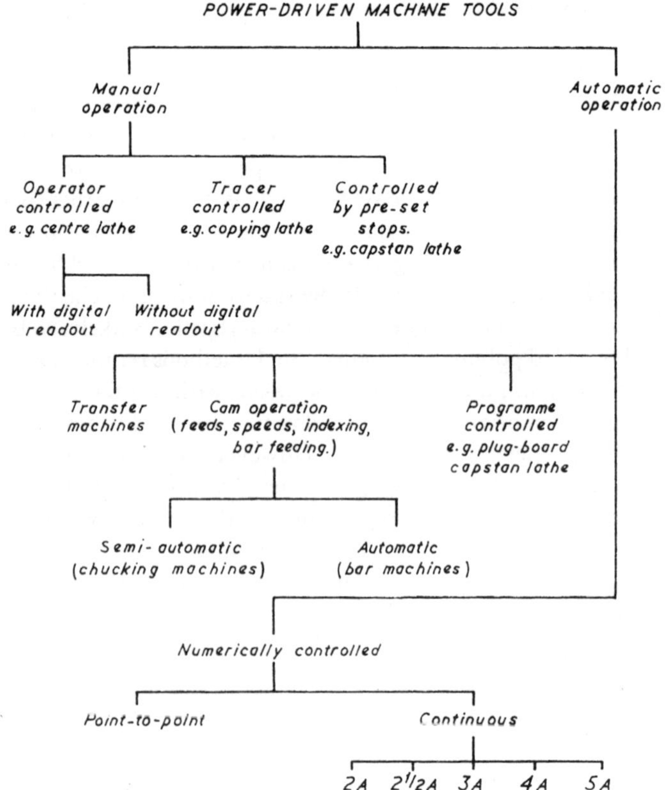

Fig. 16.1 Methods of controlling machine tools (A = axis)

greater length on those of comparatively recent introduction, whose potentialities are often not fully appreciated.

16.2 DIGITAL READOUT SYSTEMS

In many instances where full numerical control cannot be economically justified, displacement measuring devices can be fitted to slideways and connected to digital counters. The transducers sense the displacement of the sliding member in relation to the fixed slideway, and record its position on a visual display. Slide position can thus be read directly by the operator as cutting proceeds, allowing a higher degree of repeatability to be achieved and reducing production time.

Centre lathes, measuring machines and jig-boring machines are good examples of equipment in which digital readout can be used to advantage on small batches of components.

16.3 PLUGBOARD PROGRAMMING

This method of control does not require a machine operator. It is claimed that accuracy and finish of the workpiece are improved and tool life is increased because of the improved control over speeds and pressures. So far, the main application has been to capstan lathes.

The stops are set as for a normal capstan lathe, but they operate microswitches instead of stopping against a solid surface. Each operation is programmed by inserting plugs and wires into a plugboard which allows spindle speeds and feeds to be selected and turret indexing, bar feeding, etc. to be performed. The operation of a microswitch on a stop causes a stepping selector switch to move one position and actuate the next series of motions. The operator's movements are simulated by solenoid-operated air cylinders coupled to the existing machine controls.

There is some increase in setting time when compared with manually operated machines, but often a reduction in running time, and of course, a total saving of operators' wages. For complex components of comparatively long cycle time, plugboard controlled machines are more economical than conventional capstan lathes for all but the smallest batches. Fig. 16.2 shows a typical series of cost curves for a given part, plotted against batch size, for centre lathes and capstan lathes. These curves indicate that a substantial decrease in cost per piece can be achieved by using program-controlled capstan lathes. This, in turn, will move the break-even point between capstan lathes and automatic lathes further to the right, so that previously held views on the economic size of batches for automatic lathes will require revising upwards.

16.4 NUMERICAL CONTROL

Numerically controlled machine tools which operate from suitably coded numerical inputs have been developed for a wide range of engineering applications, although the vast majority are used for metal cutting, based either on milling or turning principles. Tools may be changed manually, or in the case of machining centres a number of tools can be fitted in a carousel and changed automatically.

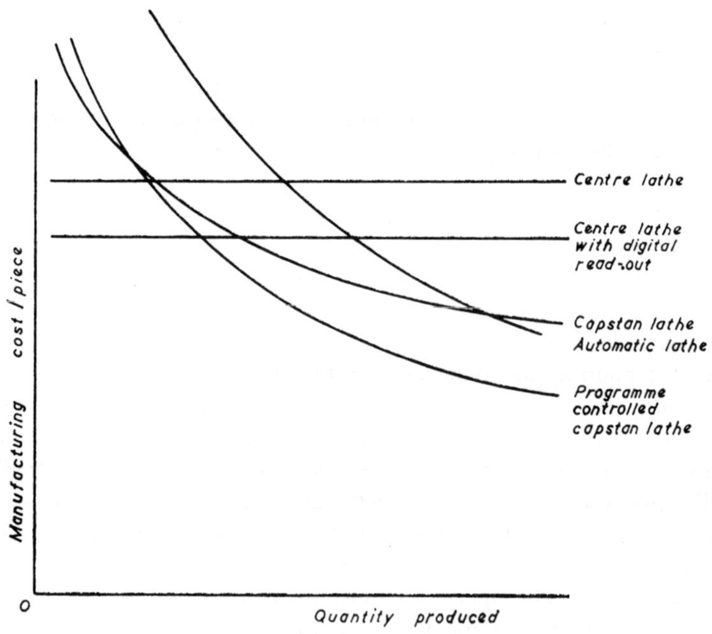

Fig. 16.2 Production cost curves using different manufacturing methods for small and medium sized batches

Basically they are of three types, point-to-point, paraxial and continuous path. Point-to-point machines are for drilling applications, where the cutter position is of importance only at specified points, and the path followed to reach these points is of little significance. Paraxial machines are capable of independent controlled motions in two or three mutually perpendicular axes, but cannot simultaneously move in more than one direction. Continuous path machines are capable of simultaneous motion in more than one direction, and can have up to three axes of translational and two axes of rotational motion.

N.C. machines can have either open-loop or closed-loop control systems. Although closed-loop systems are preferable from the viewpoints of accuracy and rapid positioning, open-loop systems are cheaper and are likely to become increasingly popular. Development of stepping motors which rotate in discrete steps when fed with electrical pulses, giving positional increments of table movement to the order of 20 μm per pulse, has done much to encourage open-loop systems. The disadvantage inherent in such systems, that of compensating for leadscrew backlash, can be overcome by using linear stepping motors which drive the machine table directly without the use of a leadscrew.

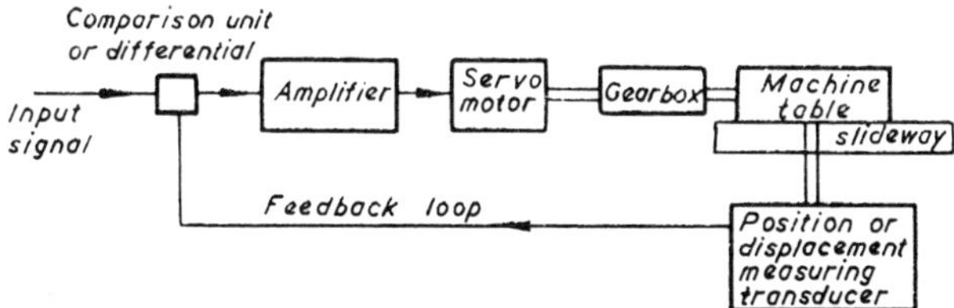

Fig. 16.3 Basic feedback loop for point-to-point control

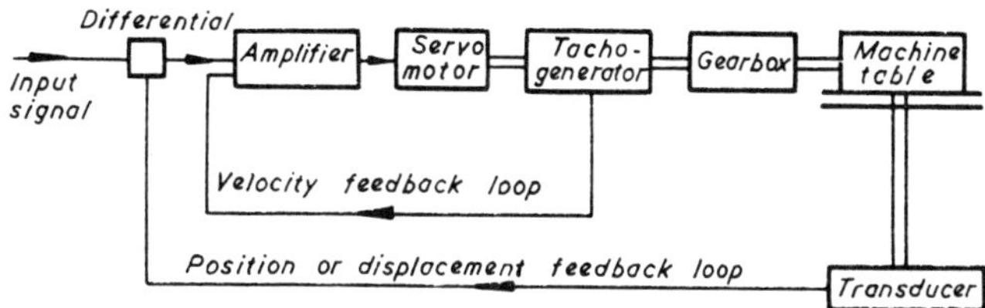

Fig. 16.4 Servo system for continuous control of one axis

A basic feed-back circuit for a closed-loop point-to-point machine is shown in Fig. 16.3. A transducer fitted to the slideway or the drive screw senses the position or displacement of the table. This information is fed back to the comparison unit or differential which compares the desired position or displacement with that achieved and generates an error signal which is amplified and used to drive a servomotor. The servomotor responds by moving the table until the error disappears. For continuous control a further feedback loop is incorporated; this measures the velocity or the derivative of the error with respect to the time (Fig. 16.4). The velocity or error rate feedback loop is used for stabilizing purposes. With continuous control the error on each axis controlled must be kept as small as possible, otherwise an inaccurate contour will be generated at the tool point.

16.4.1 Methods of data processing. Most early NC machine tools had 'hard wired' systems which were specifically designed to read an input tape and, by means of their inbuilt logic, translate the instructions on the tape into table motion signals. On point-to-point and paraxial machines this is relatively simple, but on continuous path machines, where the cutter may have to generate a curve, the input tape must either explicitly define the cutter path in very small increments of motion or alternatively the machine tool control system must be capable of interpolating between relatively coarsely spaced points along the curve. Non-interpolating systems use a magnetic input tape which is read by the tape reader in the machine console at constant speed, the increments being defined by pulses on the tape. The interpolation with such systems is performed entirely by means of a program on a digital computer. Interpolating systems use a punched tape, interpolation being either linear, circular or parabolic. Linear interpolation is obviously simpler, but requires the points on the input tape to be closer if the resulting accuracy is to compare with systems which interpolate by fitting an arc or parabola through adjacent points.

A number of different types of paper tape input were used for early systems, but the current international standards relate only to the use of 1 in wide tape with eight tracks of punched holes, using either the ISO or EIA code. Each row of holes across the tape represents a binary coded digit or a non-numerical symbol. Information is read in discrete blocks, which represent either table positions or machine instructions.

Some machines allow the operator to generate his own input tape by manually programming the first component on the machine console, but such methods of tape preparation are suitable only for relatively simple point-to-point or paraxial operations. Alternatively simple operations can be punched directly on tape by the programmer, provided he knows the co-ordinates of all the relevant points. However, even for point-to-point applications, computer programming languages have been developed which allow programming time to be reduced.

For contouring operations the magnitude of the task of manual programming demands that computer assistance is required in punching the tape. The most comprehensive of such part-programming languages is APT (Automatically Programmed Tools), which is a free format language using the nomenclature of FORTRAN. APT can control up to five axes, but requires a large computer store. For this reason a number of languages have developed, frequently based on APT, but requiring smaller computer storage; these are suitable for less sophisticated applications such as two-axis control.

The part program output defines the co-ordinates of the cutter centre line in cutting sequence, and in this form is of little practical use. To be acceptable to the given machine tool control system allowance must be made for the mechanical characteristics of the machine. For instance, limitations as regards maximum accelerations and maximum velocities must be observed. Also, in the case of machining centres, information must be transmitted to effect tool changes and table indexing. This information is handled subsequent to part programming by a post-processing program. Post processors must be tailored to suit the computer, the part programming language and the machine tool, and thus constitute a major software service provided by machine tool manufacturers.

Recently an alternative has emerged in the form of shop floor based minicomputers which either accept a partly processed input or a part program input and then perform all the necessary computation, feeding the data on-line to the machine control system, thereby eliminating the tape reader.

Computer numerical control (CNC) is rapidly replacing hard-wired systems, and offers many advantages. The buffer storage facilities of the computer enable blocks of information to be processed in advance, so that the speed limitations of tape readers no longer apply. The computer keyboard can be used to edit part programs which are in store, and machine functions can be modified if necessary by altering the executive programs. Similarly, a range of diagnostic facilities can be provided. A further development made possible by CNC is axis calibration, which increases the accuracy of machine tools by modifying motion commands, to correct for dimensional errors in the drive system and slideways.

Microprocessors, having large-scale-integration circuits (LSI) on silicon chips, have enabled machine tool designers to reduce the cost of controllers and yet retain the advantages of minicomputer CNC by using read–write semiconductor memories (RAM). Alternatively, it is possible to further reduce cost by using a dedicated read-only memory (ROM) which performs the same functions as a hard-wired system.

It is logical that the use of computer control should be extended to multi-machine systems linked with a central computer (DNC) which monitors the status of each machine as a basis for scheduling. This concept can be further extended by using a standardized method of work-holding and having automatic workpiece transportation between machines to provide a flexible machining system (FMC) which combines the advantages of NC and transfer lines. In practice, DNC is a costly investment and very few companies have yet found economic use for it.

Numerically controlled machine tools of the sorts discussed operate on pre-determined values of cutting speed and feed. Inevitably, these parameters are set so as not to overload the machine under the most arduous conditions experienced during the operation sequence. During less arduous parts of the operation sequence the machine is therefore performing at a lower production rate than is economically viable. The computing power of CNC can be used to adapt feeds and possibly speeds in-process to give near-optimal metal removal conditions. If the machine tool is provided with appropriate sensors to measure, for instance, electrical power supplied, tool wear (or a parameter analogous to it) and surface finish, it is conceptually possible to develop an adaptive control system to operate under minimum cost conditions subject to constraints due to power and surface finish. In practice, adaptive control is difficult to program and the results are generally far below the optimum, but it is an area in which many improvements can be expected in the future.

16.4.2 Measuring transducers. A large number of different types of measuring transducer have been developed, operating either on analogue or digital principles.

Analogue transducers are invariably 'absolute' in that the signal is related to a zero position or to a suitable datum, thereby enabling linear displacements or rotations to be explicitly sensed. When considering the large table traverse distances and the accuracy required, it is not surprising that most systems incorporate at least two transducers, one to give a coarse positioning accuracy over the entire traverse and the other a much finer accuracy over a short distance. This is a similar principle to that used on a micrometer, where the divisions on the barrel occur at one pitch intervals, and the divisions on the thimble sub-divide the pitch.

Fig. 16.5 Binary encoder

Synchros are widely used for this purpose, where the amplitudes of alternating currents supplied to the two stator windings are proportional to the sine and cosine of the desired angle. When the rotor is at this angle the output from the rotor winding is zero. As the rotor passes through the null position a change of frequency occurs with respect to the master oscillator, so that both the magnitude and sense of the error are detected. Inductosyns work on similar principles except that accurately produced hairpin windings on a linear scale are used. The measuring scale consists of a winding extending along the slideway, and the slider has windings displaced electrically by 90° from each other which are separated from the scale by a gap of about 0·25 mm. For measuring rotary axes, a circular configuration can be used. Although many other types of analogue transducer are in use, synchros and inductosyns are undoubtedly the most popular.

Digital transducers may be either absolute or incremental in operation. Encoders give an absolute reading, consisting of discs having circular tracks which are marked in binary, cyclical binary (Gray) or cyclic–binary–decimal codes. Fig. 16.5 shows a seven track binary encoder which sub-divides the rotation into 128 (2^7) segments. Normally, a greater number of tracks are used, giving higher resolution. They are sensed either electromechanically, using brushes, or more usually optically, using photoelectric sensors.

Pulse digitizers are incremental in operation, generating signals only when moving. The simplest form uses a quantizer, which is essentially a disc with radial slots through which light pulses pass to a photocell as the disc rotates. More popular are systems based on moiré fringe techniques, where two finely divided optical gratings, which are mutually inclined at a small angle, move relative to each other. The resulting fringes can be sensed and, by having two sensors in phase quadrature, the direction can be sensed to find whether the pulses are adding or subtracting. By making use of the fact that the fringe intensity is approximately sinusoidal, the resolution of the grating can be increased.

16.4.3 The drive system. Drive motors can be of two types, electric or hydraulic. Hydraulic motors, by virtue of their high power/inertia characteristics, respond rapidly to changes in demand. They can be used to drive rams on small machine tools where compressibility effects of the hydraulic fluid are not significant, but generally they operate through leadscrews or racks and pinions. Leadscrews are suitable for medium sized machines, but for long beds, where the added inertia of a large diameter screw is undesirable, the rack and pinion drive is favoured.

Constant speed electric motors, driving through variable clutch couplings, are capable of providing high torques, although heat generation due to clutch operation can pose problems. Alternatively variable speed a.c. induction motors or rectifier controlled d.c. motors can be used. These eliminate the heating problems associated with variable clutch systems, but in general are slower to respond to variable demands.

Stepping motors, which rotate in discrete intervals in response to electrical pulses, provide another alternative which is relatively cheap but they have a severe limitation when positioning at high traverse speeds. When operated within the speed limitations, stepping motors permit an exact number of screw revolutions to be programmed, so they are ideally suited to open-loop NC systems which are cheaper than comparable closed-loop systems when a high degree of positional accuracy is not demanded.

16.4.4 Machine tool design. Numerical control emphasizes three problems of design which are relatively unimportant with other types of machine tool. These are: backlash in leadscrews and gearing, friction between the moving parts, and lack of stiffness of the table and drive unit.

Backlash can be eliminated and friction reduced very considerably by replacing conventional leadscrews by preloaded recirculating ball leadscrews and nuts. This is of particular importance with point-to-point systems where positioning accuracy of $\pm 2{\cdot}5\,\mu$m ($\pm 0{\cdot}0001$ in) is frequently required. Many of the high accuracy systems avoid backlash errors by ensuring that the tool always approaches the required position from the same direction. Attempts have been made to allow for backlash by programming, but this is complicated and is applicable only to a particular machine tool.

Friction between metal surfaces presents particular difficulties at low traverse speeds, at which the coefficient of friction rises rapidly (Fig. 16.6). This leads to friction forces which vary rapidly and non-uniformly, and imposes a lower limit to the distance a table can be moved between starting and stopping. Although this distance may be small, it can prevent the positioning of tables to very high orders of accuracy. Two approaches have been adopted to reduce this undesirable friction effect. One is to move the table on recirculating rollers; this does not prevent friction but reduces it by replacing sliding with rolling friction. The second approach is to use hydrostatic or aerostatic bearings, where the table and slideway are separated by a thin fluid film provided by forcing oil or air under high pressures through small orifices in the bearing surfaces. It is claimed that by these means friction is practically eliminated.

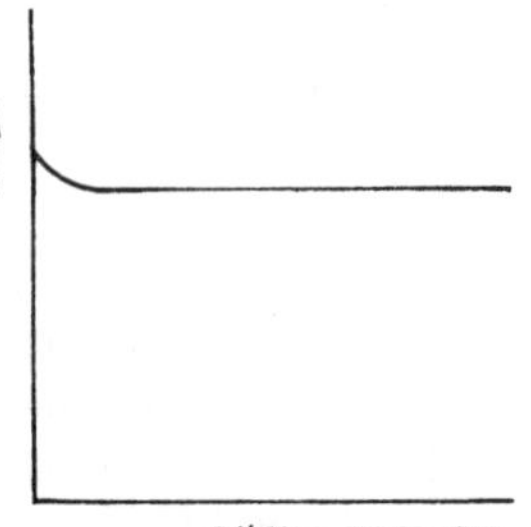

Fig. 16.6 Variation of coefficient of friction with sliding velocity

The static stiffness of a machine tool determines the amount of deflexion and gives rise to component inaccuracy. This can be allowed for with a manually operated machine by making small adjustments to the size of the cut, but it may lead to significant errors in a component made on a numerically controlled machine. Numerically controlled machines for continuous contouring should therefore be designed with greater static stiffness than comparable manually operated machines.

For fast response to changing motion demands in contouring applications, i.e. to prevent excessive overshoot, a high natural frequency is desirable. This can be achieved only by having a system with high stiffness or low mass. The traditional way of improving the stiffness of structural components, by increasing their mass, will do little to increase natural frequency and may in fact aggravate the situation. Natural frequencies should if possible exceed 50 Hz, and the most promising approach is probably to concentrate on improving the drive stiffness while keeping the mass of the moving structure as low as possible.

16.4.5 Economics of numerical control. The high cost of numerically controlled machine tools has no doubt been the main deterrent to many companies when considering their purchase. Unfortunately there is no readily available rule-of-thumb from which N C machines can be evaluated. In many companies a massive investment in N C, necessitating a complete re-organization of manufacture may result in large cost savings. Unfortunately, the sporadic purchase of single machines, which is more likely to occur, frequently results in inefficient and costly operation and the full benefits are not realized.

Substantial reductions in work-in-progress can be achieved, particularly by using machining centres where a number of operations are performed at one setting. This saving is important especially when the parts have a high value.

Fewer jigs and fixtures are required and those used are relatively simple. When design modifications are made they can be readily implemented by modifying the control tape and usually without incurring the cost of new fixtures.

The effective machine utilization is increased as the setting time is less than for manually operated machines, and the non-cutting time is minimized. The benefits of this advantage can be lost, however, if the N C machines are inefficiently programmed by the selection of conservative cutting speeds and feed rates.

Changing technology increasingly demands that the capital investment should be amortized in a short time. For this reason the manufacturing overhead due to depreciation can add considerably to unit costs and should be reduced whenever possible by multi-shift operation. For example, a machine tool costing £200 000, amortized over 10 years, incurs an hourly depreciation of £10.00 if operated on a single-shift basis, £5.00 if operated on a double-shift basis and £3.33 if operated on a three-shift basis.

17 Metrology

17.1 Metrology is a branch of engineering concerned with accurate measurement and is of considerable importance to production engineers. It is the purpose of this chapter to indicate briefly some of the ways in which length, angular displacement, shape and surface roughness can be measured. The reader will probably be conversant with some of the simpler measuring equipment, and for this reason the applications of instruments such as the micrometer and vernier caliper will not be discussed.

17.2 MEASUREMENT OF LENGTH

Length can be measured by the use of light rays, by the rotation of a screwed shaft, or by comparison with a known length frequently in the form of a calibrated scale or a block gauge.

17.2.1 Light rays. The use of light rays as a standard of length was envisaged by Babinet in 1829 and first used by Michelson in 1893. Since 1960 the metre has been defined as 1 650 763·73 wavelengths of orange radiation of krypton 86 under reference conditions. The yard is defined as 0·9144 m, from which it follows that 1 in. is exactly equal to 25·4 mm. The apparatus used to determine length standards will not be described as it is seldom used in industry, although the application of interferometry to the examination of surfaces is described in section 17.7.4.

17.2.2 Block gauges. These blocks of metal, often called slip gauges, are the usual standards of length in factories, where they are frequently used to check and calibrate length measuring equipment. They are known as end standards, because the measurement is taken over their two parallel end faces, which are a known distance apart. These measuring faces are very smooth and will wring to corresponding faces of adjacent slip gauges. By wringing together appropriate gauges, a wide range of sizes can be built up; in fact a typical set can be used to produce sizes from 2 mm to over 250 mm, in steps of 0·0025 mm. There are four grades of accuracy; they are, in descending order, reference, calibration, inspection and workshop. Permissible tolerances for gauges up to 25 mm in length are of the order of $\pm 0{\cdot}05\ \mu m$ for reference gauges and between $+0{\cdot}25\ \mu m$ and $-0{\cdot}10\ \mu m$ for workshop gauges; details of these and of permissible errors in parallelism and flatness are listed in BS 888.[73] The workshop grade is adequate for most factory purposes, although the inspection grade is frequently used for setting gauges and instruments. Calibration sets can be used for the highest grade of work, with a reference set perhaps retained as the ultimate factory standard of length. Reference and calibration sets are normally used with a certificate showing the deviation of each block from its nominal size, thereby obtaining greater accuracy. Workshop and inspection grade slip gauges can be checked against calibration gauges by using a high magnification comparator.

Slip gauges are cumbersome in combinations exceeding about 10 in. and for greater lengths are superseded by length bars, another type of end standard. These bars are wrung together, enabling lengths of several feet to be obtained. Loose fitting studs join the bars to prevent the combination from breaking up should the wringing be accidentally broken. Full details of length bars and their accessories can be found in BS 1790.[74]

The usefulness of slip gauges and length bars can be extended by accessories which, for example, enable gap gauges and marking-out tools to be made up.

17.2.3 Comparison with known lengths. Comparators are instruments enabling the size of a part to be compared with a setting of known dimensions.

As was indicated in section 17.2.2, high-sensitivity comparators are used to check inspection and workshop grade slip gauges. One of these, the level comparator (Fig. 17.1), uses a sensitive spirit level by which

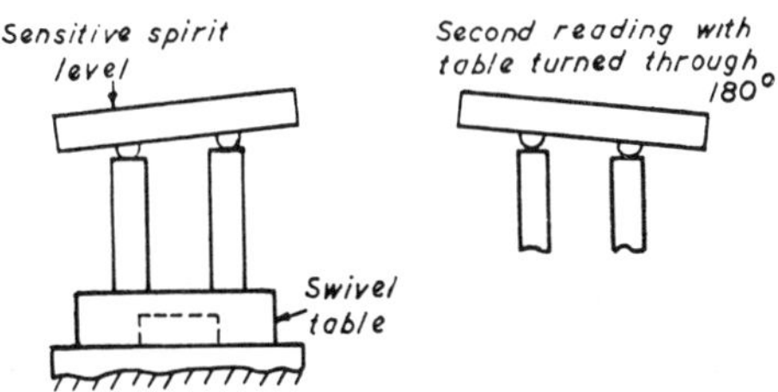

Fig. 17.1 High magnification comparator

height differences of 0·025 μm (0·000001 in) can be detected. The gauges are wrung to the base of the instrument and two readings are taken. The table is turned through 180° between readings, so that the gauges are reversed under the level, and the height difference is found by averaging the first and second spirit level readings.

High-sensitivity comparators are limited in their application, usually being confined to the standards room. Other types are in general factory use to inspect batches of close-limit parts, and some may be fitted to machine tools to control the size of parts being produced. Most comparators are set by using slip gauges or setting pieces, and adjusted until a zero reading is obtained on the measuring scale. The scale reading then records the size difference between the part being measured and the initial setting, provided the difference in size is small.

Many systems of magnifying size are available; the most common are mechanical, mechanical-optical, electrical and pneumatic. These are shown diagrammatically in Fig. 17.2.

The functioning of the mechanical and the mechanical-optical types can easily be deduced from Figs. 17.2 (*a*) and (*b*) respectively. The electrical system, illustrated in Fig. 17.2 (*c*), operates by the displacement of an armature which affects the balance of a bridge circuit. The out-of-balance current is amplified, and a suitably graduated galvonometer is used to indicate displacement.

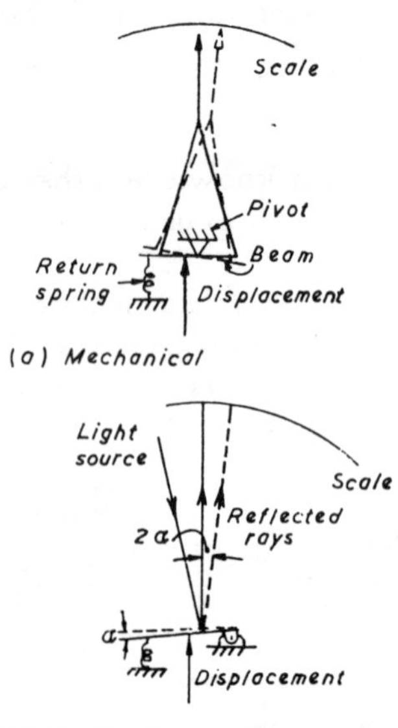

Fig. 17.2 Methods of magnifying size in comparators

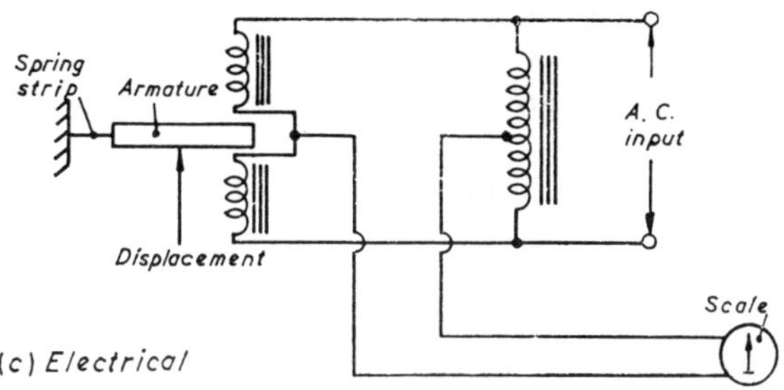

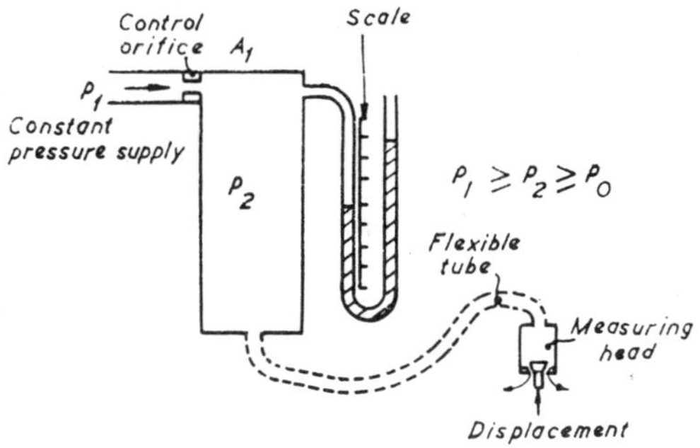

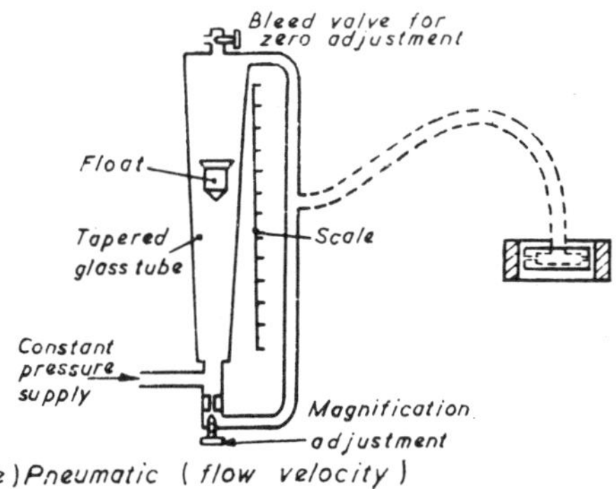

Fig. 17.2 (*contd.*)

The use of electrical type comparators has increased in recent years due to their low cost, small size and the dependability of their solid state circuitry. Magnification of displacement can be very high indeed, some comparators have meter graduations of 0·025 μm (1 μin), enabling block gauges to be checked.

Capacitive type gauge heads are also used in electrical instruments of the comparator type. These are operated by the mechanical displacement of one of the plates of a metal plate capacitor which alters its capacitance and upsets the balance of a bridge circuit. The relatively high change in voltage produced eliminates the need for further amplification. Capacitive type gauge heads are unaffected by magnetic fields but have limited measuring range.

Two types of pneumatic comparator are illustrated: Fig. 17.2 (*d*) shows a back pressure design, and Fig. 17.2 (*e*) a velocity flow type. In the back pressure comparator, the magnitude of the back pressure p_2 is an indication of the ease with which air can escape to atmosphere. Various methods of restricting this escape are available so that p_2 reflects the dimension being measured. A simple application of this type of comparator is in the measurement of small holes, such as those in carburettor jets, the scale being calibrated from holes of known sizes. A second application utilizes an escape orifice of fixed size, the effective area of which is

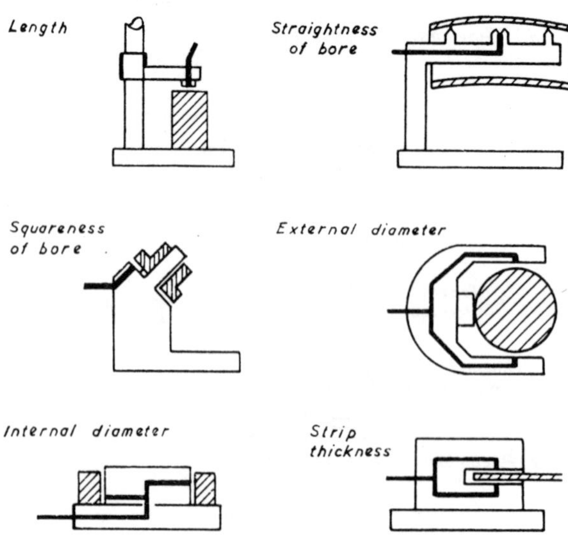

Fig. 17.3 Application of pneumatic gauging (*after Jeglum*)

varied by restricting the air flow caused by the proximity of the part being measured. Settings for flat parts can be obtained from two packs of slip gauges representing the upper and lower limits of size. Most other applications require two setting pieces machined to the appropriate limits. A third method of varying air flow to atmosphere is by displacement of a plunger which varies the size of the escape orifice when in contact with the component. Further examples of component measurement are shown in Fig. 17.3. Where more than one orifice is used the diameters must be matched to avoid errors due to misalignment of workpiece and gauge. The relationship between back pressure p_2 and effective area of measuring orifice a_2 is not linear. This can be seen from Fig. 17.4, where p_2/p_1 has been plotted against a_2/a_1, p_1 and a_1 being the supply pressure and control orifice area respectively. The characteristic can however be assumed linear for values of p_2/p_1 between 0·6 and 0·8, and within this range

$$\frac{p_2}{p_1} = -K\frac{a_2}{a_1} + C.$$

The sensitivity of the instrument is represented by the change in back pressure p_2 for a given change in the measuring orifice effective area a_2. Assuming p_1 and a_1 remain constant, then:

$$\text{Sensitivity} = \frac{dp_2}{da_2} = -K\frac{p_1}{a_1} \qquad (17.1)$$

It will be seen from this expression that the system can be made more sensitive by reducing a_1, the control orifice area, although this will reduce the range of the instrument. Sensitivity can also be increased by raising supply pressure p_1.

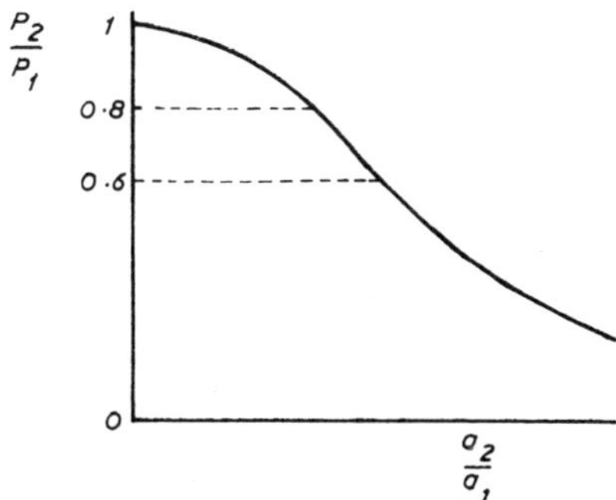

Fig. 17.4 Characteristic of pneumatic comparator

The flow type pneumatic comparator also operates on the restriction of the flow of pressurized air to atmosphere. Again the supply pressure is constant, but there is no intermediate pressure, size being measured by the velocity of air in a tapered glass cylinder (Fig. 17.2(*e*)). The air flow lifts a float in this cylinder, and the faster the flow the higher the lift. The float height is measured on a scale behind the cylinder and a bleed valve is used to zero the float. Magnification can be changed by by-passing some of the air supply, using a screw at the inlet to the glass tube. Flow type pneumatic comparators have a more rapid response than the back pressure variety, particularly when the latter has a long tube to the measuring orifice.

17.2.4 Graduated scales. Apart from the use of graduated scales for general purpose length measurement on rules and vernier calipers, they are used in conjunction with a micrometer microscope to obtain machine and instrument settings to $\pm 0{\cdot}25\ \mu$m ($\pm 0{\cdot}00001$ in). The main scale is accurately graduated at intervals of, say, 1 mm. Intermediate settings or readings are made by adjustment of the micrometer microscope so that the two setting lines symmetrically straddle the image main scale divisions in the field of view of the microscope (Fig. 17.5 (*a*)). There are two

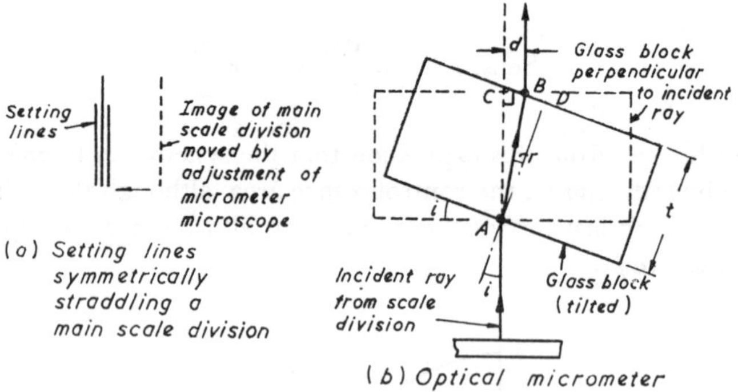

Fig. 17.5 Micrometer microscope

main types of micrometer microscope, one in which the setting lines are moved by a micrometer arrangement and the other, called the optical micrometer, where the image of the scale is moved optically. The principle of the optical micrometer is shown in Fig. 17.5 (*b*). It consists of a glass block mounted between the eyepiece and the main scale; a secondary scale, with linear graduations, is revolved as the block is rotated. When the block is perpendicular to the optical axis there is no shift in the incident ray from the main scale division, but a linear shift d is produced by a block rotation of i.

It will be seen that when the block of thickness t is rotated

$$d = BC = AB \sin (i - r)$$

where i is angle of incidence of ray from scale
r is angle of refraction of glass block

$$d = (AD/\cos r) \sin (i - r)$$

$$d = t \sin (i - r)/\cos r \qquad (17.2)$$

Hence a circular scale indicating linear displacement can be fitted to the block.

The use of an accurately divided scale enables high resolution to be obtained over a long length, and avoids the cumulative errors which occur with some other systems of length measurement.

17.2.5 Rotation of screwed shaft. This principle is employed in the well-known micrometer caliper; it is also used in a variety of other micrometers, such as those used to measure depth of hole and internal diameter. The thimbles of metric micrometer calipers are graduated in 0·01 mm divisions, those of inch instruments have 0·001 in divisions. Skill is needed to obtain accurate hand micrometer readings, largely because of the difficulty of obtaining consistent measuring pressures. Bench micrometers are, however, available for more accurate work; these have large-diameter thimbles graduated in 0·002 mm with the inch version graduated in 0·0001 in divisions. They are also fitted with fiducial indicators which enable consistent measuring pressures to be employed.

17.3 ANGULAR MEASUREMENT

A large number of methods of measuring angles is available. Some of the instruments and gauges associated with angular measurement are: combination angle gauges, sine bars, auto-collimators, angle dekkors, and instruments incorporating circular divided scales.

17.3.1 Combination angle gauges. These are similar in principle to slip gauges and allow a very large number of angles to be built up by wringing together gauges which have their measuring faces inclined at known angles. Angles between 0° and 360° can be built up in steps of 3 seconds by combining gauges from a full set. Since angle gauges can be wrung so that they subtract as well as add, only thirteen and a square block are needed for a full set.

Combination angle gauges can be used directly on a horizontal datum surface so that they incline the component until one of its faces is horizontal. Alternatively, they can be employed as standards for comparison with an angular face, using instruments such as the angle dekkor (section 17.3.4) as the comparator.

17.3.2 Sine bars. This instrument (Fig. 17.6) enables accurate angles to be produced from a flat datum surface using slip gauges to construct a right-angled triangle of known dimensions. The sine bar consists of a bar on which two rollers of equal diameter are mounted at known centre distances apart. The roller diameters and centre distances must be accurate, and the axes of the rollers must be mutually parallel and equidistant from the top surface of the bar which itself must be flat. Limits of accuracy and other details of sine bar design can be found in BS 3064.[75] The use of sine bars to create angles greater than 45° should be avoided, as the larger the angle the less sensitive the value of the sine function to a given change of angle (at 1° a difference of 6 minutes represents a vertical difference of 0·436 mm on a 250 mm sine bar, whereas at 89° the corresponding vertical difference is only 0·007 mm). When angles greater than 45° have to be measured, it is usually possible to hold the part against a vertical surface such as an angle plate, and measure the complement of the angle from the horizontal datum surface.

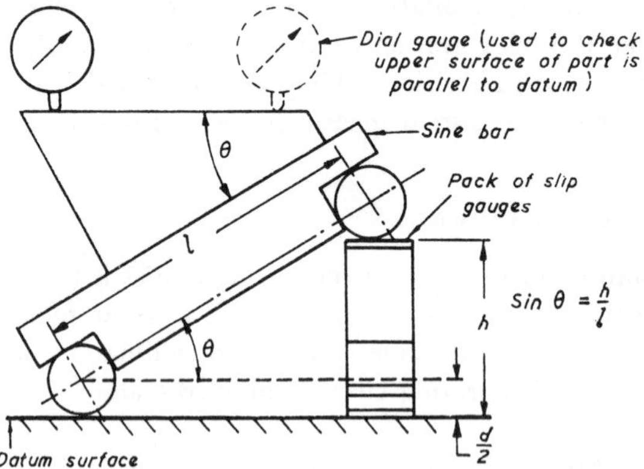

Fig. 17.6 Sine bar used to measure angle on component

Adaptations of the sine bar principle are the sine table (simple or compound) which is used to tilt large parts, and sine centres which can be used to incline conical surfaces, provided they have been centred at each end or fitted on a mandrel.

17.3.3 Auto-collimator. This instrument measures small angular displacements, and its operating principles are shown in Fig. 17.7. It will be seen from this diagram that the longer the focal length f of the collimating lens, the greater the sensitivity of the instrument. The distance of the reflecting surface from the instrument is not critical, as the light emitted from the collimating lens is a parallel beam. However, if the part being examined is too far away, the reflected light may miss the lens altogether. As it is impractical to project the image of a point source in an actual auto-collimator, a wire is placed in the focal plane and illuminated from behind. On looking through the eyepiece, both the wire and its reflected image are seen against a graduated scale (range 10 minutes, divided in 0·5 minute divisions). The displacement of the image is measured by rotating a micrometer drum until two setting lines symmetrically straddle the reflected image, the graduations on the micrometer enable readings to be taken to 1 second of arc.

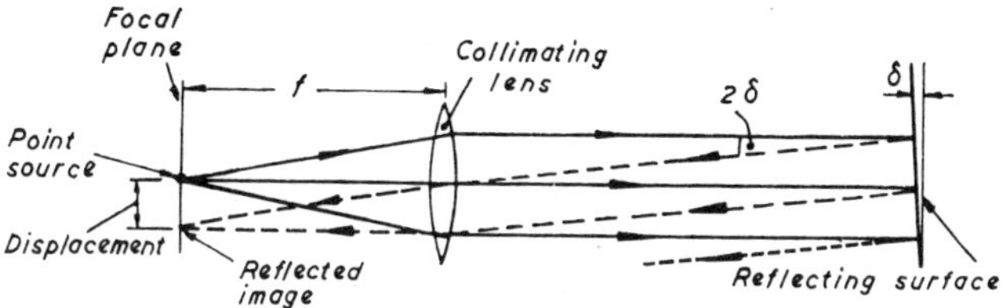

Fig. 17.7 Principle of auto-collimator

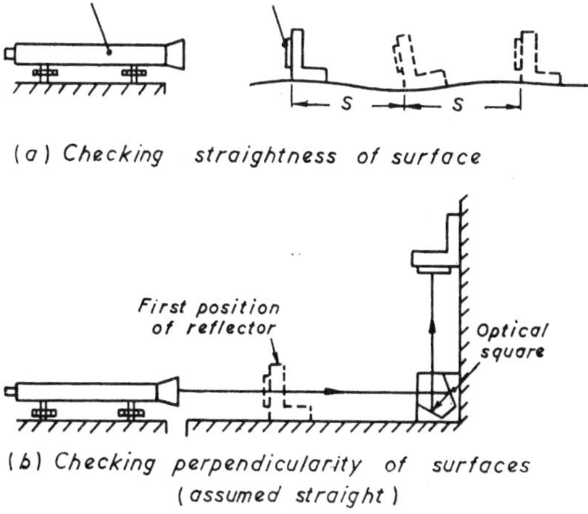

Fig. 17.8 Applications of auto-collimator

Auto-collimators can be used for a variety of measurements involving small angular deviations; two examples are shown in Fig. 17.8 (*a*) and (*b*). In the first, a perpendicular reflecting surface is used with the auto-collimator to check the straightness of a surface, and in the second an optical square, which turns a ray of light through 90°, is introduced to check the perpendicularity of one surface to another.

17.3.4 Angle dekkor. Although often less sensitive than the auto-collimator, this instrument can be used for a wide variety of angular measurements. Typically it has a range of 50 minutes, and readings can be estimated to about 0·2 minute on a scale graduated in 0·5 minute intervals. The principle of the angle dekkor is shown in Fig. 17.9 (*a*). A scale (engraved on a glass screen out of the field of vision) is illuminated and projected by the collimating lens as a parallel beam on to the surface being checked. The image of the scale is reflected from this surface and focused by the collimating lens in the field of view of the glass screen. This screen also has an engraved datum scale at right-angles to the axis of the reflected scale; the view through the eyepiece is shown in Fig. 17.9 (*b*). The purpose of the datum scale is to enable both the standard and the object being examined to be positioned in the same transverse plane. The standard angle used can be made up from combination angle gauges or from a sine bar. If the surface being checked is insufficiently reflective, a slip gauge can be placed on it, secured if necessary by an elastic band.

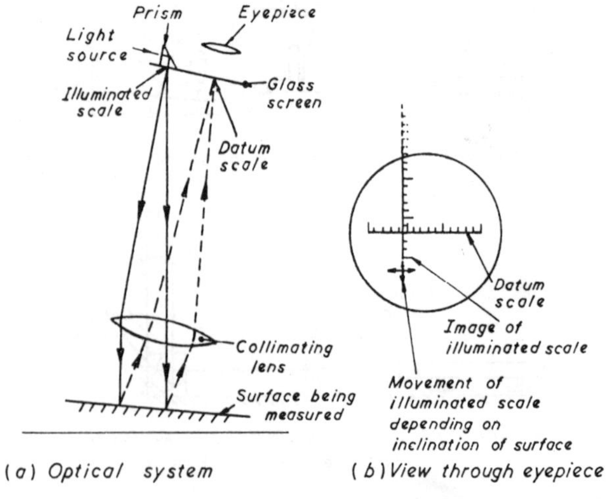

Fig. 17.9 Angle dekkor

17.3.5 Precision level. This simple instrument measures small angular deviations from the horizontal; it can be used on its own or incorporated into other instruments such as the clinometer described in Section 17.3.7. Liquid containing an air bubble is held in a vial (a sealed length of curved glass tube) which is mounted in a suitable base. It will be seen from Fig. 17.10 (*a*) that if the base of the level is tilted through a small angle $\delta\theta$ the bubble will incur a corresponding angular displacement measured from the centre of curvature of the vial. The distance the bubble moves along the vial δl will be equal to $\delta\theta R$, where R is the radius of the vial, and its sensitivity is therefore directly proportional to R. The sensitivity of a level is expressed as the angle of tilt which causes the bubble to move through one scale division; BS 958[76] recommends that these should be spaced 0·1 in (2·5 mm) apart. If a typical precision level indicates a 10 second inclination by a 2·5 mm (0·1 in) bubble movement, in this instance the vial radius $R = \delta l/\delta\theta \simeq 51{\cdot}5$ m (170 ft). Recommendations for the design and testing of precision levels can be found in BS 958.

An engineer's precision level is termed the block level. A useful variant of this instrument is the square block level (Fig. 17.10 (*b*)) which is in the form of a hollow square with a precision level set in the lower horizontal section. Each of the four sides of the square is accurately machined, so that apart from normal use it may be used to check the perpendicularity of vertical surfaces and the undersides of overhanging horizontal surfaces.

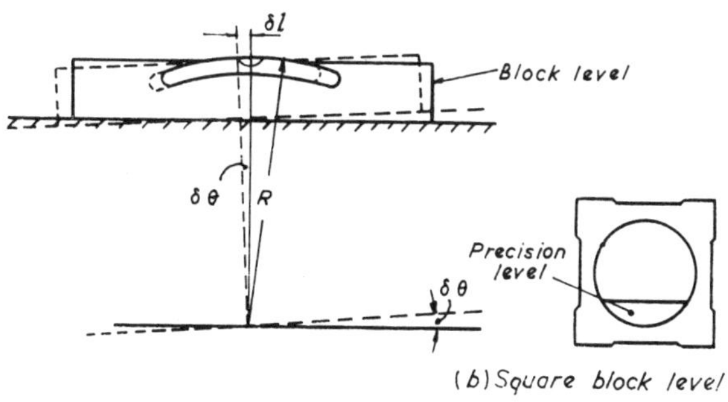

Fig. 17.10 Precision levels

17.3.6 Determination of straightness and flatness of surfaces. Auto-collimators and precision levels can be used to find the straightness and flatness of metal surfaces, and Fig. 17.8 (*a*) shows how an auto-collimator is employed to find the straightness of a horizontal surface.

When readings have been taken at intervals along a line, as in the checking of a straight edge, they are first converted to give cumulative height differences from the starting end. The best line through the points considered can then be found by applying the method of least squares, which is explained below. The best line will satisfy the following conditions:

(*a*) the sum of the deviations of the points from the line is zero;

(*b*) the sum of the squares of the deviations from the line is a minimum.

Since the sum of the deviations from the mean line is zero it follows that the centroid of the points $(\bar{X}, \bar{Y})$, must lie on this line (Fig. 17.11 (*a*)). It is convenient to retabulate the points taking $\bar{X}$ and $\bar{Y}$ as the zero values (Fig. 17.11 (*b*)), so that

$$x_1 = X_1 - \bar{X}, x_2 = X_2 - \bar{X} \ldots x_n = X_n - \bar{X}$$

$$y_1 = Y_1 - \bar{Y}, y_2 = Y_2 - \bar{Y} \ldots y_n = Y_n - \bar{Y}$$

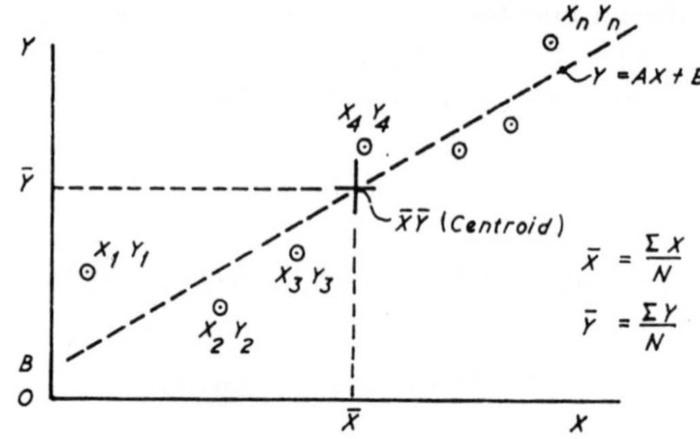

(*a*) Centroid of measured points

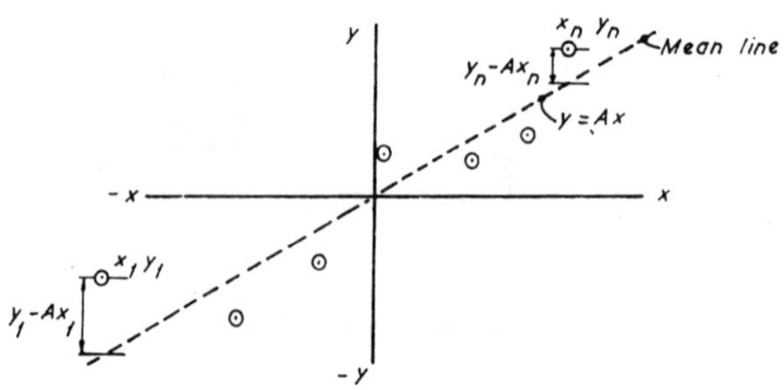

(*b*) Reference of points to axes through centroid

Fig. 17.11 Mean true line of straight edge

Assume A to be the slope of the best line through the points $y = Ax$. Deviation of a point $x_i y_i = l_i = y_i - Ax_i$

$$\sum_{i=1}^{n} e_i^2 = \sum_{i=1}^{n} (y_i - Ax_i)^2$$

To find slope of the best straight line through the points

$$\frac{d\Sigma e_i^2}{dA} = \sum_{i=1}^{n} x_i\{2(Ax_i - y_i)\}$$

$$= 2A \sum_{i=1}^{n} x_i^2 - 2 \sum_{i=1}^{n} x_i y_i$$

For $\sum_{i=1}^{n} e_i^2$ to be a minimum $\frac{d\Sigma e_i^2}{dA} = 0$

$$A = \sum_{i=1}^{n} x_i y_i \Big/ \sum_{i=1}^{n} x_i^2 \qquad (17.3)$$

Hence the slope of the best line through $\bar{X}\bar{Y}$ can be calculated and this gives the mean true plane of the straight edge, from which errors in straightness can be determined. Details of other recommended tests for straight edges are set out in BS 5204.[77]

In a similar manner, when checking the flatness of a surface such as a surface table, it is possible to calculate the equation of a mean

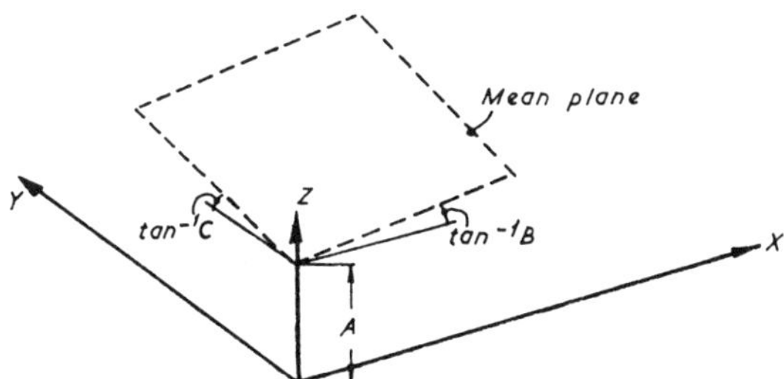

Fig. 17.12 Mean plane relative to XY and Z axes

plane through a number of points on the surface. Let the surface being measured lie approximately on the $X - Y$ plane (Fig. 17.12). Then the equation of the mean plane is

$$Z = A + BX + CY$$

In a similar way to the straight edge, the mean plane of a surface is that on which the sum of deviations is zero and the sum of the squares of the deviations is a minimum. The mean plane must pass through the centroid of the measured points ($\bar{X}$, $\bar{Y}$, $\bar{Z}$). Again for convenience the centroid is made the origin of all three axes. This is done by subtracting the centroid value $\bar{X}$, $\bar{Y}$ and $\bar{Z}$ from the individual X, Y and Z values, i.e. $x_1 = X_1 - \bar{X}$, $y_1 = Y_1 - \bar{Y}$, $z_1 = Z_1 - \bar{Z}$ etc.

The equation of the plane is now $z = Bx + Cy$ and the deviation from the mean plane at a point $x_i y_i z_i$ is e_i

where $e_i = z_i - Bx_i - Cy_i$

$$\Sigma e_i^2 = \sum_{i=1}^{n} (z_i - Bx_i - Cy_i)^2$$

$$\frac{\partial \Sigma e_i^2}{\partial B} = \Sigma 2x_i(Bx_i + Cy_i - z_i)$$

$$\frac{\partial \Sigma e_i^2}{\partial B} = 2B\Sigma x_i^2 + 2C\Sigma x_i y_i - 2\Sigma x_i z_i \tag{17.4}$$

also

$$\frac{\partial \Sigma e_i^2}{\partial C} = 2B\Sigma x_i y_i + 2C\Sigma y_i^2 - 2\Sigma y_i z_i \tag{17.5}$$

For Σe_i^2 to be a minimum, $\partial\Sigma e_i^2/\partial B = 0$ and $\partial\Sigma e_i^2/\partial C = 0$
Substituting for B in equation (19.5) it can be shown that

$$C = \frac{\Sigma x_i^2 \Sigma y_i z_i - \Sigma x_i y_i \Sigma x_i z_i}{\Sigma x_i^2 \Sigma y_i^2 : - (\Sigma x_i y_i)^2} \tag{17.6}$$

and

$$B = \frac{\Sigma y_i^2 \Sigma x_i z_i - \Sigma x_i y_i \Sigma y_i z_i}{\Sigma x_i^2 \Sigma y_i^2 - (\Sigma x_i y_i)^2} \tag{17.7}$$

By marking off the surface in a uniform square grid and measuring the height of each intersection relative to a datum point, the inclination of the mean true plane can be found by substitution into equations

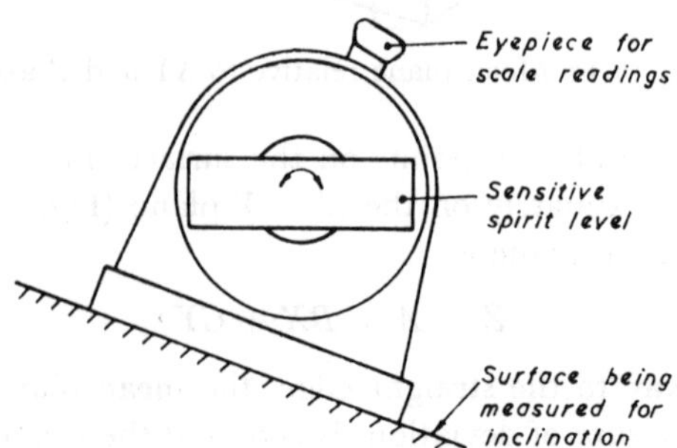

Fig. 17.13 Level type clinometer

(17.6) and (17.7). Once the mean true plane is known, the difference between the largest positive and the largest negative deviations gives the maximum error of the surface. Details of specifications and tests for surface plates are given in BS 817.[78]

17.3.7 Circular divided scales. It is possible to produce very accurately divided circular scales which can, for instance, be directly calibrated in 5-minute intervals, the scale being read with the aid of a microscope. Another arrangement uses a scale calibrated in 10-minute intervals, with an optical micrometer providing direct readings to 3 seconds of arc. Circular divided scales are incorporated in rotary tables on machines such as jig borers to obtain accurate angular displacement of work. A further application of circular scales is in clinometers, which can be used to measure the inclination of surfaces on which they are placed (Fig. 17.13). The sensitive spirit level in the clinometer illustrated enables the central portion to be rotated to a horizontal position, so that the inclination of the base can be read on an accurately divided circular scale.

The circular scale should be centrally mounted, as any eccentricity introduces an error. If a scale of radius R is mounted eccentrically by an amount e the maximum error in angular reading will be $\pm e/R$ radians (Fig. 17.14).

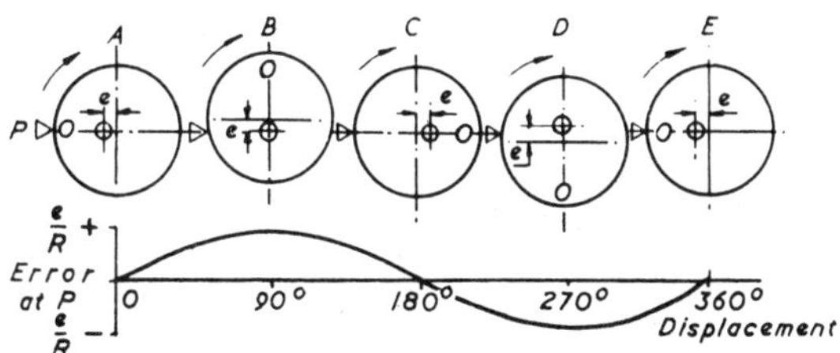

Fig. 17.14 Errors due to eccentric mounting of scale

17.3.8 Measurement of tapers. Tapers can be measured using standard diameter rollers and balls (supplied in sets by bearing manufacturers), and slip gauges.

An arrangement shown in Fig. 17.15 is suitable for measuring male tapers. The measurements l_1 and l_2 are obtained by using a micrometer caliper and h is the height of the two equal packs of slip gauges.

$$\text{Half angle of taper } \frac{\theta}{2} = \tan^{-1} \frac{l_2 - l_1}{2h} \qquad (17.8)$$

An arrangement for measuring female tapers using a ball and roller combination is shown in Fig. 17.16.

$$\tan\frac{\theta}{2} = \frac{l_2 - l_1}{2h}$$

$$l_2 = \frac{d_2 + 2r_2}{\cos(\theta/2)} \text{ and } l_1 = \frac{d_1 + 2r_1}{\cos(\theta/2)}$$

$$\cos\frac{\theta}{2}\tan\frac{\theta}{2} = \frac{(d_2 - d_1) + 2(r_2 - r_1)}{2h}$$

$$\frac{\theta}{2} = \sin^{-1}\frac{(d_2 - d_1) + 2(r_2 - r_1)}{2h} \tag{17.9}$$

The value of h can be obtained with packs of slip gauges and a dial test indicator mounted on a stand.

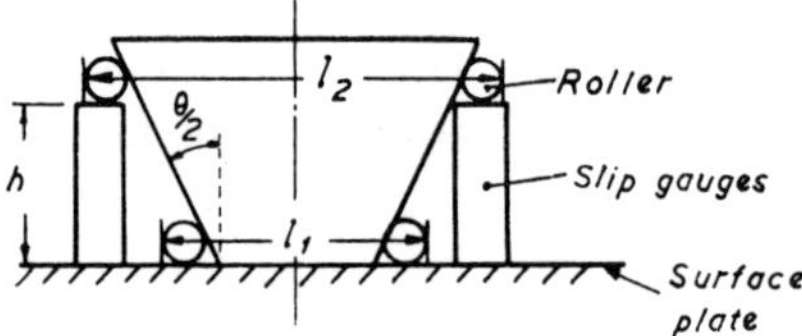

Fig. 17.15 Measurement of male taper

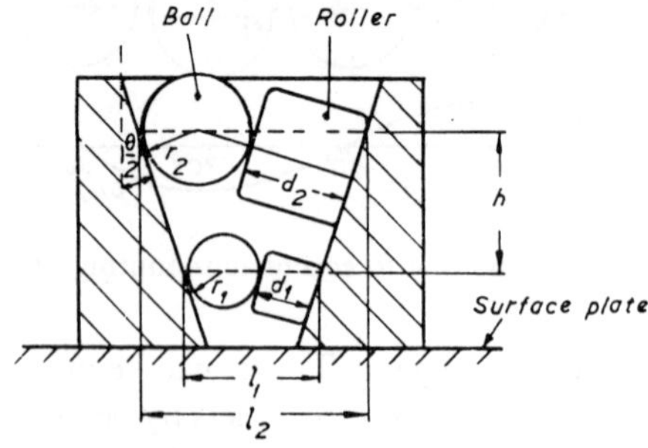

Fig. 17.16 Measurement of female taper

17.4 MEASUREMENT OF FORM

This will be considered in two parts, firstly the measurement of shape by optical projection and then that of two specific forms, screw threads and gear teeth (two important and specialized forms of measurement).

17.4.1 Optical projector. This type of instrument is used to project a magnified silhouette of the object being examined; the diagram in

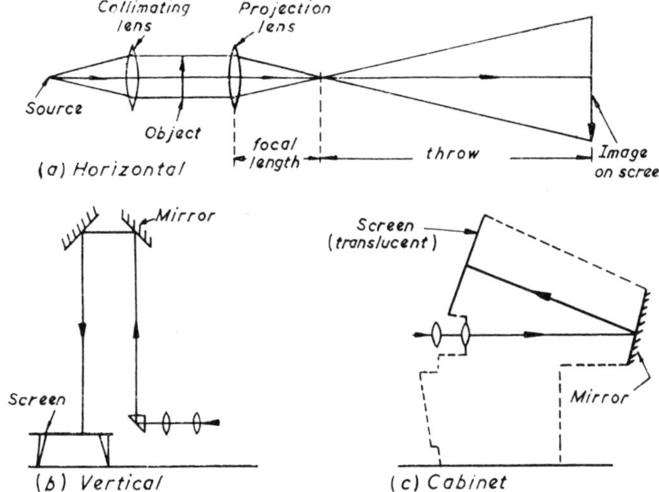

Fig. 17.17 Optical projectors

Fig. 17.17 (*a*) shows the optical principles involved. It will be seen that the magnification is equal to throw/focal length of projection lens. Magnifications of the order of 50:1 and 100:1 are obtainable and the shape of the object can be checked by using an accurately enlarged tracing or template on a screen. The floor area occupied by the projector may be reduced by using a vertical arrangement or a cabinet projector (Figs. 17.17 (*b*) and (*c*)).

A measuring system fitted to the projector enables the work to be accurately displaced both vertically and horizontally. Linear measurements can be obtained by measuring the image on the screen with a precision glass scale, or by displacing the work so that co-ordinate distances can be obtained by use of datum lines on the screen and the measuring system fitted to the instrument.

17.5 MEASUREMENT OF SCREW THREADS

A large variety of screw threads is used to transmit motion or force, to act as fastenings and to measure length. The most common is the V-form thread, the elements of which are shown in Fig. 17.18. The measurement of these elements is now described.

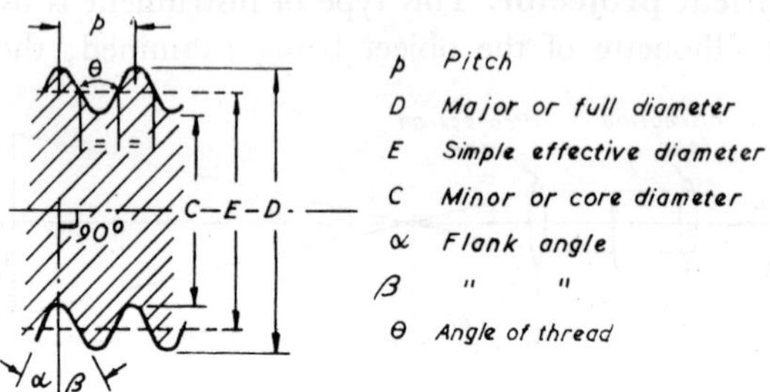

Fig. 17.18 Elements of typical vee form thread

17.5.1 Major diameter of thread. This is best found by using a bench micrometer. A number of measurements along the thread should be taken to check for taper, with different angular readings at each longitudinal position to check for ovality.

17.5.2 Simple effective diameter of thread. As can be seen from Fig. 17.18, the simple effective diameter is the diameter of an imaginary cylinder which is coaxial with the thread axis, and which cuts the thread so the distance between adjacent intercepts is half the pitch of the thread. It should be measured by floating carriage bench micrometer, provided that the thread can be held between centres; if this is impossible, a standard bench micrometer or hand micrometer can be used with a third measuring cylinder. The floating carriage micrometer consists of a bed with centres which hold the work longitudinally over the bed. A carriage can be freely moved along the bed, and there is a second slide on the carriage over which a bench micrometer can be moved perpendicular to the longitudinal axis of the bed. The ease of movement of the slideways is responsible for the name 'floating micrometer'. Two measuring cylinders of equal diameter, often called wires, are supported above the end of the micrometer anvils so that the measuring cylinder rests in Vs on opposite sides of the thread, enabling a measurement to be taken across them (Fig. 17.19 (*a*)).

By referring to Fig. 17.19 (*b*) an expression for the effective diameter of the thread can be obtained:

$$E = T + P \tag{17.10}$$

or

$$E = M - 2w + P \tag{17.11}$$

where E is effective diameter;

T is diameter under the wires;

M is measurement over the wires;
w is wire diameter;
P is 2 × distance from underside of wire to effective diameter.

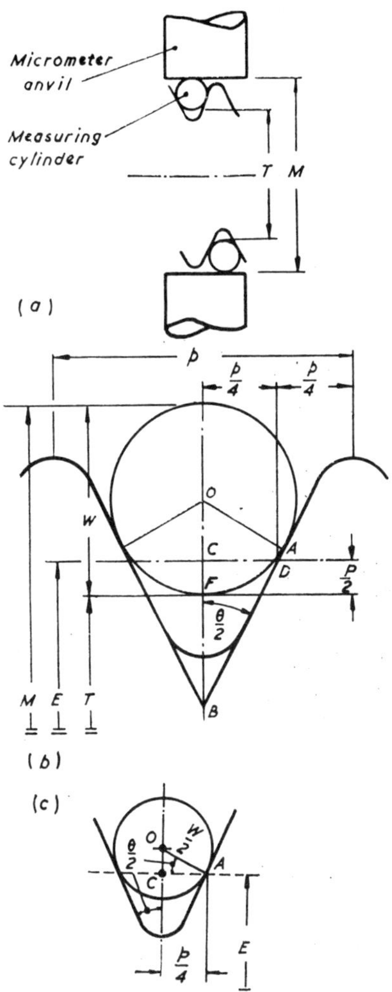

Fig. 17.19 Measurement of effective diameter (both flank angles assumed equal to $\frac{\theta}{2}$)

The value of T can be found by measuring the wires over a cylinder of known size and then measuring with the screw thread in position. Reference to equation (17.10) will show that the value of P is required in addition to that of T before the effective diameter can be found. P can be obtained by considering Fig. 17.19 (*b*).

In triangle BCD

$$\text{BC} = \text{CD} \cot \frac{\theta}{2}$$

$$\text{BC} = \frac{p}{4} \cot \frac{\theta}{2} \tag{17.12}$$

where p is thread pitch.

In triangle ABO

$$\text{BO} = \text{AO} \operatorname{cosec} \frac{\theta}{2}$$

$$\text{BO} = \frac{w}{2} \operatorname{cosec} \frac{\theta}{2}$$

$$\text{BF} = \text{BO} - \frac{w}{2}$$

$$\text{BF} = \frac{w}{2}\left(\operatorname{cosec} \frac{\theta}{2} - 1\right) \tag{17.13}$$

but

$$\frac{P}{2} = \text{BC} - \text{BF}$$

From equations (17.12) and (17.13)

$$\frac{P}{2} = \frac{p}{4} \cot \frac{\theta}{2} - \frac{w}{2}\left(\operatorname{cosec} \frac{\theta}{2} - 1\right)$$

$$P = \frac{p}{2} \cot \frac{\theta}{2} - w\left(\operatorname{cosec} \frac{\theta}{2} - 1\right) \tag{17.14}$$

Hence

$$E = T + \frac{p}{2} \cot \frac{\theta}{2} - w\left(\operatorname{cosec} \frac{\theta}{2} - 1\right) \tag{17.15}$$

If the two-wire method just described cannot be used, the three-wire method is employed. Two wires are placed on one side of the screw in adjacent threads and a third wire is located between them on the opposite side of the screw. The value of E can be obtained from equation (17.11), as in this instance a direct measurement M over the top of the wires is obtained.

Although the wires must always contact the straight portions of the thread flanks, it is preferable that they should touch as close as possible to the effective diameter; in this way inaccurate flank angles do not introduce serious errors in the measurement over the cylinder. The usual tolerance requires contact within $\pm 1/20$ of the straight flank length from the effective diameter.

The 'best' wire size can be simply calculated from consideration of Fig. 17.19 (*c*).

In triangle ACO

$$\frac{w}{2} = \frac{p}{4 \cos \theta/2}$$

$$w = \frac{p}{2} \sec \frac{\theta}{2} \qquad (17.16)$$

If cylinders of the 'best' size are used, p and θ can be taken as their nominal values and the P value (equation (17.14)) for a given pair of wires can be treated as a constant when used to measure threads of the appropriate pitch and flank angle.

The formula for effective diameter (equation (17.15)) assumes a zero helix angle and no deformation of the measuring cylinders between the thread and the micrometer anvils. With the cylinders lying over at the helix angle, the points of contact between them and the thread flanks do not lie on a plane parallel to the thread axis, as shown in Fig. 17.19. One contact point is now just above and the other just below this plane, causing a radially outward displacement of the cylinders and slightly increasing the micrometer reading. The elastic compression of the cylinders, however, produces a small reduction in the micrometer reading. As both corrections are of small magnitude and opposite direction they are usually neglected; but details of their calculation can be found if required in NPL *Notes on applied science*, 'Measurement of screw threads.'[76]

17.5.3 Minor diameter of thread. Two triangular prisms and a floating carriage micrometer are used to measure the minor diameter (Fig. 17.20 (*a*)). The size of the prisms themselves does not have to be taken

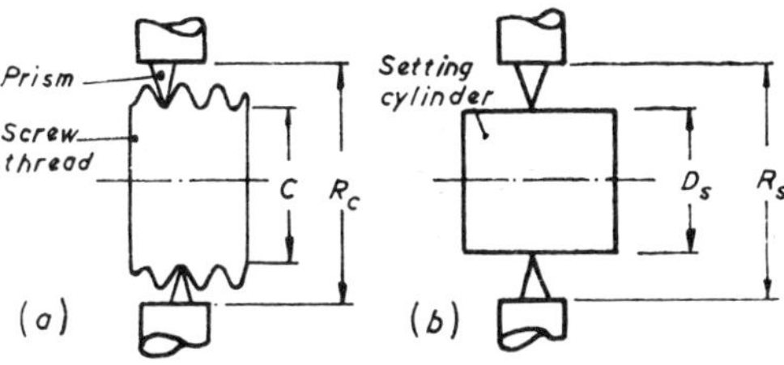

Fig. 17.20 Measurement of minor diameter

into account, as an initial micrometer reading R_s is taken with the prisms in position over a cylindrical setting piece of known diameter D_s (Fig. 17.20 (*b*)). If the measurement across the minor diameter of the thread using prisms is R_c

Minor diameter $C = D_s + (R_c - R_s)$.

17.5.4 Thread pitch. A specially designed instrument is used to measure the pitch of external and internal threads. A stylus is traversed along the thread by means of a micrometer unit which indicates the distance travelled. When the stylus is in contact with the thread flanks, a pointer is moved to either one side or the other of a scale, depending on which flank is in contact. However, when the stylus is bedded in at the root of the thread the pointer is at the centre of the scale. By taking successive micrometer readings when the scale is centred, the pitch of the thread can be obtained.

Several types of pitch error may be present; these can be cumulative, periodic or irregular. Cumulative errors are shown by a cumulative lengthening or shortening of the thread and can, for instance, occur when a thread changes length during heat treatment. Periodic errors repeat at regular intervals, resulting in successive shortening and lengthening of the pitch. Drunkenness is another type of periodic error but one which recurs at each revolution of the thread. Although drunken threads have correct pitch, the helix when developed is a curve and not a straight line. Periodic errors are likely to be caused by a lathe having an inaccurate leadscrew thrust race. As the name implies, irregular errors follow no set pattern and result from disturbances during machining.

17.5.5 Thread form. Optical projectors described in section 17.4.1 are used for the measurement of thread form. Templates or enlarged drawings can be used for comparison with the shadow image on the screen. Flank angles can be measured by protractors fitted to the screen or by rotating the screen and using its datum lines and the angular scale

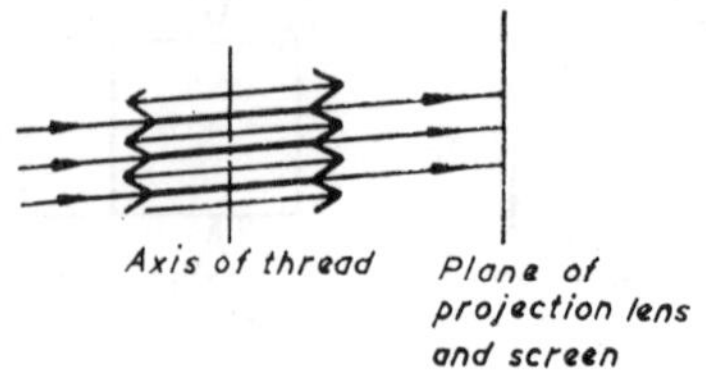

Fig. 17.21 Projection of thread form

at its periphery. To avoid interference, the lamp holder is swung through the helix angle of the thread so that the light rays travel along the helix. The thread form is, however, measured in a plane containing the thread centre line and not in a plane perpendicular to the thread helix. The foreshortening effect of light rays travelling along the helix is compensated for and a centre line image obtained, because the projection lens and the screen are parallel to the axis of the thread (Fig. 17.21). A second method of measuring flank angle is to use a microscope fitted with a goniometric head.

17.5.6 Virtual effective diameter of thread. If a perfect male thread is fitted to the smallest perfect nut, the effective diameter of each will be the same. If, however, the male thread has pitch or flank angle errors, a larger perfect nut will be needed before the two can be assembled. The virtual effective diameter of a screw is defined as the effective diameter of the smallest nut of perfect form or pitch which can be assembled to the screw.

The effects of both pitch and flank angle errors will have to be considered to obtain an expression for virtual effective diameter. It can be seen from Fig. 17.22 that with a pitch error of δp_s over the screw length,

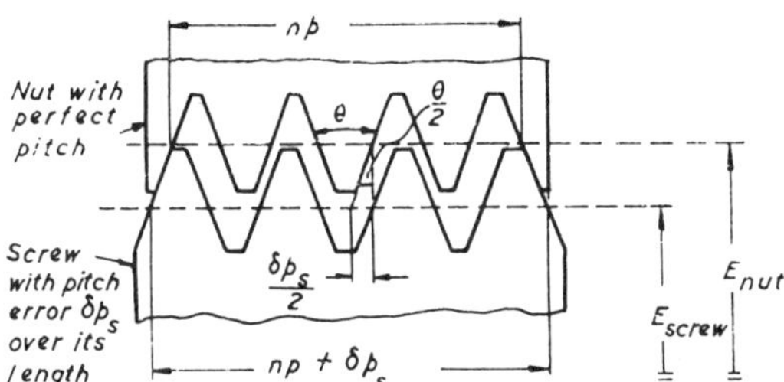

Fig. 17.22 Effect of pitch error on effective diameter

the increase in effective diameter δE_p is found by substituting into the following expression.

$$\delta E_p = \cot\frac{\theta}{2}\,\delta p_s \qquad (17.17)$$

Flank angle errors, assuming they are constant over the flank length, also have the effect of increasing the simple effective diameter of male threads. Consider Fig. 17.23 (*a*) showing a nut and screw with V threads. There are no pitch errors, but there is a screw flank angle error of $\delta\theta$ on one flank. The increase in effective diameter of the screw can be obtained by consideration of Figs. 17.23 (*a*) and (*b*), from which it can be

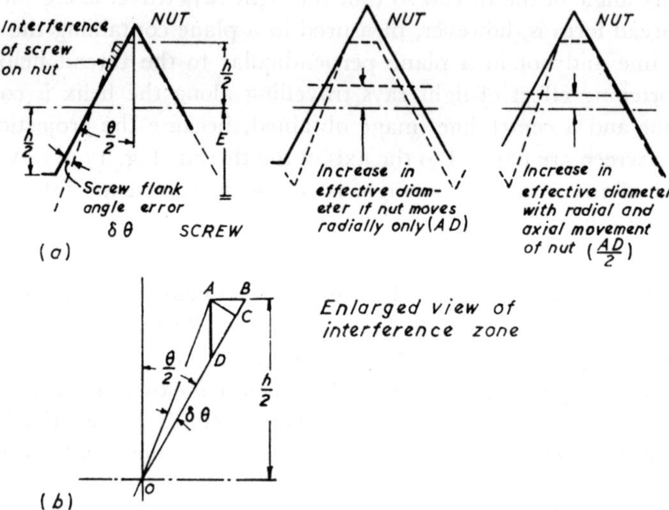

Fig. 17.23 Effect of flank angle error on effective diameter of screw

seen that the increase in effective diameter is not $2AD$ but AD, due to an axial movement of the nut, which is possible when its effective diameter is increased.

$$AD = \frac{AC}{\sin(\theta/2)}$$

If $\delta\theta$ is small

$$\delta\theta \simeq \frac{\mathrm{AC}}{\mathrm{BO}}$$

$$\mathrm{AC} = \mathrm{BO}\delta\theta$$

$$\mathrm{AD} = \frac{\mathrm{BO}\delta\theta}{\sin(\theta/2)}$$

but $\mathrm{BO} = \dfrac{h}{2}\dfrac{1}{\cos\theta/2}$ where h is the depth of the thread.

$$\therefore \qquad \mathrm{AD} = \frac{h\delta\theta}{2\sin(\theta/2)\cos(\theta/2)} = \frac{h\delta\theta}{\sin\theta}$$

If there are angular errors $\delta\theta_1$ and $\delta\theta_2$ on each of the flanks of the screw, the increase in effective diameter δE_θ is given by

$$\delta E_\theta = \frac{h}{\sin\theta}(\delta\theta_1 + \delta\theta_2) \qquad (17.18)$$

The virtual effective diameter of a screw can be found by adding δE_p and δE_θ to the simple effective diameter E.

$$\text{Virtual effective diameter} = E + \cot\frac{\theta}{2}\,\delta p_s + \frac{h}{\sin\theta}(\delta\theta_1 + \delta\theta_2) \qquad (17.19)$$

17.6 MEASUREMENT OF GEARS

Gears are a standard method of transmitting torque. The gear form almost universally used is the involute; it has the advantages that it can be cut if required by a straight-sided cutting tool of rack form, and that it gives a constant velocity ratio even though there may be errors in the centre distances of mating gears. A large variety of gears is in use, but space will enable only the measurement of straight tooth involute gears to be described.

17.6.1 The involute form. Before gear measurement is discussed, something should be known of the properties of the involute.

The involute curve can be defined as the shape traced out by the end of a taut string unwound from a circle known as the base circle. Another method of producing an involute is to roll a straight-edge round the base circle (Fig. 17.24). Two properties of the involute worthy of note are that the tangent to any point on the involute is perpendicular to the line from that point to the base circle, and that the length of the line is equal to its corresponding arc on the base circle, i.e. CB = arc AB in Fig. 17.24.

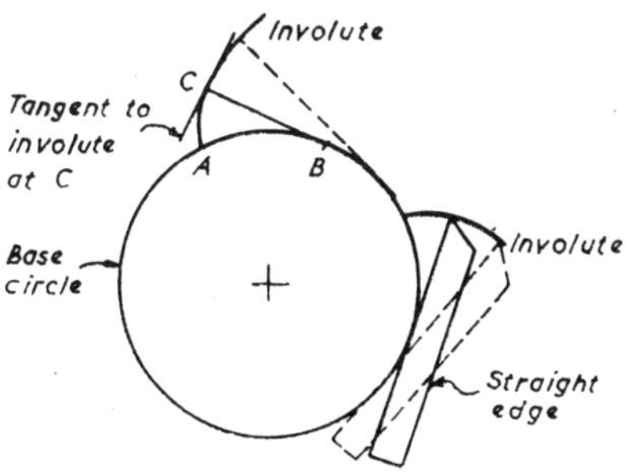

Fig. 17.24 Generation of involute curve

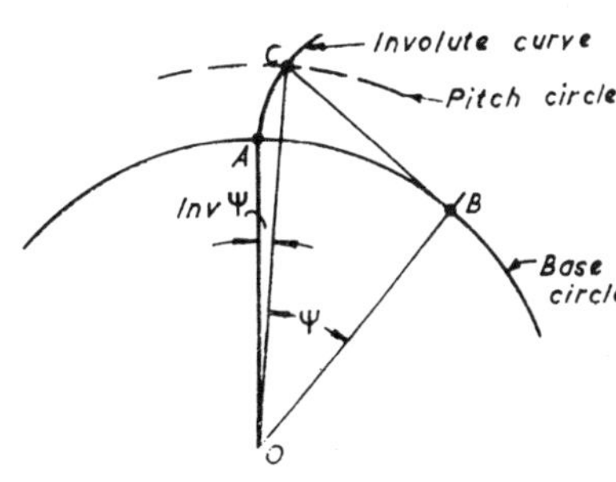

Fig. 17.25 The involute function

The involute function is a property useful in gear measurement. For a given point on the involute curve the involute function is the angle made at the gear centre between that point and the origin of the involute. In Fig. 17.25 the involute function of $C\hat{O}B$ (ψ) is $A\hat{O}C$ (inv ψ). It will be seen from this figure that

$$\tan \psi = \frac{BC}{BO} = \frac{\text{arc AB}}{BO}$$

but

$$\frac{\text{arc AB}}{BO} = (\psi + \text{inv } \psi) \text{ radians}$$

$$\therefore \tan \psi = (\psi + \text{inv } \psi) \text{ radians}$$

$$\text{and inv } \psi = (\tan \psi - \psi) \text{ radians.} \qquad (17.20)$$

17.6.2 Gear tooth elements. The terms used to describe gear tooth elements are shown in Fig. 17.26. The pitch and base circles of two

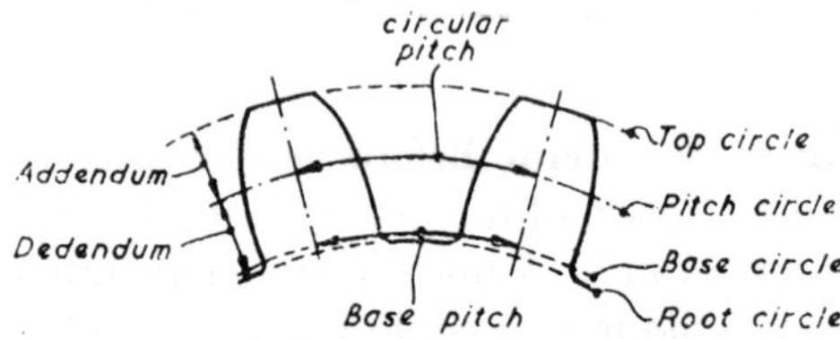

Fig. 17.26 Gear tooth elements

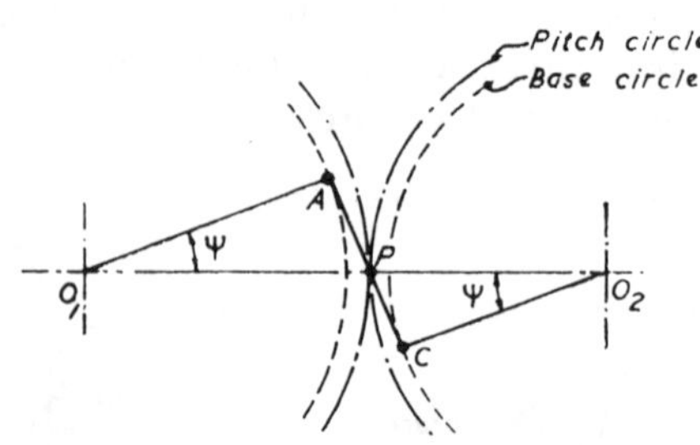

Fig. 17.27 Pressure angle

spur gears in contact are shown in Fig. 17.27. The pitch circles are coincident at point P, the pitch point, and the involutes are in contact along APC, the line of action. Angles $A\hat{O}_1P$ and $C\hat{O}_2P$ are equal, and are termed the pressure angle ψ, which is usually 20° but occasionally $14\frac{1}{2}$°. Apart from the number of teeth N and the pressure angle, a third term is necessary to specify a spur gear; this is the diametral pitch P, or its inverse the module M. Diametral pitch is defined as the number of teeth/pitch diameter ($P = N/D$). The base diameter can be deduced from the pitch diameter and the pressure angle; reference to Fig. 17.27 shows that diameter of base circle $= D \cos \psi$.

17.6.3 Rolling gear test. This is a simple test in which the gear to be checked is rotated against an accurate master gear. The gears are kept in close mesh by spring pressure, and a dial gauge is used to measure any change in centre distance as the gears revolve.

Some rolling gear testing machines produce a trace record (Fig. 17.28) of the variation in centre distances. If a perfect gear is tested, the trace will be a horizontal line, but if the pitch circle is eccentric to the centre of rotation this is indicated by a sine wave with a period of one revolution of the faulty gear. Individual tooth errors are shown by undulations in the trace which have an appropriately shorter period.

Backlash can also be measured by these testing machines. The gear and the master gear are adjusted to their correct centre distance and the change in centre distance δx is noted when the gears are pressed into close mesh. As the master gear is cut so that it has no backlash δx must be doubled, since in service both meshing gears will have backlash. The

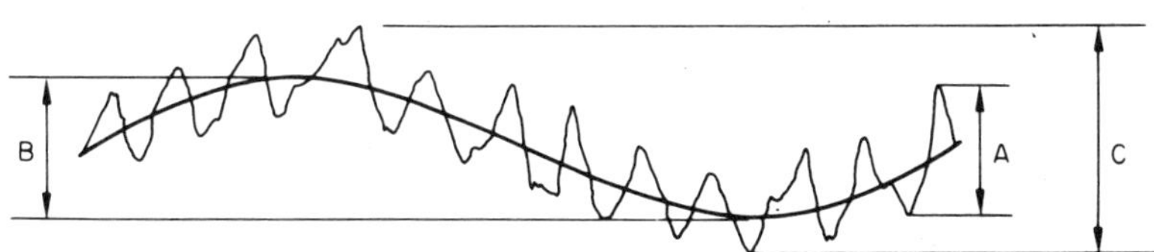

Fig. 17.28 Recording from rolling gear test

backlash along the line of action, known as the normal backlash, can be obtained by considering the gear geometry.

$$\text{Normal backlash} = 2\delta x \sin \psi.$$

Circumferential backlash, i.e. the total free movement in the direction of the circumference $=$ normal backlash/$\cos \psi$.

17.6.4 Checking gear tooth profile. The involute form of a gear tooth can be tested by generating a perfect involute and comparing it

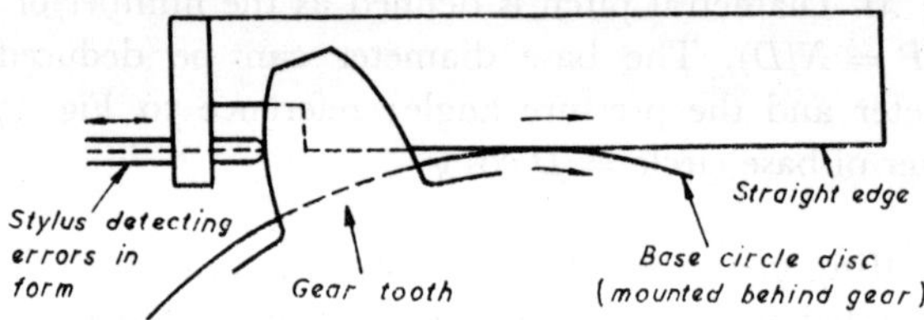

Fig. 17.29 Principle of involute testing machine

with the tooth profile of the gear being checked. The perfect profile is produced by rolling a straight edge along a disc with an appropriate base circle diameter. The principle of an involute testing machine is shown in Fig. 17.29. Simple models use dial gauges to detect errors in profile but more elaborate designs record errors on a trace.

17.6.5 Measurement of gear tooth thickness. The thickness of a gear tooth is defined as the length of the arc of the pitch circle between opposite faces of the same tooth.

Gear tooth vernier method. Gear teeth can be measured by a gear tooth vernier, which is essentially a vernier caliper combined with a vernier depth gauge. The depth gauge is set so that the jaws of the caliper lie on the pitch circle (Fig. 17.30). It should be noted that this method measures the chordal thickness of the tooth and not the distance along the pitch circle; however, it is usual to neglect this difference.

The tooth thickness can be found by considering triangle ABO in Fig. 17.30.

$$\theta = \frac{360^\circ}{4N}$$

where N is number of gear teeth.

$$AB = AO \sin\left(\frac{90^\circ}{N}\right)$$

$$AO = \frac{D}{2} = \frac{N}{2P} \quad (P = N/D, \text{ from the definition of diametral pitch})$$

$$AB = \frac{N}{2P}\sin\left(\frac{90^\circ}{N}\right)$$

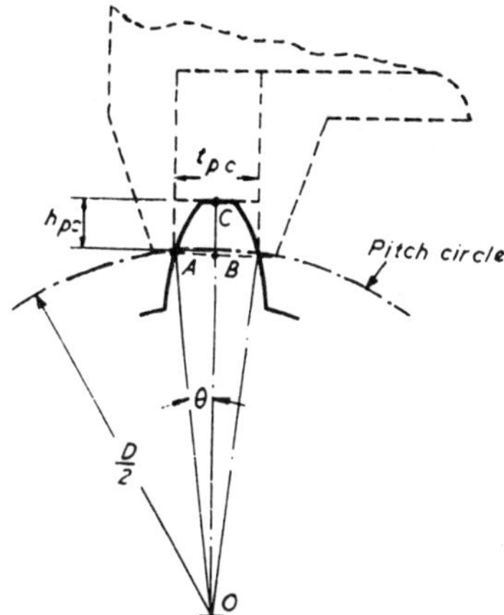

Fig. 17.30 Use of gear tooth vernier to measure tooth thickness

$$\text{But tooth thickness} = 2\text{AB} = \frac{N}{P}\sin\left(\frac{90^\circ}{N}\right)$$

$$\therefore \qquad t_{pc} = \frac{N}{P}\sin\left(\frac{90^\circ}{N}\right) \tag{17.21}$$

To find depth of setting, Fig. 17.30 is again considered.

$$h_{pc} = \text{CO} - \text{BO}$$

$$h_{pc} = \left(\frac{D}{2} + \text{addendum}\right) - \text{AO}\cos\theta$$

Assuming no correction is needed to the teeth, addendum $= 1/P$

$$h_{pc} = \frac{N}{2P} + \frac{1}{P} - \frac{N}{2P}\cos\left(\frac{90^\circ}{N}\right)$$

$$h_{pc} = \frac{N}{2P}\left[1 - \cos\left(\frac{90^\circ}{N}\right)\right] + \frac{1}{P} \tag{17.22}$$

The calculated value of tooth thickness will have to be adjusted for the required circumferential backlash and the depth setting will require modification if a non-standard addendum is used.

As the gear tooth vernier can at best be read to only $\pm 0{\cdot}001$ in., and the thickness measurement is made with the edges rather than the faces of the measuring jaws, some inaccuracy results.

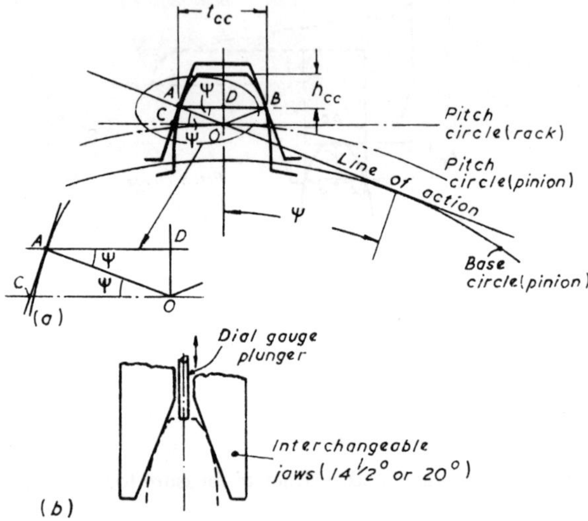

Fig. 17.31 Constant chord measurement of tooth thickness

Constant chord method. This method of checking gear tooth thickness measures the tooth width at the points of contact of a symmetrically placed close meshed rack (Fig. 17.31 (*a*)). The length of chord AB is constant for all gears of the same diametral pitch and pressure angle irrespective of the number of teeth; for this reason a single calculation and comparator setting will suffice when checking a set of meshing gears with different numbers of teeth.

In triangle ACO (Fig. 17.31 (*a*))

$$\text{AO} = \text{CO} \cos \psi$$

but $\text{CO} = \dfrac{\pi}{4P}$ ($\frac{1}{4}$ circular pitch, where circular pitch $= \pi D/N = \pi/P$)

In triangle ADO

$$\text{AD} = \text{AO} \cos \psi = \frac{\pi}{4P} \cos^2 \psi$$

and

$$\text{AB} = 2\text{AD} = \frac{\pi}{2P} \cos^2 \psi$$

$\therefore$

$$t_{cc} = \frac{\pi}{2P} \cos^2 \psi \qquad (17.23)$$

The distance h_{cc} between constant chord and top of tooth has now to be found:

$$\mathrm{DO} = \mathrm{AO} \sin \psi = \frac{\pi}{4P} \sin \psi \cos \psi$$

$$h_{cc} = \text{addendum} - \mathrm{DO}$$

$$h_{cc} = \frac{1}{P} - \frac{\pi}{4P} \sin \psi \cos \psi \qquad (17.24)$$

The expressions for t_{cc} and h_{cc} will have to be modified for any tooth correction and backlash.

Gear tooth comparators are set with master plug gauges for $14\frac{1}{2}°$ or $20°$ pressure angles and have a dial gauge which contacts the top of the gear tooth (Fig. 17.31 (*b*)).

Base tangent method. In this method a measurement is taken over a number of teeth. A vernier caliper can be used to obtain the reading, but greater accuracy is obtainable by a base tangent comparator. A portable version is shown in Fig. 17.32 (*a*). The inclination of the instrument in the plane of the base circle is not critical, as the distance between a pair of opposed involutes is constant. It is, however, desirable that the

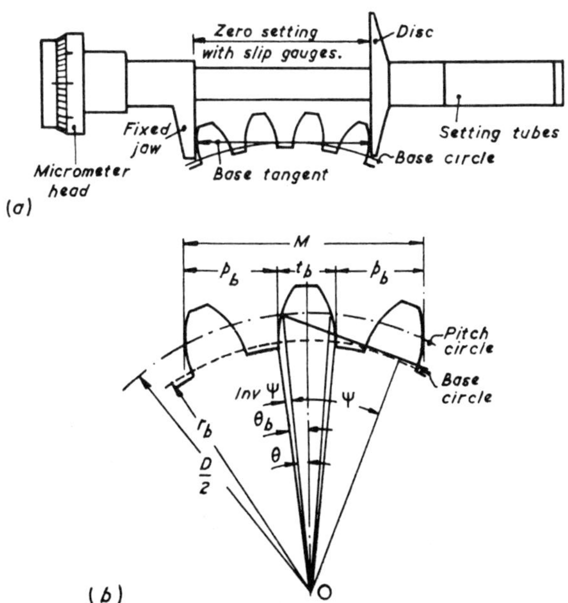

Fig. 17.32 Base tangent method of checking tooth thickness

number of teeth chosen should enable the measuring faces to contact the involute near to the pitch circle because the shape here is likely to be most accurate.

It will be seen from Fig. 17.32 (*b*) that the base pitch measurement M is made up of a number of elements:

$$M = Sp_b + t_b$$

where S is number of tooth spaces,
p_b is base pitch of the gear,
t_b is tooth thickness at the base circle

$$p_b = \frac{2\pi}{N} r_b$$

where r_b is radius of the base circle

$$t_b = 2\theta_b r_b$$

where θ_b is angle subtended at the centre of the gear by a $\frac{1}{2}$ tooth at the base circle.

$$\theta_b = \theta + \text{inv}\ \psi$$

where θ is angle subtended at the centre of the gear by a $\frac{1}{2}$ tooth at the pitch circle $= 90°/N = \pi/2N$ radians, and inv. ψ is the involute function of the pressure angle ψ

$$r_b = \frac{D}{2}\cos\psi = \frac{N}{2P}\cos\psi\left(\text{as } D = \frac{N}{P}\right)$$

hence
$$M = \frac{N}{P}\cos\psi\left(\frac{\pi S}{N} + \frac{\pi}{2N} + \text{inv}\ \psi\right) \tag{17.25}$$

A correction to M will have to be made for backlash.

17.6.6 Measurement of gear pitch. The term pitch of a gear, when used unqualified, is taken to mean the circular pitch p. This is the distance along the pitch circle between similar faces of successive involutes, and is equal to $\pi D/N$. The base pitch of a gear is a similar measurement taken around the base circle and related to the circular pitch by the pressure angle, i.e. base pitch $= p \cos \psi$.

Specially developed hand-held instruments, incorporating a dial gauge and locating and measuring fingers, are available to measure base pitch over a single tooth.

Variations in circular pitch can be measured by mounting the gear on a mandrel and indexing it through an angle equal to 360°/number of teeth. A dial gauge in contact with the involute at the pitch circle can be used

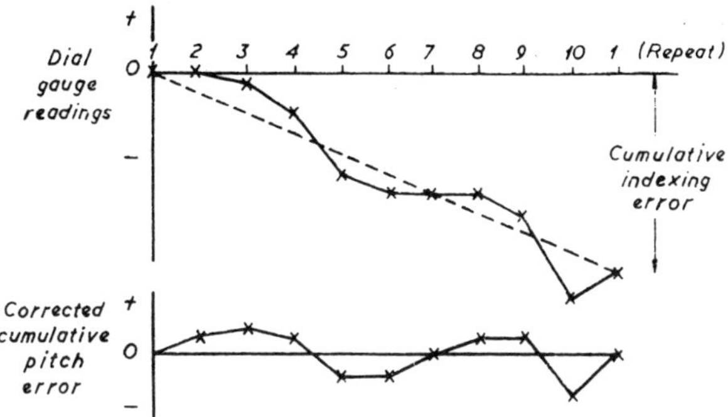

Fig. 17.33 Effect of indexing error on cumulative pitch error of 10-tooth gear

to detect cumulative variations in pitch. If indexing has been correct, the dial gauge should repeat its initial reading when back on the first tooth after one revolution. Should there by any cumulative indexing error, it will be revealed by this test and enable the intermediate readings to be adjusted (Fig. 17.33).

17.7 SURFACE FINISH

Specification of surface finish is of increasing importance to designers. In particular, the roughness of surfaces can influence fatigue failure and determine the rate of wear of rubbing surfaces. Surface errors may be broadly divided into two categories: waviness and roughness. Waviness consists of surface undulations of relatively long wavelength caused by errors in the manufacturing process, whereas roughness results from surface irregularities of high frequency caused by the interaction of the shaping process and the material. When wayiness and roughness are both present, they must be separated as far as possible if roughness is to be measured. Instruments which measure roughness are designed to examine comparatively short lengths of surface so that the effect of waviness is minimized. Standard meter cut-off values which are recommended in BS 1134[80] vary from 8 mm (0·31 in) to 0·08 mm (0·003 in), although a typical range for a given instrument might be 0·25 mm (0·01 in), 0·8 mm (0·03 in) and 2·5 mm (0·10 in). The length chosen should be long enough to reproduce the surface characteristics of the metal shaping process; for instance, if a turned surface is being examined it should exceed the feed/rev.

17.7.1 Measurement of surface finish. The visual assessment of surface is inadequate, although the results can be somewhat improved if touch is used. Surfaces where the lay of the machining marks is in a single direction may be highly reflective and seem to be smooth, but they may be considerably rougher than another surface having a matt appearance. To assist in the estimation of surface finish, set of standards of known roughness[81] can be compared with the surface being assessed. As a general rule only standards produced by the same machining process should be compared.

The most satisfactory method of measuring surface roughness is to use an instrument in which a stylus and skid are drawn over the surface, and where the movement of the stylus is amplified and recorded on a trace. A well-known example of this type of instrument is the Rank, Taylor Hobson 'Talysurf', which is shown in schematic form in Fig. 17.34. A carrier wave of constant frequency and magnitude is supplied to a bridge circuit, the inductance of which is altered by the vertical movement of the stylus relative to the skid as it moves over the work surface. This results in a modulation of the carrier wave, which is then amplified and demodulated, leaving a current which represents the vertical displacement of the stylus. The displacement current is made to operate a

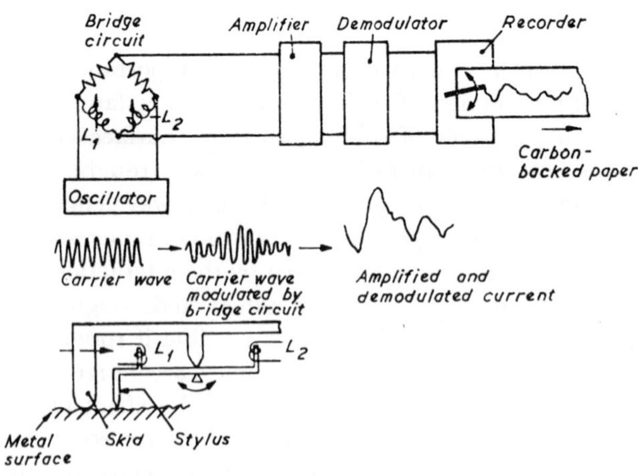

Fig. 17.34 Principle of operation of Talysurf

recorder which produces a trace of surface finish on carbon backed paper. Suitable vertical magnifications can be selected, a typical range being from x1000 to x50 000. Horizontal magnifications are smaller, for instance x20 and x100, so that the trace exaggerates the true profile in the vertical direction.

17.7.2 Analysis of surface finish. Inspection of the trace provides an excellent guide to the surface condition, although it should be appreciated that a stylus with an end radius of even 2·5 μm (0·0001 in) acts as a filter, passing over and filtering out many of the small valleys which can be detected by an electron microscope.

If surface finish is to be specified it must be given a numerical value. Two systems are in common use, the arithmetical mean deviation (R_a), previously known as the centre line average method, and the ten point height of irregularities (R_z), previously the peak-to-valley height method.

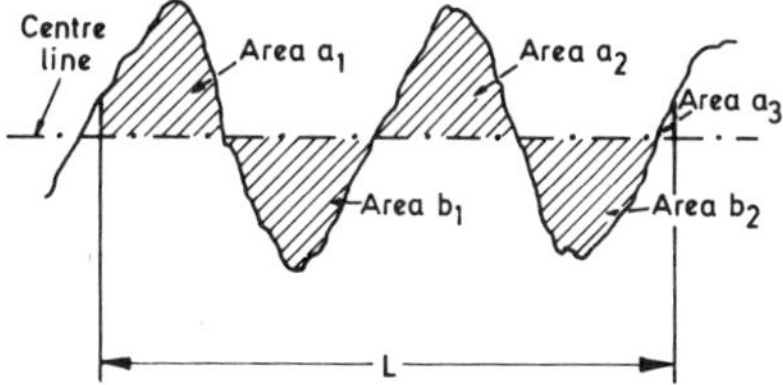

Fig. 17.35 Arithmetic mean deviation method (R_a)

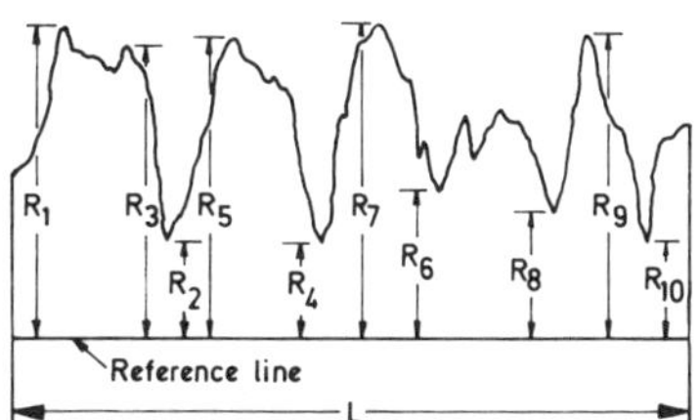

Fig. 17.36 Ten point height method (R_z)

The arithmetical mean deviation method, which is widely used in the U.K., is a measurement of the departure of the surface profile from a centre line drawn throughout the sampling length. The ten point height method is the average distance between the five deepest valleys and the five highest peaks within the sampling length. The peak and valley heights are measured from a base line drawn below the profile of the surface and parallel to the centre line of the profile. The two methods are illustrated in Figs. 17.35 and 17.36, from which it can be seen that:

$$R_a = \frac{\text{sum of areas } (a) + \text{sum of areas } (b)}{L}$$

$$R_z = \frac{(R_1 + R_3 + \ldots + R_9) - (R_2 + R_4 + R_{10})}{5}$$

Although values of R_a and R_z can be obtained graphically from surface profile traces, they are usually read directly. Talysurf instruments have both a direct reading and a profile trace facility.

The values of R_z are generally from four to seven times the corresponding R_a values depending on the surface profile being measured.

17.7.3 Bearing curves. When two surfaces rub, wear occurs. Bearing curves enable probable rates of wear to be estimated, and can be constructed from surface finish traces. The curve is plotted by selecting a

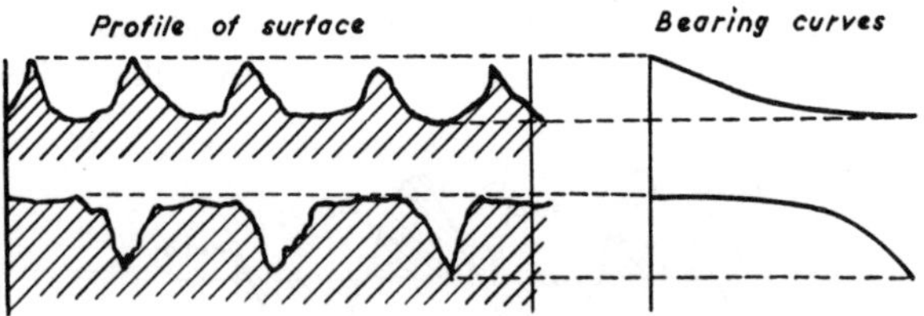

Fig. 17.37 Bearing curves

number of depths and then finding the lengths of metal surface available at these depths. This has been done for two traces in Fig. 17.37, where it can be seen that the trace with the smaller peak-to-valley roughness is, in this instance, likely to have the higher initial rate of wear.

17.7.4 Optical assessment of surface finish. By placing optical flats on a comparatively smooth surface it is possible to obtain contour lines in the form of interference fringes, from which the shape of the surface can be deduced. Optical flats are pieces of glass or quartz, usually in the form of a disc, which have been ground so that their two circular surfaces are flat and smooth to very close limits. Fig. 17.38 represents an optical flat lying on a metal surface with a minute air gap at its underside. If it is assumed that the difference in length of ray path between $B_1C_1D_1$ and $B_2C_2D_2$ is $\lambda/2$, and the light rays emanating from F_1 and E_1 are in phase, then the light emitted from F_2 will be optically displaced by 180° from that from E_2. This will produce alternate bright and dark bands on the surface of the flat. If the light falling on the flat approaches the normal to its surface, then $B\hat{C}D \rightarrow 0$ and $B_1C_1 \rightarrow t_1 \rightarrow C_1D_1$ where t_1 is the perpendicular distance from the underside of the flat to the metal surface. A similar result is obtained by considering triangle $B_2C_2 \rightarrow t_2 \rightarrow C_2D_2$. As has been stated, the change from a bright to dark zone is caused by differences in the ray path length of $\lambda/2$,

$$\text{hence } 2(t_2 - t_1) = 2(t_3 - t_2) \simeq \frac{\lambda}{2}$$

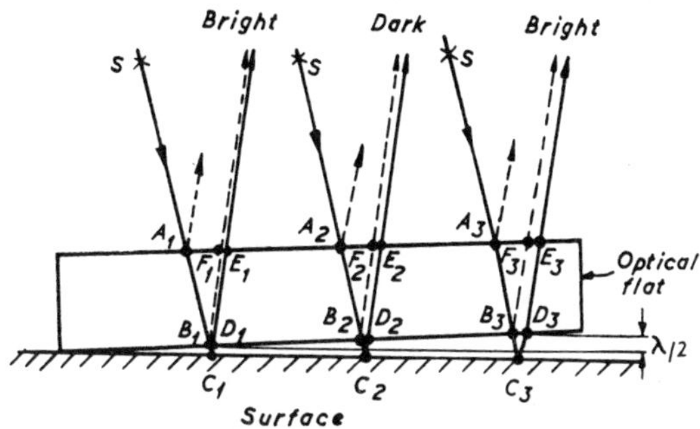

Fig. 17.38 Examination of surface using optical flat (inclination greatly exaggerated)

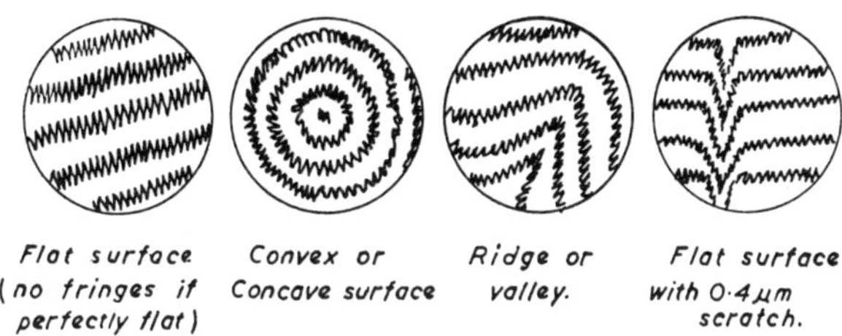

Fig. 17.39 Fringe patterns

If the average wave length of white light in air is taken as 0·5 μm (0·00002 in), then $\lambda/2 = 0{\cdot}25$ μm (0·00001 in) and $2(t_2 - t_1) = 2(t_3 - t_2) = 0{\cdot}25$ μm (0·00001 in). Hence the air gap difference between adjacent light and dark bands represents 0·125 μm (0·000005 in) and that between adjacent pairs of bright or dark bands 0·25 μm (0·00001 in).

Some of the basic patterns obtained by optical flats are shown in Fig. 17.39. These can be interpreted in the same way as contour lines on a map. The pattern will not indicate whether the surface is convex or concave, but if pressure is applied near the centre of the pattern to reduce the air gap, the lines will move towards the pressure point if the surface is concave, and away if it is convex.

17.8 MEASUREMENT OF ROUNDNESS

The simplest way of checking the roundness of a part is by placing it in a V-block and rotating it under a dial gauge.

Instruments measuring roundness are of two types: the first rotates the work against an accurate indicator, and the second holds the work stationary while the indicator is rotated. The latter method is used by the Rank, Taylor Hobson 'Talyrond', where the rotating stylus is carried on a spindle with a run out of as little as 0·025 μm (0·000001 in). Polar diagrams produced on carbon-backed paper discs indicate departures from roundness, with errors magnified radially from 50 to 10 000 times (Fig. 17.40).

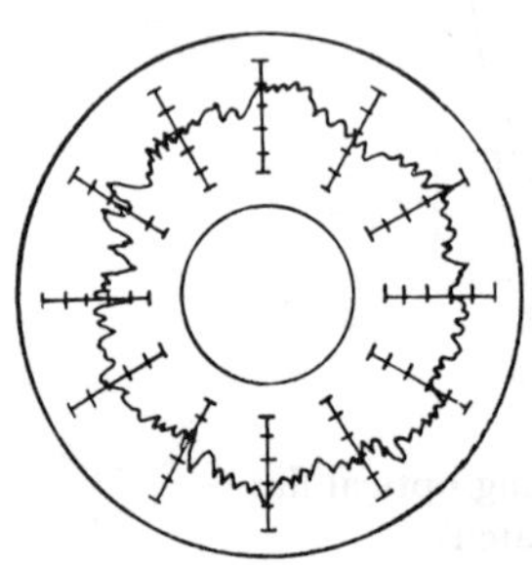

Fig. 17.40 Talyrond polar diagram showing departures from roundness

Three methods can be used to obtain a numerical assessment of roundness. They are all of the peak-to-valley type, but differ in the way centres of measurement are chosen. The first method varies depending on whether a hole or a shaft is being measured. If the part is a shaft, a circumscribing circle with minimum diameter is drawn round the trace; then, using the same centre, a second circle is drawn which just fits inside the trace. The difference in radii represents the maximum inward departure of the metal from the smallest possible ring gauge which would fit over the shaft. If a hole is being examined, the maximum inscribed circle is drawn to the trace. In this way the maximum outward radial departure of the largest possible plug gauge which will just fit the hole can be found. A second approach is to draw two concentric circles which just enclose the trace and have minimum radial separation. Thirdly, a least squares mean line can be drawn through the trace; the line must be a defined shape, normally a circle. Using the centre of this least squares circle, a minimum circumscribing circle and a maximum inscribed circle can be drawn. The least squares circle can be computed electrically or determined graphically. Further details of methods of roundness measurement are given in BS 3730.[82]

17.9 LIMITS AND GAUGING

It is impossible to manufacture two parts of exactly the same size. Therefore on most production drawings the designer specifies an upper and lower limit of size for important dimensions in the form of a nominal size and tolerances, e.g. 10·00 $^{+\cdot 000}_{-\cdot 025}$. The tolerances should be chosen so that they ensure satisfactory functioning over the anticipated life of the part, permit interchangeable assembly, and allow economical production.

17.9.1 Limit gauging. In order to ensure that the size limits specified are maintained in production, inspection of work is necessary. For one-off and small batch production, standard measuring equipment such as

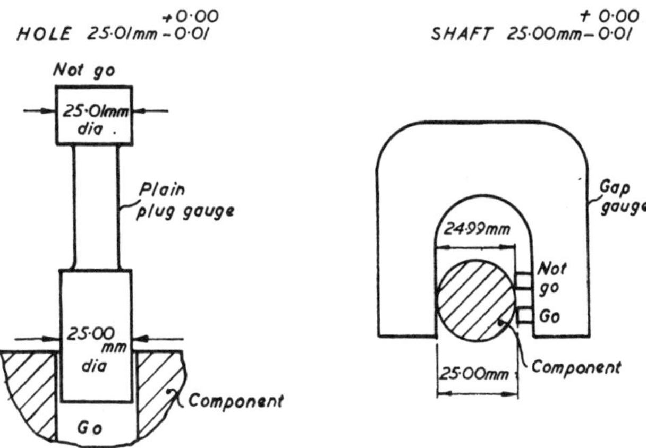

Fig. 17.41 Application of limit gauging to holes and shafts

micrometers and verniers is used. However, when large quantities of similar parts have to be inspected, gauges set to upper and lower limits of size provide a better method of inspection. Not only are these 'go' and 'not go' gauges faster in use, but they can also be used by less skilled workers. The gauging arrangements for a mating hole and shaft are shown in Fig. 17.41.

17.9.2 Taylor's principle of gauging. Taylor's principle (patent granted to W. Taylor of Taylor, Taylor and Hobson, 1905), states that the 'go' gauge should check maximum metal conditions on as many dimensions as possible, but that the 'not go' gauge should check one dimension only at the minimum metal conditions. As a result a number of 'not go' gauges may be required to check a component completely.

An example of how an unsatisfactory part can be accepted if Taylor's principle is ignored is illustrated in Fig. 17.42. If the 'not go' portion of the gauge had been designed as a pin gauge with radiused ends, so that it checked one dimension only, an oversize hole would have been detected, provided the gauge had been tried in a sufficient number of angular positions. Taylor's principle as applied to the 'not go' portion of a gauge is frequently ignored when the manufacturing process ensures adequate geometric accuracy; in fact the 'not go' portion of most hole gauges is cylindrical, as shown in Fig. 17.42. To satisfy Taylor's principle, the 'go'

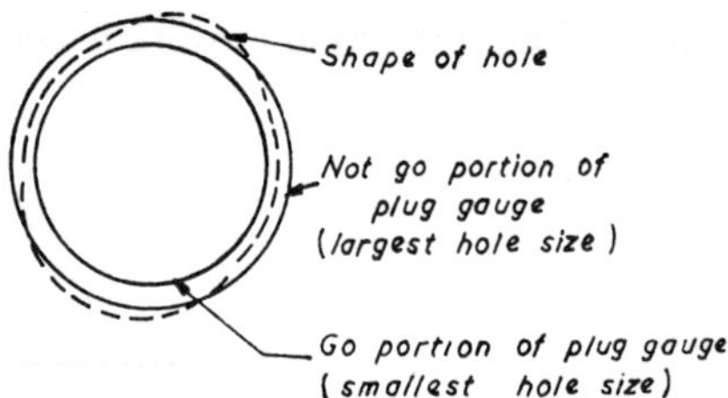

Fig. 17.42 Acceptance of incorrect hole by plug gauge not designed to Taylor's principle

portion of a plug gauge should be as long as the hole it is checking, in order to detect any lack of straightness. Although this is not always practicable, the 'not go' ends of plug gauges are normally made shorter than the 'go' ends to facilitate easy identification.

17.9.3 Gauge tolerances. As with components, gauges cannot be manufactured to an exact size, and the gaugemaker must be provided a small tolerance, usually 10% of the component tolerance. It is recommended in BS 969[83] that for all 'go' plain plug, gap and ring gauges the gaugemaker's tolerance should be positioned within the work tolerance, but for 'not go' portions the tolerance should be outside the work tolerance. A wear allowance, amounting to 20% of the gauge tolerance, is also used on the 'go' portion of the gauge provided the component tolerance is $\nless$0·09 mm (0·0035 in). The application of gaugemaker's and wear allowances to a plug gauge is shown in Fig. 17.43.

17.9.4 Gauging screw threads. If Taylor's principle was applied to the gauging of a male thread, it would be necessary to provide a 'go' gauge in the form of a ring gauge of the same length as the thread and of the full thread form, and two 'not go' gauges, one to measure major diameter and another to measure the effective diameter. Ring gauges, which are occasionally used for reference purposes, are expensive, wear quickly and are slow in use. They have been largely superseded as 'go' gauges by adjustable gap gauges having anvils of full thread form, which, as far as practicable, are a similar length to the thread being checked. The 'not go' gauge is also in the form of a gap gauge with the thread form truncated at the top and relieved at the bottom, so that contact is made with the work only on the thread flanks. The number of teeth is reduced, often to one on one side and two on the other, so that as far as possible only effective diameter is gauged.

Similarly, when gauging female threads, a compromise is made with Taylor's principle. Here the 'go' gauge is often a full length, full form screw gauge, with a 'not go' plug gauge to ensure that the minor diameter is not oversize. Recommendations on gaugemakers' tolerances and wear allowances for screw gauges are made in BS 919.[84]

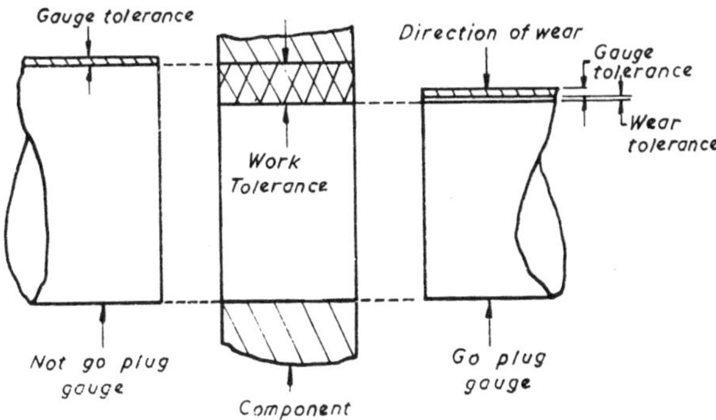

Fig. 17.43 Application of gauge tolerances and wear allowance to a plug gauge

17.10 LARGE-SCALE ALIGNMENT TESTING

The alignment testing of machine tools has been outlined in Chapter 8.11. In these tests use is made of instruments such as precision levels, auto-collimators and dial gauges to check alignment. Large-scale alignment checks, often over distances of 30 m (100 ft) or more, are, however, required in the shipbuilding and aircraft industries. For this work metrologists adapt surveying instruments and techniques.

17.10.1 Collimators and telescopes. A collimator and telescope can be used to determine both the angular and linear alignment of widely separated surfaces. The collimator carries two illuminated graticules: the first, at the focal plane of the collimating lens, has a central reference cross and two scales set at right angles to each other so that angular displacement can be measured; the second graticule is in front of the collimating lens, and is marked with a small central star, with horizontal and vertical scales from which any linear misalignment can be determined. Both collimator and telescope are housed in accurately ground tubes, which are concentric with their optical axes to within very close limits. The telescope and the collimator are located on the two surfaces to be checked; if bearing alignment is being checked each is held in specially machined bushes.

Angular and linear alignment is checked by focusing the telescope on the appropriate collimator scale (Fig. 17.44). Accuracy of angular readings does not depend on the separation of telescope and collimator and

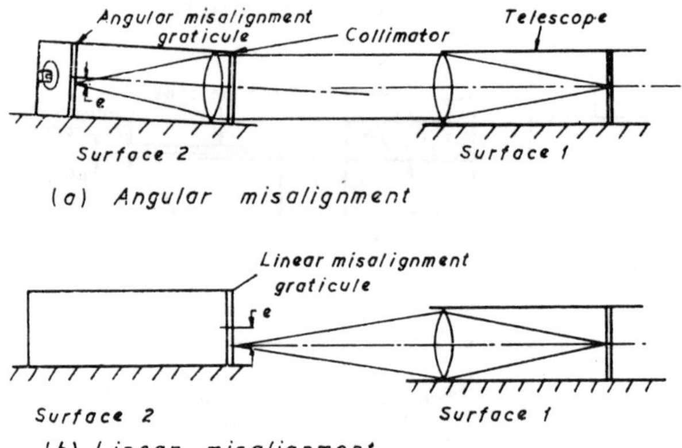

Fig. 17.44 Measurement of misalignment using collimator and telescope

can, in still air, be estimated to six seconds of arc. Readings of linear misalignment are, however, dependent on separation, e.g. if one graticule graduation represents 0·01 mm at 2 m it becomes 0·10 mm at 20 m.

17.10.2 Optical tooling level. This instrument enables a horizontal plane to be established and the vertical distances of widely separated points above or below it to be measured. The optical tooling level consists of a horizontally mounted telescope which can be rotated about a vertical axis. An optical micrometer is fitted to the telescope, and this enables the line of sight to be adjusted upward or downwards by known small amounts. When using scales to determine the heights of points relative to the level plane, graduated scales are held vertically against the points and the telescope is focused on each in turn (Fig. 17.45 (*a*)). The optical micrometer (Fig. 17.45 (*b*)), enables the scale to be accurately sub-divided, for instance, if the scale divisions are 1 mm, the optical micrometer enables readings to 0·01 mm to be made.

17.10.3 Transit level. This instrument, which is sometimes called a jig transit or an optical tooling square, is similar to a surveyor's theodolite. It has a telescope which can be inclined in the vertical plane, and this with its mounting can also be swivelled in the horizontal plane. Transit

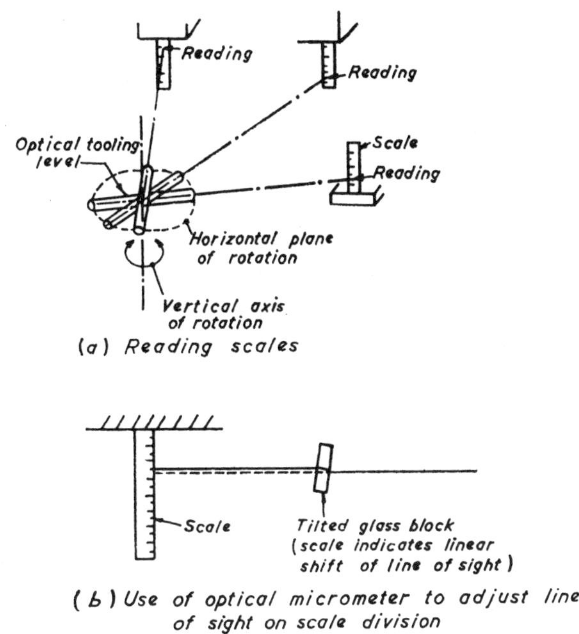

Fig. 17.45 Use of optical tooling level to determine heights

levels are used with scales to measure horizontal distances from a vertical reference plane. Firstly, a horizontal master reference line is established by sighting an optical tooling level on the intersection of a suitably positioned illuminated target. The optical tooling level is then replaced by a transit level. The vertical plane, which is swept out from the master reference line by inclining the telescope of this instrument, is called the width plane.

Horizontal distances from the width plane can be determined by using the telescope to read horizontally held scales (Fig. 17.46). If necessary, additional transit level positions can be set up at known distances along the master reference line. These enable secondary reference planes, called station planes, to be established perpendicular to the width plane and the position of additional features to be measured.

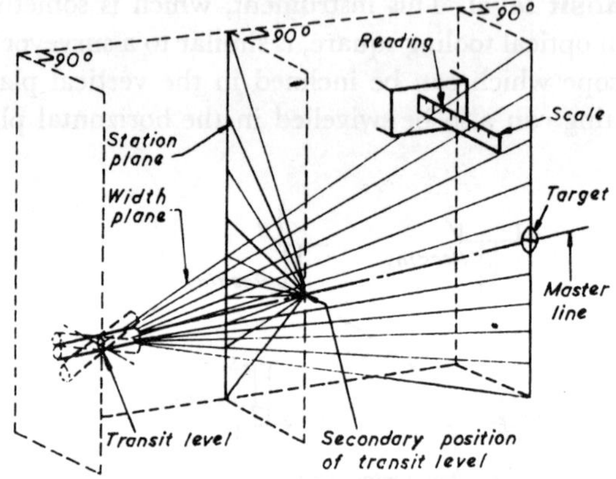

Fig. 17.46 Width and station planes

17.11 LASER INTERFEROMETRY

Lasers provide coherent highly monochromatic light which is particularly suitable for interferometry. The light intensity is such that it can be easily detected by photocells and the noise-to-signal ratio is low enough for extremely rapid fringe counting (up to 10^8 Hz). One of numerous applications in the field of metrology is checking the positional accuracy of tables on N.C. machine tools. Automatic compensation can be arranged for variations in ambient temperatures and pressures, both of which will alter the light wavelength. Accuracies in the order of one part in a million, $\pm$ one count, can be obtained over distances of up to 50 m.

Helium–neon lasers are normally used, having a wavelength of 0·6328 μm (25 μin). A simple system using a laser to measure displacement is shown in Fig. 17.47.

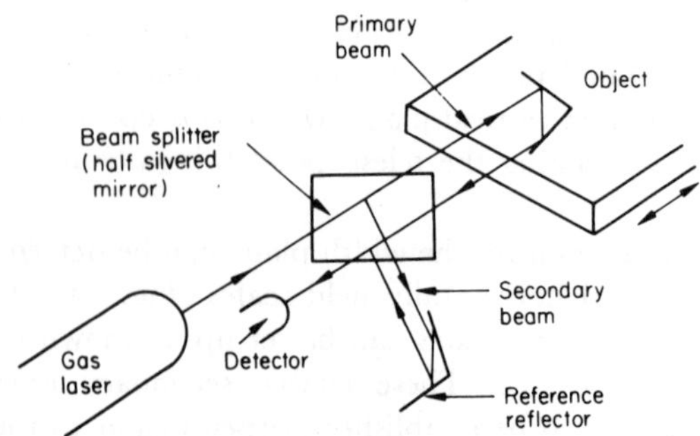

Fig. 17.47 Measurement of linear displacement using laser

The beam from the laser impinges on a semi-reflecting beam splitter which provides a reference from which displacement is measured. The beam splitter passes part of the incident light, the primary beam, but the rest, the secondary beam, is reflected by the splitter. The primary beam is reflected back by a triangular prism or a cube-corner reflector mounted on the object the displacement of which is to be measured. On its return journey the primary beam again passes through the beam splitter, here it combines with the secondary beam which has been reflected by a second reflector mounted at a fixed distance from the beam splitter. These two beams form an interference pattern which changes with the displacement of the object being measured. As the object moves through half a wavelength 0·3164 μm (12·5 μin) there is one complete change in the interference pattern. Changes in the interference pattern can be monitored by photodetectors linked with digital displays which provide a read out of displacement, those calibrated in Imperial units reading to 1 μin.

17.12 MULTIPLE AND AUTOMATIC GAUGING

17.12.1 Multiple gauging. Components which have several dimensions to be measured are expensive to inspect by conventional means. However, if the quantity of parts justifies the cost of multiple dimension gauging, a very large number of dimensions can be simultaneously measured in a specially designed fixture fitted with pre-positioned pneumatic or electrical measuring heads. The dimensions are seen at a glance on multi-tube or multi-dial displays which can be arranged to indicate upper and lower drawing limits.

17.12.2 Pre-process gauging. This type of gauging is performed on parts to detect either excessive stock, which could damage pre-set tooling, or insufficient stock, which would result in the part not cleaning up during the machining operation. Manually loaded gauging fixtures can be used, or the parts can be moved through an automatic measuring station.

17.12.3 In-process gauging. With this type of gauging the part is measured as it is being machined and the cutting tool is retracted when the desired size has been reached. Most systems use pneumatic gauging heads, as these are superior to electrically operated types in a metal cutting environment.

17.12.4 Automatic post-process gauging. This type of gauging falls into two categories. The first is an automatic measuring machine in which work is automatically positioned and several dimensions are simultaneously measured, after which there is automatic segregation into correct, oversize and undersize categories. The second combines automatic gauging and sorting with machine control, see Fig. 17.48.

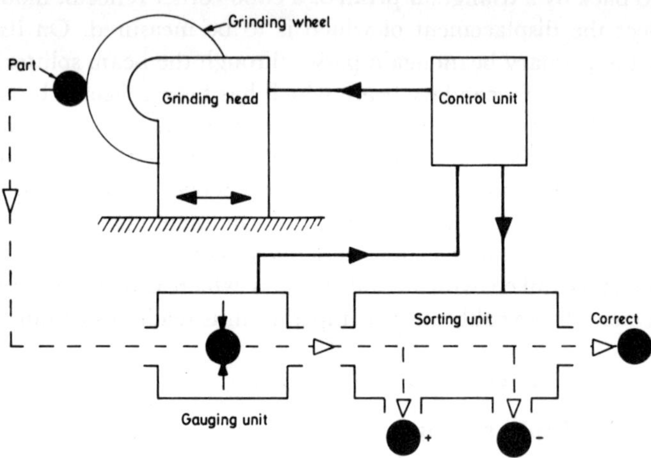

Fig. 17.48 Automatic gauging and sorting unit with automatic size compensation

17.13 ACCURACY OF MEASURING SYSTEMS

Measuring systems are available to meet most product tolerances, and a frequently used guide is that the maximum error in measurement should not exceed $\pm 10\%$ of the product tolerance. As might be expected, the closer the tolerance, the greater the precautions necessary to minimize errors. Some of the major sources of error in metrology are now discussed.

Temperature can produce serious errors in length measurement, particularly if the scale or end standard has a different coefficient of expansion from the part being measured. To avoid this type of error, accurate measurements should be made at a temperature of 20°C, this being the temperature at which scales and standards are calibrated. Even though the temperature in the metrology laboratory may be kept at 20°C, temperature errors will still occur unless care is taken in the handling of equipment. For instance, slip gauges should be handled with a piece of chamois leather or left on a surface table to assume room temperature, otherwise body heat will produce expansion.

Elastic deformation can also cause errors when measuring objects which lack rigidity. Long thin parts therefore should be supported at points of minimum deflexion. For parts of uniform section, these are two symmetrically spaced points 0·554 l apart, where l is the component length. It is of interest that a 1·83 m (6 ft) straight edge, 100 mm (4 in) wide and 12·7 mm ($\frac{1}{2}$ in) thick, will sag 0·061 mm (0·0024 in) when supported at its extreme ends with the 100 mm side vertical. Sag is reduced to 0·00125 mm if it is supported at its points of minimum deflexion. Elastic deformation of a comparator stylus and the work under it caused by measuring pressure can produce a significant error in very accurate measurement. These have been calculated by Hertz.[85]

Instrument errors can result from a variety of sources. Frequently manufacturers quote a maximum error for their equipment. There are also British Standards specifying maximum errors for many instruments. For instance, BS 907[86] indicates that maximum error of a dial gauge graduated in 0·01 mm divisions should not exceed 0·010 mm in any half revolution of the pointer, and 0·020 mm for the total travel of the plunger. Additional sources of instrument error are wear, excessive friction and backlash in mechanisms.

The designer, by adopting certain design principles, can do much to increase accuracy. For instance, the scale or other dimensional reference should coincide as far as possible with the axis of measurement. In a vernier caliper, the scale is about 50 mm (2 in) from the ends of the measuring jaws, and any lack of straightness in the beam on which the scale is engraved will produce a measuring error. Instrument accuracy deteriorates with time, due to wear or possibly misuse. Periodic checking recalibration and maintenance are therefore important if instrument accuracy is to be maintained.

Human errors are a frequent source of inaccuracy. They can result from a lack of discrimination or from carelessness. Where feel is involved, as in applying the correct measuring pressure to the thimble of a micrometer, the operator can be assisted by fitting a fiducial indicator which shows when a standardized measuring pressure has been reached. With instruments incorporating dials, the ability of the operator to discriminate and interpolate determines reading accuracy. Here, good pointer and dial design minimizes parallax errors as well as gross reading errors.

Appendix 1

HEAT DIFFUSION IN SOLID BODIES

Consider an elemental parallelepiped AA′BB′CC′DD′ (Fig. A.1) in a homogeneous material subjected to a temperature gradient in the x, y and z directions, where material is flowing with a velocity V in the x direction. Let the temperature at the centre of the parallelepiped be θ.

Then the temperature along face AA′DD′ $= \theta - \frac{1}{2}\frac{\partial\theta}{\partial x}\,\delta x$

and the temperature along face BB′CC′ $= \theta + \frac{1}{2}\frac{\partial\theta}{\partial x}\,\delta x$

Rate of heat flow into element through

$$\text{AA}'\text{DD}' = -K\,\delta y\,\delta z\,\frac{\partial}{\partial x}\left(\theta - \frac{1}{2}\frac{\partial\theta}{\partial x}\,\delta x\right)$$

where K = thermal conductivity

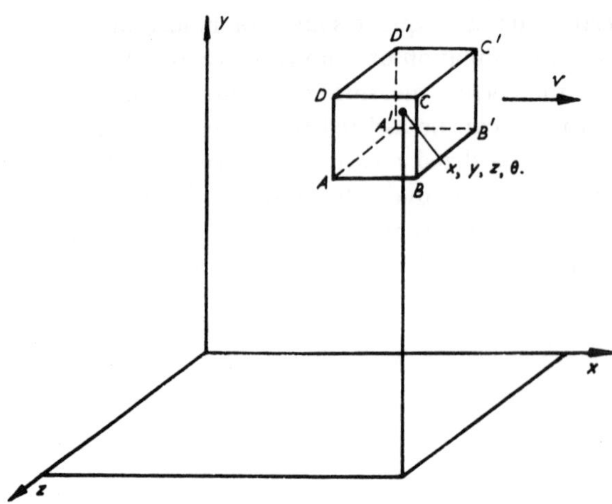

Fig. A.1 Parallelepiped referred to x, y, z axes, showing direction of mass transfer

Rate of heat flow into element through

$$\mathrm{BB'CC'} = +K\,\delta y\,\delta z\,\frac{\partial}{\partial x}\left(\theta + \tfrac{1}{2}\frac{\partial\theta}{\partial x}\,\delta x\right)$$

Rate of heat gain through conduction in x direction $= K\,\dfrac{\partial^2\theta}{\partial x^2}\,(\delta x\,\delta y\,\delta z)$

Similarly, rate of heat gain to the element through conduction in y and z directions $= K(\delta x\,\delta y\,\delta z)\left(\dfrac{\partial^2\theta}{\partial y^2} + \dfrac{\partial^2\theta}{\partial z^2}\right)$

Quantity of heat carried into the element through AA′DD′ due to mass transfer/unit time $= V\delta y\,\delta z\,\rho C\left(\theta - \tfrac{1}{2}\dfrac{\partial\theta}{\partial x}\,\delta x\right)$

where ρ = density of material and C = specific heat.

Heat carried out of element through BB′CC′ due to mass transfer/unit time $= V\delta y\,\delta z\,\rho C\left(\theta + \tfrac{1}{2}\dfrac{\partial\theta}{\partial x}\,\delta x\right)$

Rate of heat gain due to mass transfer $= -V\rho C\,\dfrac{\partial\theta}{\partial x}\,(\delta x\,\delta y\,\delta z)$

Total rate of heat gain $= C\rho\,\dfrac{\partial\theta}{\partial t}\,(\delta x\,\delta y\,\delta z)$

The general heat flow equation in three dimensions with a moving heat source is thus,

$$\frac{\partial^2\theta}{\partial x^2} + \frac{\partial^2\theta}{\partial y^2} + \frac{\partial^2\theta}{\partial z^2} - V\frac{\rho C}{K}\frac{\partial\theta}{\partial x} = \frac{\rho C}{K}\frac{\partial\theta}{\partial t} \tag{A.1}$$

$$\frac{\partial^2\theta}{\partial x^2} + \frac{\partial^2\theta}{\partial y^2} + \frac{\partial^2\theta}{\partial z^2} - \frac{R_T}{t}\frac{\partial\theta}{\partial x} = \frac{R_T}{Vt}\frac{\partial\theta}{\partial t} \tag{A.2}$$

where R_T, the thermal number, $= (\rho VCt)/K$

For metal cutting, V is the cutting speed

and t is the uncut chip thickness.

To obtain a mathematical solution, several simplifying assumptions are usually necessary. Since metal cutting operations such as turning rapidly attain steady state conditions, the time-dependent term can be taken as zero, and for orthogonal cutting, heat conduction across the chip in a direction parallel to the cutting edge is also zero, i.e. $\partial^2\theta/\partial z^2 = 0$, so equation (A.2) reduces to

$$\frac{\partial^2\theta}{\partial x^2} + \frac{\partial^2\theta}{\partial y^2} - \frac{R_T}{t}\frac{\partial\theta}{\partial x} = 0$$

Dutt and Brewer[87] attempted to find the temperature distribution in the chip, workpiece and tool by relaxation, assuming the heat sources to be plane and of uniform intensity along the shear plane and the rake face. The temperatures calculated for the workpiece were appreciably higher than those obtained experimentally by Hollander.[88] These discrepancies were probably due mainly to the assumptions that the primary deformation occurred along a shear line, and the frictional work done along the rake face ignored a secondary deformation zone extending into the chip (Fig. A.2).

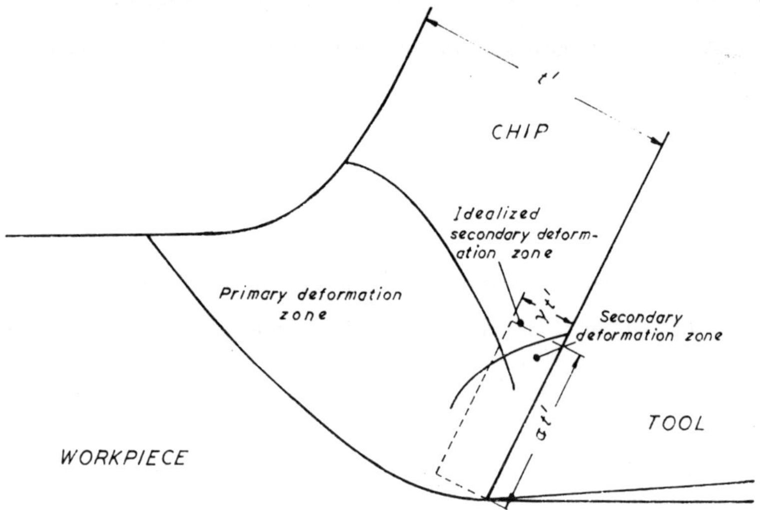

Fig. A.2 Primary and secondary deformation zones in metal cutting

Several attempts have been made to evolve a generalized formula for calculating temperatures in the shear zone and along the rake face in orthogonal cutting.[89,90] Rapier, assuming the friction heat source to be plane and uniform along the rake face and the conduction of heat in the direction of movement to be negligible compared with the heat removed by mass flow, further simplified the heat flow equation to

$$\frac{\partial^2\theta}{\partial y^2} - \frac{R_T}{t}\frac{\partial\theta}{\partial x} = 0$$

and obtained the solution $\dfrac{\theta_m}{\theta_f} = 1{\cdot}13\sqrt{\dfrac{R_T}{\alpha}}$

where θ_m = maximum temperature rise in the chip (along the rake face) due to the frictional heat source.

θ_f = average rise in temperature of the chip due to friction heat source, and

$$\alpha = \frac{\text{chip contact length along the rake face}}{\text{chip thickness}}$$

Boothroyd[91] showed that the assumption of a plane frictional heat source gave a maximum temperature of greater magnitude than that found in practice. He concluded that the secondary deformation zone could be regarded as a rectangle extending a distance $\gamma t'$ into the chip from the rake face, and if the heat source caused by this secondary deformation was regarded as having uniform intensity over the rectangular area, the value of θ_m/θ_f was reduced as shown in Fig. A.3. For steel, the value of γ appears to be about 0·2 when cutting without lubricant.

From a knowledge of the frictional work done it is possible to arrive at a value for θ_f and hence for θ_m. To obtain a value for the maximum temperature along the rake face, it is necessary to know the rate of heat generation on the shear plane

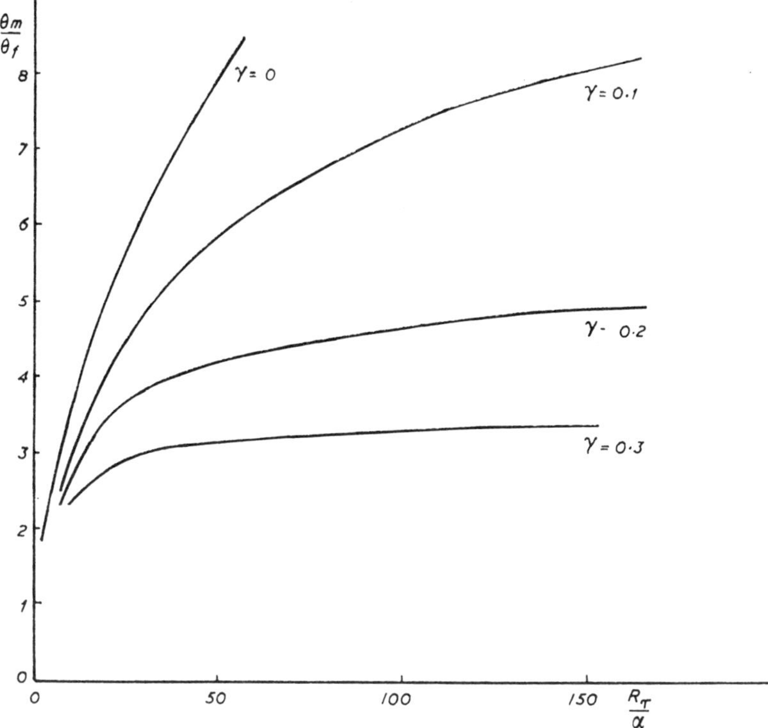

Fig. A.3 Variation of θ_m/θ_f with R_T/α for different values of γ (*adapted from Boothroyd*)

and the proportion of shearing energy which is conducted into the workpiece. Weiner showed that the ratio β, representing the shear plane heat conducted to the workpiece divided by total heat generated by shearing, bore a unique relationship to R_T . tan ϕ. Experimental work by Nakayama[92] and Boothroyd showed that such a relationship did appear to exist, but Weiner's theory considerably understated the ratio at large value of R_T . tan ϕ. From the experimental results obtained, the relationship is as shown in Fig. A.4.

When considering the temperature in a solid at a distance x from a plane of semi-infinite dimensions at which a sudden increase in temperature occurs at time $t = 0$, e.g. in a mould for a casting, the generalized heat conduction equation

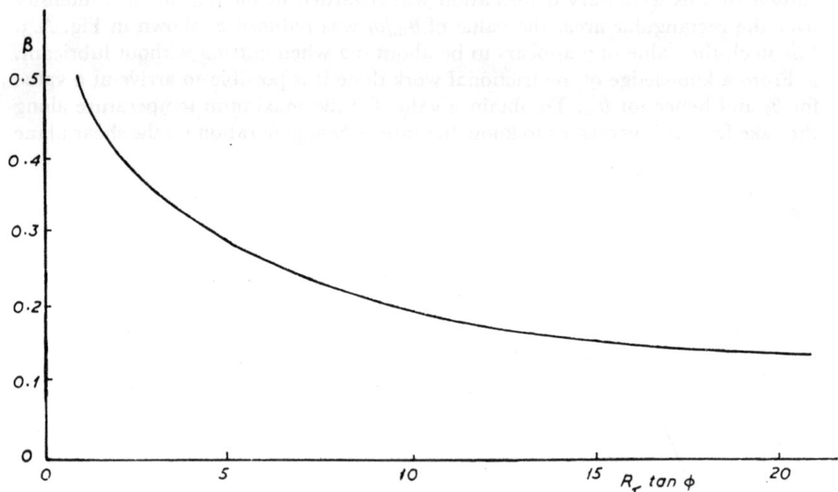

Fig. A.4 Experimental relationship between β and $R_T \tan \phi$
(from Nakayama and Boothroyd)

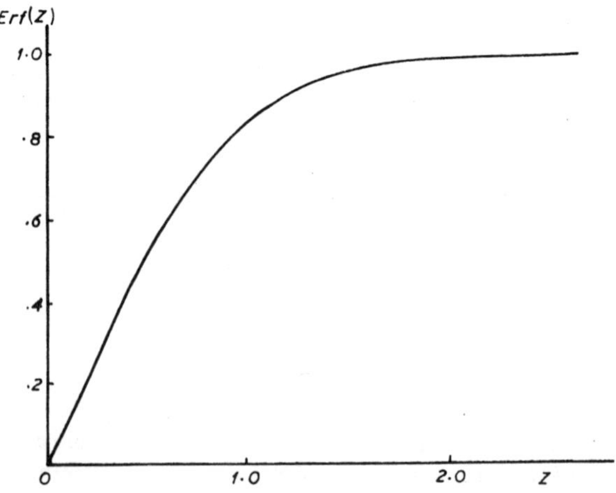

Fig. A.5 Graph of error function

simplifies to one of conduction in one dimension, and there is no heat flow due to mass transfer.

$$\frac{\partial^2 \theta}{\partial x^2} = \frac{1}{\kappa}\frac{\partial \theta}{\partial t}$$

where $\kappa = \frac{K}{\rho C}$, the thermal diffusivity.

The solution to this equation is given by:

$$\theta_x - \theta_0 = (\theta_1 - \theta_0)\left(1 - \operatorname{erf}\frac{x}{2\sqrt{(\kappa t)}}\right)$$

where θ_0 is the initial temperature of the solid and θ_1 is the temperature to which the plane is raised, e.g. by pouring the molten metal in the case of a mould.

Assuming the thermal diffusivity to remain constant over the temperature range considered, θ_x can be calculated from the graph of error function (Fig. A.5). If the thermal diffusivity varies, a numerical solution is necessary.

Appendix 2

THE PLANE STRAIN COMPRESSION TEST

This test was devised by Watts and Ford[93] to enable uniaxial stress/strain relationships to be found for material in strip form when being deformed under plane strain conditions. The main advantage of the test is that large strains (of the order of 200%) can be achieved, and hence the forces involved in the forming of work-hardened material can be more accurately predicted.

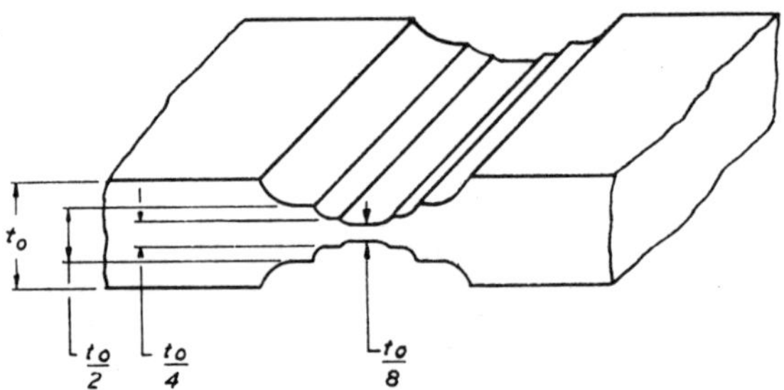

Fig. A.6 Specimen at completion of test

Strips of material are prepared with a width/thickness ratio of at least 10:1, and are squeezed between flat rectangular platens whose length exceeds the strip width. An effective lubricant, such as molybdenum disulphide, is used between the platens and the specimen, so that friction is almost eliminated.

When the width of the platen b is an exact multiple of the thickness t, the slip-line field will consist of a number of right-angled isosceles triangles, as shown in Fig. 6.25, in which case the stress on the platens, $p = 2k$, but for non-integral values of b/t, the value of p is greater than $2k$. However, for values of b/t between 2 and 4, the increase in p is so small that it can be ignored.

The test is usually carried out in the following manner. A pair of platens of width b, such that $b/t = 2$, are accurately aligned, usually in a press die set, between the anvils of a compression testing machine. Load is applied in small steps until the thickness of the specimen is reduced to half its initial value, values of p and t being recorded at each thickness reduction.

A second pair of platens, half the width of the original pair, is fitted, and the thickness of the reduced section of the specimen is again squeezed to half its original value. The procedure is then repeated a third time using platens of half the width of the second pair (Fig. A.6).

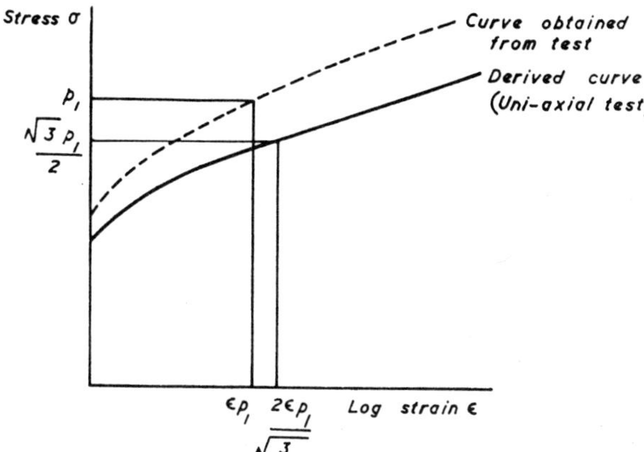

Fig. A.7 Derived curve obtained from plane strain test

Strain ε_p in the direction of p, the applied stress, is calculated for each value of p. If p is plotted against ε_p, a stress/strain curve is obtained; but since $p = 2k$, and (using the von Mises criterion) $Y = \sqrt{(3)}k$, it follows that the equivalent uniaxial stress is found by multiplying p by $\sqrt{(3)}/2$. From consideration of work done, the areas under the plane strain and the equivalent stress/strain curves are the same, so the effective strain is found by multiplying ε_p by $2/\sqrt{3}$, (Fig. A.7).

Appendix 3

Calculation of economic turning speed in Imperial units (see Section 10.1.3 for similar example in SI units).

Example. A positive rake throw-away carbide tip is used to turn EN35B. The values of the various parameters are:

tool approach angle $\psi_r = 30°$
nose radius $r = 0{\cdot}100$ in
depth of cut $d = 0{\cdot}200$ in
feed $f = 0{\cdot}030$ in

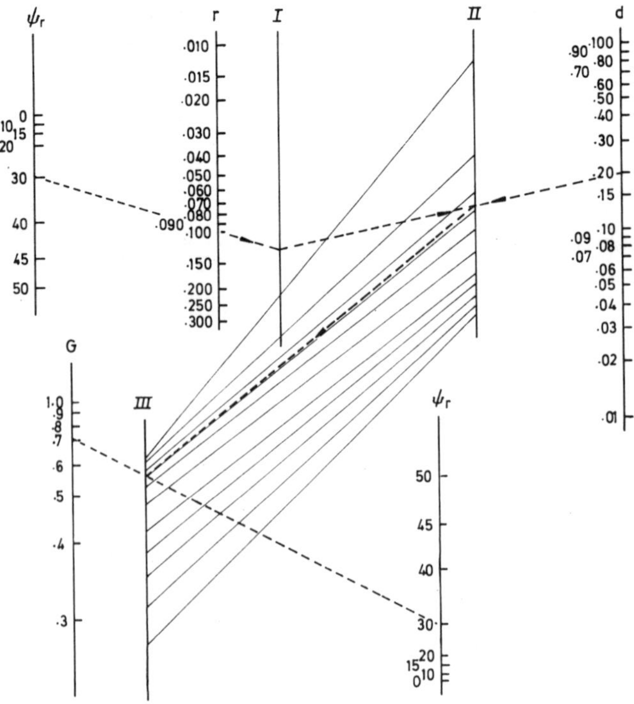

Fig. A.8 Nomogram for determining value of G

From Fig. A.8 G is found to be 0·72 and from Fig. 10.5, λ is 235. Using these values Fig. A.9(a) gives C as 900. The value of X is obtained from Fig. 10.6 using a value of 0·25 for n ($X = 0{\cdot}54$.)

Optimum cutting speed $V = XC = 486$ ft/min.

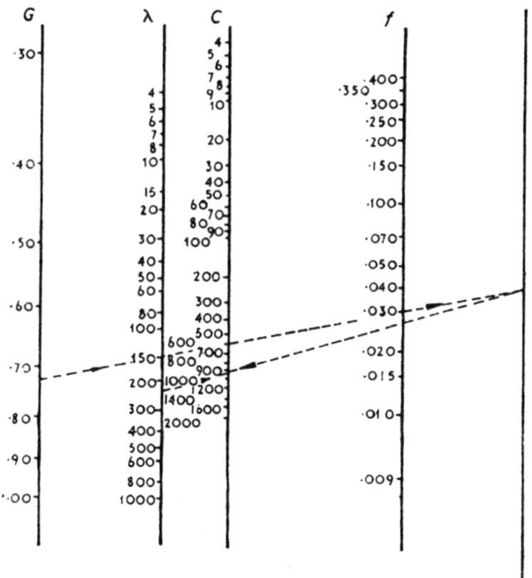

Fig. A.9 (a) Nomogram for determining tool life constant C for steel

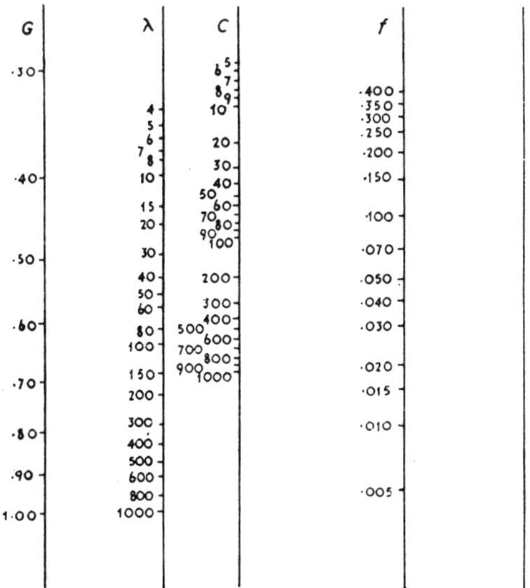

Fig. A.9 (b) Nomogram for determing tool life constant C for cast iron

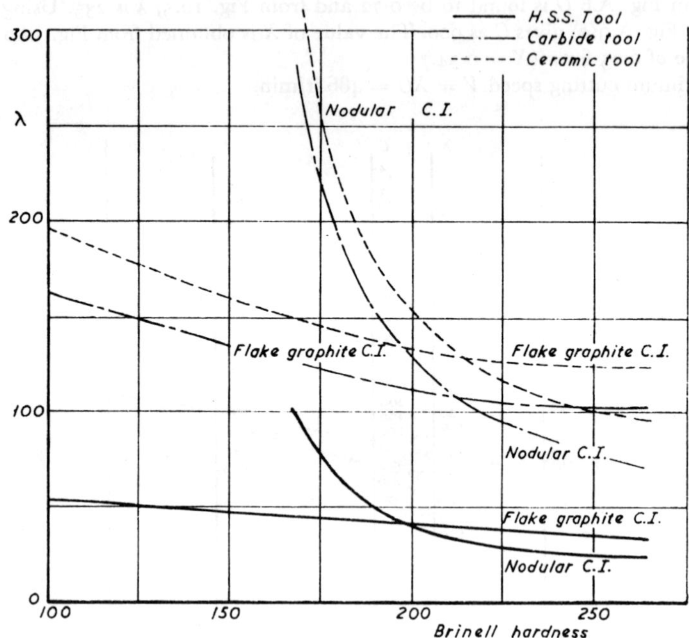

Fig. A.10 Variation of λ with hardness for cast iron (*after Brewer and Rueda*)

Examination Questions

1. The lower half of the slip-line field for a 4 to 1 perfectly lubricated plane extrusion is shown below. Calculate the extrusion pressure using the upper bound method, when $k = 200\ \text{N mm}^{-2}$.

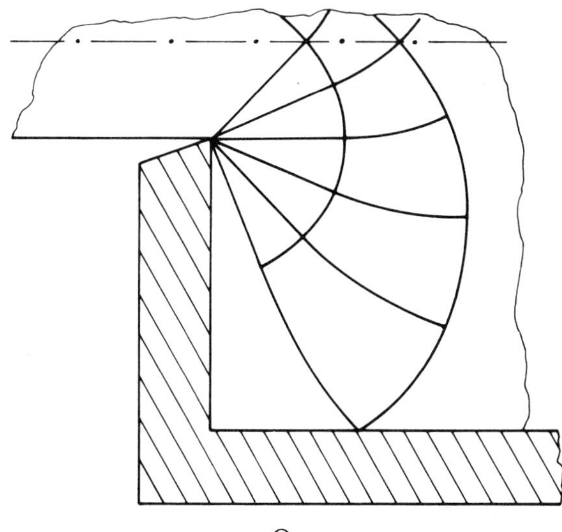

Q. 1

Find the extrusion pressure using the work formula and comment on the result.

Explain how Hencky's equations for a rigid plastic material can be modified to allow for strain hardening effects.

[Answer: $1000\ \text{N mm}^{-2}$ (upper bound), $480\ \text{N mm}^{-2}$ (work formula)]

2. Comment briefly on the usefulness of the three main methods by which deforming force can be found in metal-forming operations. The slip-line field below is that used for the cold drawing of wide strip when the drawing ratio is 1·2 and the half angle of the die is 20°.

   Calculate, using the upper bound method, the drawing tension if the value of k is $160\ \text{N mm}^{-2}$. What condition does the slip-line field indicate along the work/die interface? How can friction be minimized in cold drawing?

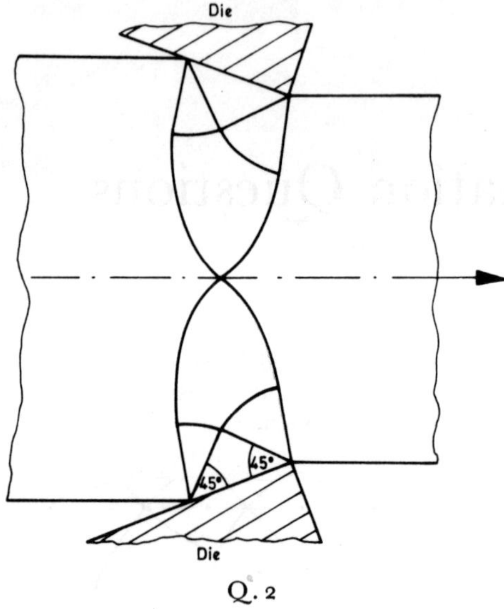

Q. 2

[Answer: 110 N mm^{-2}]

3. Discuss the difficulties and limitations of slip-line field theory when applied to practical problems in metal forming.

   The slip-line field for the lubricated indentation of a wide thick block by a flat punch is shown below.

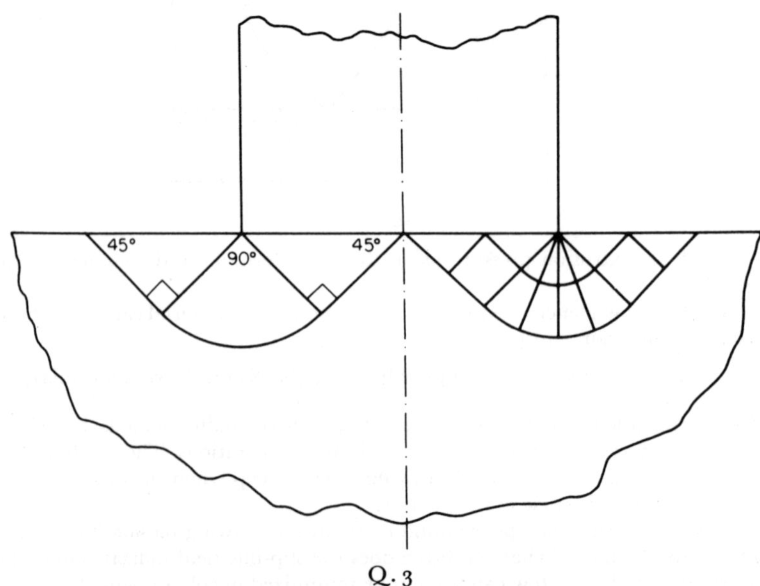

Q. 3

Calculate the indentation pressure (a) by using slip-line field theory and (b) by the upper-bound method. Comment on the difference in results obtained. Assume k for material is 300 N mm^{-2}.

[Answer: (a) 1540 N mm^{-2}, (b) 1730 N mm^{-2}]

4. In metal forming the two frictional conditions occurring at the work/die interface are either Coulomb or sticking friction. By means of diagrams indicate for both cases the directions at the interface of the slip lines and principal stresses. Construct the Mohr stress circle for the Coulomb case and derive an expression for slip line inclination to the interface.

   Show that in metal forming the coefficient of friction cannot exceed a value of 0·577.

   What frictional conditions are found in both

   (1) hot

   and (2) cold extrusion of steel. Suggest a suitable method of lubrication for each process.

5. State Hencky's first theorem and show how it can be used to sketch the slip-line field for a 4 to 1 plane lubricated extrusion (the field emanates from the mouth of the die as a fan).

   Find the magnitudes and directions of the principal stresses in the plastic zone at a point adjacent to the container wall.

   Outline a method by which the extrusion pressure can be calculated using the slip-line field (no calculations are required).

   [Answer: $\sigma_1 = 4{\cdot}7\,k$, $\sigma_2 = 5{\cdot}7\,k$, $\sigma_3 = 6{\cdot}7\,k$ (all compressive)]

6. Contrast stretch and hydrostatic forming and comment on the magnitudes of deformation without fracture which are possible in both cases.

   Describe a practical method of obtaining strain measurement in sheet metal forming. Explain how these measurements can be used in conjunction with forming limit diagrams.

7. Show that for uniaxial stretch forming instability occurs when the subtangent to the stress–strain curve for the material reaches unity. Use this result to show that an annealed material with a high strain hardening rate is best suited for stretch forming. It can be assumed that $\sigma = A(B + \epsilon)^n$, where A, B and n are constant for a given material.

   Why is good quality steel necessary if deep drawing operations are to be successful and what is the most likely cause of failure when drawing deep cups?

8. Describe the mechanism of failure in (a) high-velocity and (b) conventional blanking and its effect on the quality of the blanked edge. Explain how punch shear can be used to reduce the magnitude of the blanking force?

   A press of 50 tonnes capacity is needed to produce a square blank from material 10 mm thick using a conventional blanking tool. The most powerful press available is, however, of only 25 tonnes capacity. Calculate the minimum diagonal shear needed on the punch if the part is to be manufactured on the 25 tonnes press. Assume a rigid plastic material which fails at 40% penetration.

[Answer: 8 mm diagonal shear]

9. Why is it important in hot extrusion to obtain the highest possible extrusion ratio? Explain the interaction of extrusion speed, billet pre-heat temperature and ram pressure on the magnitude of extrusion ratio. For direct lubricated extrusion show that ram pressure (p) is given by

$$p = p_0 e^{(4\mu L/D)}$$

where p_0 is the frictionless ram pressure;

μ is coefficient of friction between billet and container wall;
L is unextruded length of billet;
D is billet diameter.

Two billets of identical material one 175 mm long, the other 150 mm long are extruded in a 100 mm diameter container. If the commencing extrusion pressure is 15% greater for the longer billet, calculate the value of μ.

[Answer: $\mu = 0{\cdot}14$]

10. Describe the casting process which you consider to be the most suitable for the following applications and analyse the reasons for your choice.

(a) Main frame of a typewriter (500 per week for 5 years);
(b) Turbine blade for a gas turbine (2000 in all);
(c) Tailstock for a centre lathe (10 per week for 5 years)

11. Describe the solidification of solid solution alloy castings and discuss the effect of riser design on the quality of these castings. What criteria are available to assist in the sizing and placement of risers in steel castings?

12. Explain why grinding is a much less efficient method of metal removal than turning or milling, when judged on the basis of specific energy.

How can the undeformed chip thickness be altered by varying grinding parameters? Discuss the effect of chip thickness on grinding wheel wear.

13. Discuss the factors other than wheel grade which will influence the 'hardness' of a grinding wheel when in use. In practice almost all of the wheel loss is caused by dressing. Explain why this should be so and discuss the difficulties of designing a truly self-dressing grinding wheel.

14. Discuss the factors which impose limits on the rate of metal removal in electro-chemical and electrical discharge machining.

What action can be taken in each case to raise the limits imposed?

15. Comment on the most common modes of cutting tool failure, and explain how tool wear in each case can be reduced or eliminated.

16. Discuss the assumptions implicit in Merchant's theory of chip formation. Which factors are most likely to invalidate this theory, and why?

17. Assuming a chip to shear across a narrow rectangular zone, derive a formula for the shear strain in terms of the shear angle ϕ and the tool rake angle β_e. In an orthogonal turning operation the cutting speed is 1 m s^{-1}, the reduction in diameter is 24 mm and feed/rev is 0·5 mm. Calculate the power required to shear the material if the chip thickness is 0·8 mm, the tool rake angle is 15° and the shear flow stress is 200 N mm^{-2}. If the observed cutting power is nearly twice that calculated, suggest reasons for the discrepancy.

[Answer: $\gamma = \cot \phi + \tan (\phi - \beta_e)$, 2.1 kW]

18. In tests conducted to incorporate feed/rev as a variable in the tool life equation the following results were obtained when cutting EN 1A.

feed (mm/rev)	0·1	0·2	0·3	0·4	0·5	0·6
T (min)	200	72	40	26	19	14

where T is tool life at a constant cutting speed of 3·0 m s^{-1}. Positive rake carbide tools were used, for which the Taylor equation $VT^{0 \cdot 25} = C$ applies.

From the results evaluate the constant in the tool life equation containing feed/rev as a variable.

[Answer: $VT^{0 \cdot 25} f^{0 \cdot 37} = 4 \cdot 83$]

19. In a plain turning operation to reduce the diameter from 75 mm to 50 mm on a number of identical parts made from EN3A, the feed (f) is 0·5 mm/rev and a positive rake carbide throw-away tip is used.

Assuming a generalized tool-life equation of the form $VT^n(fG)^m = \lambda$, calculate the cutting speed to give maximum rate of production and the expected tool life.

Time to index or change tool tip = 2 minutes
λ for EN3A when using carbide tools = 7·1
n = 0·25 for positive rake carbide tools when cutting steel
m = 0·37 for steel
r, the nose radius on the tool = 1·0 mm
ψ_r, the tool approach angle = 15°
V = cutting speed (m s^{-1}) T = tool life (minutes)

$$G = \frac{d}{[d\text{-}r\,(1 - \sin \psi_r)]/\cos \psi_r + (90 - \psi_r)\pi r/180}$$

where d = depth of cut (mm).

[Answer: 6·02 m s^{-1}, 6 minutes]

20. The machining cost per piece in a turning operation can be simplified to give:

$$K = k_1 t_s + k_1 \frac{L\pi D}{fV} + \frac{L\pi D}{fC^{1/n}} V^{1/n-1}[k_1 t_i + k_g]$$

where k_1 = cost/minute of labour and overhead;
t_s = set-up and idle time/piece;
L = turned length of the component;
D = diameter of component;
f = feed/rev;
V = cutting speed;

C = material constant in Taylor's equation ($VT^n = C$);
t_i = time required to change a tool;
k_g = tool regrinding and depreciation cost per regrind.

Express the optimum cutting speed in terms of the other variables, assuming a given maximum feed/revolution (f), and hence show that:

(a) optimum tool life is independent of cutting conditions;
(b) optimum tool life using throw-away tips is appreciably less than that for tools which are re-ground.

Explain why the above analysis of cost is valid only within certain cutting speed limitations.

21. One of the primary functions of a production engineer is to specify the most appropriate method of manufacture after taking account of all the relevant factors. What are these factors and how can they be ascertained? Discuss the hazards of process specification and indicate ways in which the risk of error can be minimized.

REFERENCES

1. Christopherson, D. G., Oxley, P. L. B. and Palmer, W. B. *Orthogonal Cutting of Work-hardening Material* (1958), Engineering, **186,** 113
2. Alexander, J. M. *Deformation Modes in Metal Forming Processes* (1958), Proc. Conf. Technol. Engng. Manuf. Paper 42, (Inst. Mech. Engrs).
3. Wistreich, J. C. and Shutt, A. *Theoretical Analysis of Bloom and Billet Forging* (1959), J. Iron and Steel Institute, **193,** 163.
4. Orowan, E. and Pascoe, K. J. *A Simple Method of Calculating Roll Pressure and Power Consumption in Hot Flat Rolling* (1946), Iron and Steel Institute Special Report, **34,** 124.
5. Saxl, K. *The Pendulum Mill—A New Method of Rolling Metals* (1965), Proc. Inst. Mech. Engrs, **179,** part 1, 453.
6. Bland, D. R. and Ford, H. *The Calculation of Roll Force and Torque in Cold Strip Rolling with Tensions* (1948), Proc. Inst. Mech. Engrs, **159,** 144
7. Hitchcock, J. H. *Elastic Deformation of Rolls during Cold Rolling* (1935), ASME Research Publication, Roll Neck Bearings.
8. Haffner, E. K. L. and Séjournet, J. *The Extrusion of Steel* (1960), J. Iron and Steel Institute, **195,** 145.
9. Hirst, S. and Ursell, D. H. *Some Limiting Factors in the Extrusion of Metals* (1958), Proc. Conf. Technol. Engng. Manuf. Paper No. 32, (Inst. Mech. Engrs).
10. Johnson, W. *The Pressure for the Cold Extrusion of Lubricated Rod through Square Dies of Moderate Reduction at Slow Speed* (1957), J. Inst. of Metals, **85,** 403.
11. Metcalfe, J. and Holden, C. *Tubemaking with Continuously Cast Material* Steel International, June 1966.
12. Sachs, G. and Baldwin, W. M. *Stress Analysis of Tube-sinking* (1946), Trans. ASME, **68,** 655.
13. Siebel, E. *The Present State of Knowledge of the Mechanics of Wire Drawing* (1947), Stahl und Eisen, **66,** 142.
14. Chang, T. M. and Swift, H. W. *Shearing of Metal Bars* (1950–51), J. Inst. of Metals, **78,** 119.
15. Chang, T. M. *Shearing of Metal Blanks* (1950–51), J. Inst. of Metals, **78,** 393.
16. Chung, S. Y. and Swift, H. W. *Cup Drawing from a Flat Blank, Part I—Experimental Investigation; Part II—Analytical Investigation* (1951), Proc. Inst. Mech. Engrs., **165,** 199.
17. Veerman, C. C., Hartman, L., Peels, J. J., and Neve, P. F. *Determination of Appearing and Admissable Strains in Cold-reduced Sheets* (Sept. 1971), Sheet Metal Industries, p. 678.
18. Ogura, T. and Ueda, T. *Liquid Bulge Forming* (Aug. 1968), American Machinist.
19. Johnson, W. BISRA Report No. MW/E/55/54.
20. Kalpakeioglu, S. *On the Mechanics of Shear Spinning* (1961), Trans. ASME Series B, 83, 125.
21. Davis, R. *Some Effects of Very High Speeds in Impact Extrusion*. Paper presented at Machine Tool Design and Research Conference, Birmingham, 1966.
22. Sawie, R. *Forming Superplastic Aluminium* (1978), Chart. mech. Engr,
23. BS 1296: 1972. Specification for Single Point Cutting Tools.
24. Stabler, G. V. *The Fundamental Geometry of Cutting Tools* (1951), Proc. Inst. Mech. Engrs., **165,** 14.

25. Enahoro, A. E. and Oxley, P. L. B. *An Investigation of the Transition from Continuous to a Discontinuous Chip in Orthogonal Machining*, Int. J. of Mech. Sci., **3**, No. 3, 145.
26. Luk, W. K. and Brewer, R. C. *An Energy Approach to the Mechanics of Discontinuous Chip Formation* (1963), ASME Winter Annual Meeting.
27. Bisacre, F. F. P. and Bisacre, G. H. *The Life of Carbide-tipped Turning Tools* (1947), Proc. Inst. Mech. Engrs., **157**, 452.
28. Chao, B. T. and Bisacre, G. H. *The Effect of Feed and Speed on the Mechanics of Metal Cutting*, Proc. Inst. Mech. Engrs., **165**, 1.
29. Heginbotham, W. B. and Gogia, S. L. *Metal Cutting and the Built-up Nose* (1961), Proc. Inst. Mech. Engrs., **175**, 892.
30. Palmer, W. B. and Yeo, R. C. K. *Metal Flow Near the Tool Point during Orthogonal Cutting with a Blunt Tool* (1963), Proc. Machine Tool Design and Research Conf. (Pergamon).
31. Merchant, M. E. *Mechanics of the Metal Cutting Process* (1945), J. Appl. Phys., **16**, 267 and 318.
32. Bridgman, P. W. *On Combined Torsion and Compression* (1943), J. Appl. Phys., **14**, 273.
33. Lee, E. H. and Shaffer, B. W. *The Theory of Plasticity Applied to a Problem of Machining* (1951), J. Appl. Mech., **18**, 405.
34. Oxley, P. L. B. *A New Approach to the Mechanics of Metal Cutting* (1964), Production Engnr., **43**, 609.
35. Oxley, P. L. B. and Welsh, M. J. M. *Calculating the Shear Angle in Orthogonal Metal Cutting from Fundamental Stress-Strain-Strain Rate Properties of the Work Material* (1963), Proc. Machine Tool Design and Research Conf. (Pergamon).
36. Zorev, N. N. *Interrelation between Shear Processes Occurring along Tool Face and on Shear Plane in Metal Cutting* (1963), Proc. Int. Prod. Engng. Research Conf. Pittsburgh, p. 42.
37. Shaw, M. C. *On the Action of Metal Cutting Fluids at Low Speeds* (1958), Wear, **2**, 217.
38. Childs, T. H. C. *Rake Face Action of Cutting Lubricants: An Analysis of, and Experiments on, the Machining of Iron Lubricated by Carbon Tetrachloride* (1972), Proc. Inst. Mech. Engrs., **186**.
39. Trent, E. M. *Tool Wear and Machinability* (1959), J. Inst. Prod. Engrs., **38**, 105.
40. Taylor, F. W. *On the Art of Cutting Metals* (1907), Trans. ASME, **28**, 31.
41. Brewer, R. C. and Rueda, R. *A Simplified Approach to the Optimum Selection of Machining Parameters* (Sept. 1963), Engineers' Digest, **24**, 9.
42. *Throwaway Tip Turning Tools*, PERA Report, No. 163.
43. *Test Charts for Machine Tools* (1948), Inst. Mech. Engrs./Inst. Prod. Engrs.
44. Pearce, D. F. and Richardson, D. B. *Improved Stability in Metal Cutting by Control of Feed and Tool/Chip Contact Length* (1977), Joint Polytechnic Symposium on Manufacturing Engineering, Leicester.
45. Richardson, D. B. and Pearce, D. F. *Measurement of Dynamic Cutting Force when using Restricted Contact Tools* (1977), XVIIIth International Machine Tool Design and Research Conference, Imperial College, London.
46. Pearce, D. F. and Richardson, D. B. *Improved Machining Capability using Controlled Contact Tools* (1977), Chart. mech. Engr, **24**, 55–7.
47. Tlusty, J. *Die Berechnung des Rahmens der Werkzeugmaschine* (1955), Schwerindustrie der CSR, **1**, No. 1, 8.
48. *The PERA Face Milling Cutter*, PERA Report, No. 171.

49. Guest, J. J. *The Theory of Grinding with Reference to the Selection of Speed in Plain and Internal Work* (1915), J. Inst. Mech. Engrs., **79**, 543.
50. Backer, W. R., Marshall, E. R. and Shaw, M. C. *The Size Effect in Metal Cutting* (1952), Trans. ASME, **74**, 61.
51. Marshall, E. R. and Shaw, M. C. *Forces in Dry Surface Grinding* (1952), Trans. ASME, **74**, 51.
52. Outwater, J. O. and Shaw, M. C. *Surface Temperature in Grinding* (1952), Trans. ASME, **74**, 73.
53. Hahn, R. S. *On the Nature of the Grinding Process*, Proc. 1962 Machine Tool Design and Research Conf. (Pergamon).
54. Colwell, L. V., Lane, R. O., and Soderlund, K. N. *On Determining the Hardness of Grinding Wheels* (1962), Trans. ASME, **84**, 113.
55. Pahlitzsch, G. *Features and Effects of Novel Cooling Methods in Grinding* (1954), Microtecnic, No. 4, 119.
56. Goefert, G. J. and Williams, J. L. *Wear of Abrasives in Grinding* (1954), Mechanical Engineering, **81**, 69.
57. Yang, C. T. and Shaw, M. C. *The Grinding of Titanium Alloys* (1955), Trans. ASME, **77**, 65.
58. Bokuchava, G. V. *Cutting Temperatures in Grinding* (1963), Russian Engineering Journal, No. 11, p. 48.
59. Harada, M., and Shinozaki, N. *Effect of Grinding Fluids on Grinding* (1963), Proc. Int. Prod. Engng. Research Conf. Pittsburgh, p. 218.
60. Yosikawa, H. *Fracture Wear of Grinding Wheels* (1963), Proc. Int. Prod. Engng. Research Conf. Pittsburgh, p. 209.
61. Selby, J. S. *Dressing Abrasive Grinding Wheels with Diamond Tools*. De Beers Diamond Information L.12.
62. Voronin, A. A., and Markov, A. J. *Effects of Ultrasonic Vibrations in Machining Creep Resistant Alloys*, Machines and Tooling, **31**, No. 11, p. 15.
63. Colwell, L. V. *Vibrations, an Aid in Metal Cutting and Grinding* (1962), Paper presented at Machine Tool Design and Research Conf. Birmingham.
64. *Electrochemical Machining*, PERA Report, No. 145.
65. Waller, D. N. *A Weld-quality Monitor for Resistance Welding*, Europäisher Maschiner—Markt/Europa Industrie Revue 11/1965.
66. Carslaw, H. S., and Jaeger, J. C. *Conduction of Heat in Solids*, (1947), London U.P.
67. Chvorinov, N. *Control of the Solidification of Castings by Calculation* (1938), Proc. Inst. Brit. Found., **32**, 229.
68. Caine, J. B. *Risering of Castings* (1949), Trans. AFS, **57**, 66.
69. Bishop, H. F. and Pellini, W. S. *Contribution of Riser and Chill Edge Effects to the Soundness of Cast Steel Plates* (1950), Trans. AFS, **58**, 185.
70. Bishop, H. F., Myskowski, E. T. and Pellini, W. S. *Soundness of Cast Steel Bars* (1951), Trans. AFS, **59**, 174.
71. Hodge, E. S. *Gas Pressure Bonding of Refractory Metals* (1961), Metals Engineering Quarterly, **1**, 4, 3.
72. Thümmler, F. and Thomma, W. *The Sintering Process* (1967), Metals and Materials, **11**, No. 6, 69.
73. BS 888: 1950, Slip (or Block) Gauges and their Accessories.
74. BS 1790: 1961. Length Bars and their Accessories.

75. BS 3064: 1978, Metric Specification for Sine Bars and Sine Tables.
76. BS 958: 1968, Spirit Levels for use in Precision Engineering.
77. BS 5204: Straight Edges, Part 1 1975 Cast Iron, Part 2 1977 Steel and Granite.
78. BS 817: 1972, Surface Plates and Tables.
79. Notes of Applied Science Vol. I *Gauging and Measuring of Screw Threads* (1959), NPL.
80. BS 1134: 1972. Method for Assessment of Surface Texture.
81. BS 2634: Roughness Comparison Specimens.
82. BS 3730: 1964. Method for the Assessment of Departures from Roundness.
83. BS 969: 1953. Plain Limit Gauges Limits and Tolerances.
84. BS 919: Screw Gauges Limits and Tolerances.
85. Hertz, H. Miscellaneous Papers, Macmillan, 1896.
86. BS 907: 1965. Dial Gauges for Linear Measurement.
87. Dutt, R. P. and Brewer, R. C. *On the Theoretical Determination of the Temperature Field in Orthogonal Machining*, Int. J. of Prod. Research, **4**, No. 2.
88. Hollander, M. B. *An Infra-red Micro-radiation Pyrometer Technique Investigation of the Temperature Distribution in the Workpiece during Metal Cutting* (1959), ASTE Research Report, No. 21.
89. Weiner, J. H. *Shear Plane Temperature Distribution in Orthogonal Cutting* (1955), Trans. ASME, **77**, 1331.
90. Rapier, A. C. *A Theoretical Investigation of the Temperature Distribution in the Metal Cutting Process* (1954), Brit. J. Appl. Phys., **5**, 400.
91. Boothroyd, G. *Temperatures in Orthogonal Cutting* (1963), Proc. Inst. Mech. Engrs., **177**, 29.
92. Nakayama, K. *Temperature Rise of Workpiece During Metal Cutting* (1956), Bull. Fact. Engng., Nat. Univ. Yokohama, **5**, 1.
93. Watts, A. B. and Ford, H. *An Experimental Investigation of the Yielding of Strip between Dies* (1952), Proc.(B) Inst. Mech. Engrs., **1B**, 488.

Index